Trace Fossils

Trace Fossils

BIOLOGY, TAPHONOMY AND APPLICATIONS

Second edition

Richard G. Bromley

Geological Institute
University of Copenhagen
Denmark

CHAPMAN & HALL

London · Glasgow · Weinheim · New York · Tokyo · Melbourne · Madras

Published by Chapman & Hall, 2–6 Boundary Row, London SE1 8HN

Chapman & Hall, 2–6 Boundary Row, London SE1 8HN, UK

Blackie Academic & Professional, Wester Cleddens Road,
Bishopbriggs, Glasgow G64 2NZ, UK

Chapman & Hall GmbH, Pappelallee 3, 69469 Weinheim, Germany

Chapman & Hall USA, 115 Fifth Avenue, New York, NY 10003, USA

Chapman & Hall Japan, ITP-Japan, Kyowa Building, 3F, 2-2-1 Hirakawacho,
Chiyoda-ku, Tokyo 102, Japan

Chapman & Hall Australia, 102 Dodds Street, South Melbourne,
Victoria 3205, Australia

Chapman & Hall India, R. Seshadri, 32 Second Main Road, CIT East,
Madras 600 035, India

Second edition 1996

Typeset in Times by Acorn Bookwork, Salisbury, Wilts
Printed in Great Britain by St Edmundsbury Press Limited, Bury St Edmunds, Suffolk

ISBN 0 412 61480 4

A catalogue record for this book is available from the British Library

Library of Congress Catalog Card Number: 95-71087

∞ Printed on permanent acid-free paper, manufactured in accordance with ANSI/NISO Z39.48-1992 and ANSI/NISO Z39.48-1984 (Permance of Paper).

Contents

Preface to the second edition

Since the publication of the first edition of this book in 1990, research in ichnology has shown spectacular acceleration. In the first half of 1995, a seminal paper in ichnology was published about every two weeks. For the purposes of this book, I have said 'Stop' and drawn the line at June 1st.

Since 1990, several palaeobiology textbooks have included a chapter on ichnology. This is a major step forward. The chapter in Dodd and Stanton (1990) is a good one although, like this book, it needs revising! Several major works on ichnology have appeared, including a book on the evolution of behaviour (Boucot 1990), Goldring's book (1991) *Fossils in the Field*, a whole part in *Palaios* (edited by Ekdale and Pollard 1991), a chapter in Walker and James' *Facies Models* (Pemberton *et al.* 1992), SEPM core workshop notes (Pemberton 1992), notes for a Paleontological Society short-course (Maples and West 1992) and the 1992 Lyell Lectures (edited by Goldring and Pollard 1993).

The journal *Ichnos* has played an important part in the publication of ichnological papers. Established in 1989 by Bob Frey and George Pemberton for promoting trace fossil research, one of the journal's most important functions has been to draw together the work of specialists in invertebrate and vertebrate trace fossils, borings, coprolites, fossil insect frass in plants, termite nests, etc. This has had a unifying effect on the discipline.

During the past few years, specialists in ichnofabric have been organizing workshops. The first International Ichnofabric Workshop was held in Norway in 1991, the second in Utah in 1993. The third is to take place in 1995 in Denmark.

This book is deliberately written to function as a bridge between the living burrowing faunas and trace fossils; it is conceived in two parts, neoichnology and palaeoichnology. However, the first edition certainly failed to establish that bridge between the sister disciplines of benthic ecology and ichnology. It seems extraordinary to me that a book (an excellent one) can be published with the title *The Environmental Impact of Burrowing Animals and Animal Burrows* (Meadows and Meadows 1991) and yet somehow avoid almost all reference to ichnology. I hope the present book will help remedy this state of affairs. But I saw only one review of the first edition by a biologist. The word *Fossils* in the title seems to be intimidating.

I have referred extensively to the literature in the hope of promoting 'further reading'. The first edition referred to 493 papers. This figure has nearly doubled in the present edition (861), reflecting the enormous activity in the discipline. There are 29 papers from 1990, 48 from 1991, 31 from 1992, 32 from 1993, 45 from 1994, and already 13 from 1995 and in press.

The whole text has been up-dated and several new sections inserted. A new chapter (12) has been added and the title correspondingly extended with the addition of 'Applications'. I had intended to include an ichno-taxonomical appendix, but there is no room. Nevertheless, most of the ichnotaxa mentioned in the book are described or illustrated somewhere; use the index.

In the first edition I was taken to task for muddling my oxygen-depletion terms, 'anoxic' versus 'anaerobic', and so on. I have tried to follow the recommendations of Tyson and Pearson (1991) this time, and hope this is acceptable.

As I write this preface, Ulla Asgaard is battling with the index, for which I thank her profoundly. She has also helped me greatly with much nitty-gritty word-processing, as well as her inspiration as my closest colleague.

The international ichnological community is closely knit and functions well. I thank all members of it with whom I have contact for their input that has made this book possible. No names, you know who you are.

I am grateful to the following organizations for being able to record their permission to publish the following photographs: the Director of the Geological Survey of Greenland for Figs 3.6, 6.8, 8.3, 8.5, 8.13, 9.4, 9.9, 10.5 and 10.17; Conoco for Figs 9.2, 10.4, 10.14, 10.23–10.25, 11.1–11.3, 11.5a, 11.7, 11.14, 11.15 and 11.18; Statoil (Norway) for Figs 11.4, 11.5b, 11.6, 11.8, 11.12, 11.13, 11.16 and 11.17; and the Institute for Continental Shelf Research (Trondheim) for Figs 10.21, 11.9 and 11.10.

Preface to the first edition

This is a personal book about animals and sediments: one man's trace fossils. It is intended as a course book for advanced students rather than a general introduction to ichnology. Nevertheless, all terms are explained as they occur, or may be found in the glossary.

Most trace fossil specialists are geologists and were introduced to trace fossils in the rocks before they ever examined the sedimentary activity of tracemaking animals. Some indeed have never had the opportunity to watch endobenthic animal behaviour. Most of the existing books and chapters in text books are slanted in this way, beginning with the rocks rather than with the organisms that create the structures that become trace fossils (e.g. Frey 1975; Frey and Pemberton 1984, 1985; Ekdale *et al.* 1984a).

This geological approach may lead to an over-rigid attitude to trace fossils, a tendency to treat them like postage stamps, or to consider them as having a one-to-one relationship with particular environments or trace-making animal species.

It makes a healthy change, therefore, to remind ourselves that most trace fossils are produced in unconsolidated sediment by animal movements: sediment processing by deposit feeders, excavation of residences, or the substrate restructuring caused by locomotion. (Borings and bioerosion are not treated in this book.) As Adolf Seilacher told us, trace fossils represent 'fossil behaviour', and Hallam (1975, p. 55) emphasized that 'a thorough knowledge of the manner in which crawling and burrowing organisms produce preservable structures in soft sediments is obviously vital to a proper understanding of trace fossils'. But the many forms and fabrics that are trace fossils also are the product of the varied processes of taphonomy and diagenesis that the sediment and its structures are submitted to in the course of the history of the rock.

I have arranged this book, therefore, in two parts. We begin by examining animal–substrate relationships in a broad spectrum of different organisms and behavioural programmes. I have personal experience of some of these case histories; others are drawn from the literature. Part One closes with a chapter on bioturbation seen from the biological point of view. On the basis of this neoichnological approach, we are now in a position to take a refreshingly new look at palaeoichnology. We cross the 'fossilization barrier' in Part Two and apply the biological principles to trace fossils and biogenic textures as found in sedimentary rocks.

The biological and taphonomical approach places emphasis on the relative depths into the substrate at which different animals operate. The endobenthic community thus is said to be 'tiered', and this has absolutely fundamental significance in determining the rock fabric or the trace fossils that come to be preserved. Furthermore, the tiering principle allows an ecological grouping of trace fossils in 'ichnoguilds', a new concept introduced here. And finally, a new look is taken at the Seilacherian Ichnofacies.

A closing chapter discusses the transfer of this information into the two-dimensional format of the drilling core. Cored wells produce rock sequences in unweathered condition in wonderfully uninterrupted vertical continuity, but having a ridiculously limited lateral extent of only a few centimetres of core diameter. The study of trace fossils, on the other hand, was born on the extensive, weathered, sun-warmed bedding planes of onshore outcrops. Transferring the discipline in a meaningful way to the core-shed is stretching many well-trained minds. In this cylindrical format, biogenic textures or ichnofabrics are generally more revealing than are individual trace fossils, which usually remain anonymous in vertical section alone.

I have tried to avoid re-publishing earlier illustrations, and almost without exception, the figures are new. I have based my drawings on published figures in many cases, of course, but always changing their emphasis to my personal view. This has taken much time and energy, and would have been beyond my powers had I not had the continual encouragement and assistance of my most special colleague, Ulla Asgaard. She has searched out difficult literature, skimmed German texts for me, laboured critically and constructively through numerous early drafts and has guided my fingers on the word-processor. Henrik Egelund has helped me with some of the artwork, and Jan Asgaard made the photographic prints and some of the negatives.

The photographs derive from my field and laboratory work over many years – work which would have been impossible without the help, encouragement and guidence of very many friends and colleagues. I refrain from listing names here, as I do not know where to end the list. Laboratory photos were taken at the Marine Biological Laboratory of the University of Copenhagen at Helsingør, together with Margit Jensen, who also critically read an early draft of Part One of the book. Individuals who kindly placed photographs at my disposal are mentioned in the captions to the respective figures.

Introduction

Traces are structures produced in rocks, sediments and grains by the life processes of organisms, and their study comprises the discipline **ichnology**. **Trace fossils** are the fossilized equivalents of those structures (**palaeoichnology**).

Designed to encompass all trace fossils, these definitions are vague. However, if we subdivide the concept into component groups, the coverage of each of these becomes more precise and meaningful. Trace fossils include: (1) footprints, tracks and burrows in unconsolidated sediments; (2) raspings, borings and etchings in rigid substrates; and (3) faecal pellets, pseudofaeces and coprolites. Some geologists also include: (4) plant root penetration structures, and even (5) algal-mat laminites and stromatolites.

I restrict myself here almost entirely to the first category; thus, in the present context, trace fossil means **biogenic sediment structure**. In zoological terms, as used in the International Code of Zoological Nomenclature (ICZN), this is equivalent to **the work of an animal**. Trace fossils contrast with **body fossils**, which are biological skeletal structures or other remains of bodies of organisms.

Animals that have adopted an **endobenthic** existence, i.e. living within unconsolidated to consolidated sediment of the sea or lake floor, or subaerially, derive from virtually all the phyla. This taxonomic diversity brings with it a wide spectrum of pre-adaptational tools and strategies for dealing with and living within sediment. Each of these behavioural variants produces a correspondingly different type of sediment structure. Burrowing strategy is further modified according to the nature of the substrate that has to be penetrated or processed. Not only endobenthic, but also many **epibenthic**, animals disturb the substrate beneath them in their life processes, and thereby produce biogenic sediment structures.

As a result, a bewildering array of structures arises, ranging from those produced by stampeding dinosaurs, aestivating lungfish and egg-laying turtles to those of mantis shrimps lying in ambush, mud shrimps systematically processing sediment for food, and tiny ostracodes working their way among sand grains larger than themselves.

This wide morphological variation needs a common terminology that allows the structures to be treated within a single discipline and classifies them on the basis of comparable qualities. A host of terms has emerged with the growth of ichnology, although many are treated differently by different authors. Terms used here are collected in a Glossary.

But, let us begin by discussing terms and processes before going on to examine some examples of animal–sediment relationships as observed in living material. In a book of this size, a comprehensive treatment cannot be presented, but I have chosen a number of case histories that illustrate distinctive categories of sediment processing.

The book is arranged in two parts. In the first, processes and organisms of the present day are examined (neoichnology). In the second part we cross the fossilization barrier and examine trace fossils and the results of endobenthic activity as expressed in ancient sedimentary rocks.

PART ONE

Neoichnology

1 Animal–sediment relationships

1.1 Why do animals burrow?

Animals disturb sediments for many reasons. Some do it accidentally, as the crab or quadruped leaves surficial footprints. At the other end of the scale are animals that spend their entire life constructing intricate structures as part of their feeding processes. The specialist **endobionts** have learnt to employ various properties of their substrate to serve a wide spectrum of the essential requirements of life: respiration, feeding, reproduction and protection.

1.1.1 Protection and concealment

Perhaps the most basic advantage for an animal that can establish itself beneath the depositional interface is the protection this confers from the above-sediment environment. In turbulent water conditions, and in environments that are susceptible to periodic sub-aerial exposure, the endobenthic environment is far less rigorous than the epibenthic. A sun-baked beach exposed at low tide may seem a barren place, but a little spadework may reveal it to be richly populated beneath the surface.

In addition to physical protection, endobenthic animals are concealed from potential predators, in particular those that hunt by sight. The so-called 'resting traces' so common in the photic sea floor largely represent such hiding traces; it is safer to be just covered with sediment than exposed on the sea floor. Deeply or rapidly burrowing animals find refuge from shallowly or slowly burrowing predators. However, the tables may be turned. Many predators (fish, crustaceans, polychaetes) put concealment within a **burrow** to their advantage and wait in ambush for passing epibenthic prey.

1.1.2 Respiration

Another advantage of residence within a cylindrical burrow is the relative ease with which water may be circulated through it. Different species bring different pre-adaptational mechanisms to work for this purpose: worms send peristaltic contractions along their elongate bodies and thus squeeze rings of water past them; crustaceans and fish modify swimming organs to waft a current.

In either case, a strong current can be produced using little energy, bringing quantities of water past the respiratory organs. Outside the burrow, the same activity would pass hardly any water over the gills. Thus, although penetration into the sediment brings the burrower into regions of lowered oxygen tension, irrigation using bottom water removes this problem. Indeed, the hydraulic efficiency of canalization allows burrowers to live at lower oxygen levels than most epibenthic animals (Sassaman and Mangum 1972).

1.1.3 Suspension feeding

The current produced so easily within a burrow can also be used for gathering suspended food particles. Many different devices have been developed by suspension-feeding burrowers to intercept the **seston** suspended in the water current, including tentacles, sieves created by hairs on limbs, and nets of slime (section 3.3). Many epibenthic suspension feeders depend on natural currents to bring them food, and spread delicate catchment devices that are exposed and vulnerable. Burrow dwellers control their own current and have protection for their filtering equipment. They may thus colonize environments that are unsuitable for epibenthic suspension feeders.

1.1.4 Deposit feeding

By far the most **endobenthos** are deposit feeders, working the sediment for its contained nutrients. A host of diverse niches is available to multifarious specializations, and strategies are numerous within this way of life (sections 3.5, 4.1.1, 4.2.2 and 4.3.2). Few organisms engulf the sediment whole and uncritically, and digest what they can of it. Almost all are selective, extracting those grains that carry special organic components. Many animals digest microalgae and the bacteria that are taking part in decomposing the organic material (Taghon *et al.* 1978; Newell 1979; Hauksson 1979). Together with organic exopolymeric substances, these organisms form a biofilm covering grain surfaces (Westall and Rincé 1994). Several of these deposit feeders find such material well below the redox level (**RPD**), deep beneath the depositional interface. Unlike seston and surface detritus, supplies of which fluctuate temporally, the organic content of the sediment constitutes a relatively predictable food source (Levinton 1972).

The enzymes available to deposit feeders are limited, and the animals may depend to some extent on gut and sediment microbes for the conversion of particulate organic matter into simpler, assimilable units. In a sense, the substrate can function as an 'external rumen' for deposit feeders (Rice and Rhoads 1989; Levinton 1989).

1.1.5 Surface detritus feeding

In aquatic environments, the richest supply of organic matter is that at the depositional interface – immature, copious and continually renewed. Many vagile epibenthic animals exploit this food resource. If supplies are renewed rapidly, however, they can also support stationary detritus feeders, and these generally avail themselves of the advantages of an endobenthic domicile. Many species of sipunculan, echiuran and annelid worms, bivalves and crustaceans have adopted this mode of life, protected beneath the sediment surface and feeding radially around the burrow aperture or trapping detritus in a funnel (section 3.4.3).

Surface detritus feeding is not a universally recognized category. Ecologists tend to view detritus feeders as either collecting their food *in situ* as deposit feeders, or resuspending it and catching it as suspension feeders. Moreover, many ecologists use detritus as a synonym for particulate organic matter, whether at the surface or buried (de Wilde 1976). Most phytodetritus arrives on the ocean floor pelagically as a surface layer (Billett *et al.* 1983; Suchanek *et al.* 1985; Rice *et al.* 1986; Carney 1989). Much less commonly it arrives in a turbidity current and is immediately buried (Reichardt 1987).

From a sedimentological point of view, feeders on the organic accumulations at the depositional interface represent a specialist group that is important in community and trace fossil analysis. Thus, the biologically impure group is separated here as a trophic unit. In this book, 'detritus feeders' are 'surface detritus feeders'.

1.1.6 Gardening

The search for suitable bacteria may be alleviated by actually culturing them (or other food organisms) on burrow walls or on deliberately accumulated organic matter in chambers within the burrow. The activity of burrowers within the sea floor in any case increases the microbial activity in the sediment (e.g. Andersen and Kristensen 1991). Gardening is said to take place if the activities of a burrowing animal promote the production of microbes in such a way that this positively contributes to the food resources of the animal (de Wilde 1991).

As leaf-cutter ants culture fungi on leaf clippings in the terrestrial realm, so do some shrimps apparently cultivate bacteria on sea-grass cuttings (section 4.3.4) and some annelids and echiurans may culture organisms on their burrow walls (section 3.3.3).

1.1.7 and the opposite

A thin scattering of endobenthic animals, distributed over many phyla, produce highly toxic halogenated compounds (Woodin *et al.* 1987). These compounds occur in the body and are also commonly concentrated in the burrow wall. The toxins reduce or prevent the development of microbes and other fauna in the burrow walls, as well as acting as a deterrent to predators. One group, the enteropneusts, have taken this poisoning seriously (section 3.4.5).

1.1.8 Chemosymbiosis

Chemosymbiosis was first detected in deep-sea vent habitats, where communities of pogonophore worms and large bivalves have formed an association with autolithotrophic bacteria. These bacteria can oxidize HS^- from the vents, thereby allowing the fixation of carbon and production of carbohydrates and enzymes for the growth and metabolism of the host (e.g. Hand and Somero 1983; Felbeck *et al.* 1984; Jannasch 1984). Bacterial chemosymbiosis has since been recognized in other habitats where HS^- or methane occurs in close proximity with oxygenated water, as it is in the soft sea floor generally, and particularly around natural gas and petroleum seeps (Paull *et al.* 1984; Hovland and Thomsen 1989).

Some host animals carry the bacteria within their bodies, such as the lucinacean bivalves (Fisher and Hand 1984; Reid and Brand 1986). Others use a burrow to house the bacteria, usually in the wall, or they dig a deep 'sulphide well' into the anoxic sediment. These burrow modifications are exercising the imagination of ichnologists, and the trace fossil *Chondrites* has been convincingly reinterpreted as the work of a chemosymbiont (Seilacher 1990; Fu 1991).

Oxidation of such sulphide exploits energy that would otherwise be lost from the system since it occurs beneath the **RPD**. The symbiosis with invertebrates extends the habitat available for bacterial oxidation of sulphide, thus benefiting the bacterial side of the association.

1.1.9 Predation

Endobionts themselves constitute a food source. Some fossorial predators trap vagile endobenthos as the mole does earthworms in an extensive open-burrow system; others actively burrow and seek less mobile prey, as do priapulid worms and some carnivorous snails. Many predators find it more effective to search epibenthically, and then burrow down to their hidden prey as they find them, as do some rays and starfish (Smith 1961; Mauzey *et al.* 1968; Veldhuizen and Phillips 1978). Such epibenthic predators may cause considerably more disturbance of the substrate than do their truly endobenthic prey. As mentioned before, yet other predators produce permanent burrows in which to lie in ambush (section 4.4).

Finally, many sedentary burrowing animals are subject to **browsing predation**, whereby upper ends are browsed off or harvested by epibenthic predators without killing the prey animal. The lost parts are later regenerated (de Wilde 1976; Woodin 1982; section 4.1.2).

1.1.10 Reproduction and trauma

Special chambers for brooding young have been described, especially in the case of insects (e.g. Wohlenberg 1939, fig. 34; Frey and Pemberton 1985, fig. 10); their morphology is extremely variable. For example, the nests of termites make fine trace fossils (section 9.2.10).

Panic structures may be produced by animals escaping from predators or from sudden burial (section 9.2.9). In other cases, an animal may quietly shift its burrow upwards, adjusting its position in response to a gradually accreting sea floor (section 9.2.8).

1.1.11 Other behaviour

Endobenthic animals create sediment structures through still other behavioural traits than these, and the picture is further complicated by the fact that many broadly specialized burrowers exhibit several different feeding styles, adapting to changing environmental conditions (Cadée 1984). Indeed, individual burrow systems may comprise different parts that relate to different life activities. Thus, although it is not always possible to relate trace fossil morphology directly to animal behaviour, it is nevertheless possible to classify trace fossils in broad behavioural categories (section 9.2).

1.2 How do animals burrow?

As stated previously, endobenthic animals have approached the problem of substrate penetration along many different pre-adaptational pathways. Many studies have been made of individual groups or species of burrowing organisms, but little general work has been done, and nothing approaching a universal terminology has been developed. This is despite the excellent general works of Schäfer (1956, 1962), Trueman and Ansell (1969) and Trueman (1975). Several authors used terms such as 'excavate' and 'push' in order to differentiate between different means of digging, but the terms are vaguely defined (e.g. Carney 1981).

Thayer's (1979, 1983) term 'bulldozing' for all sediment processing is inappropriate, except for the work of some lobsters. Likewise, Thayer's (1983) subdivisions of bulldozing, i.e. gliding, handling, digging and insinuating, do not cover the general needs of the ichnologist.

But we need a set of terms to define general categories of bioturbation processes. On the basis of whether the sediment is transported and redeposited during the burrowing process, or whether it is merely pushed aside, the following penetration styles can be distinguished: **intrusion**, **compression**, **excavation** and **backfilling**. These terms relate to substrate properties rather than to the muscular activity of the burrowers; they are ichnological rather than biological terms.

1.2.1 Intrusion

An animal may intrude into its substrate by pushing sediment aside and temporarily displacing it with its body. The medium closes behind as the animal moves on. A process of **eddy diffusion** (Fig. 1.1) mixes and tends to homogenize the sediment. Wetzel (1983b) followed Schäfer (1956) in calling the result a **biodeformational structure**.

In an extremely fluid sediment containing much water, some animals proceed effectively through it by using modified swimming technique (section 2.4). For example, some polychaete worms use sinuous wave movement instead of a peristaltic behaviour which the same worm would find necessary to penetrate firmer sediments.

The activity of **meiofauna** and very small macrofauna may also be considered as intrusion. Active among grains little smaller or even larger than themselves, these organisms constantly dislodge and minutely displace the particles, tending to blur and ultimately to obliterate earlier structures and thereby homogenize the sediment (section 2.1).

A burrower working a firmer substrate closely beneath the sediment surface may lift this as a ridge as it bodily displaces sediment upwards. The ridge generally collapses in disorder behind as the animal burrows forwards. This is typical behaviour for several terrestrial vertebrates such as moles (Fig. 1.2a), amphisbaenid reptiles (Gans 1960) and insects (Frey and Howard 1969, pl. 4).

Numerous animals enter loose sediment in order to hide or to catch prey by merely thrusting themselves or a part of their anatomy into the

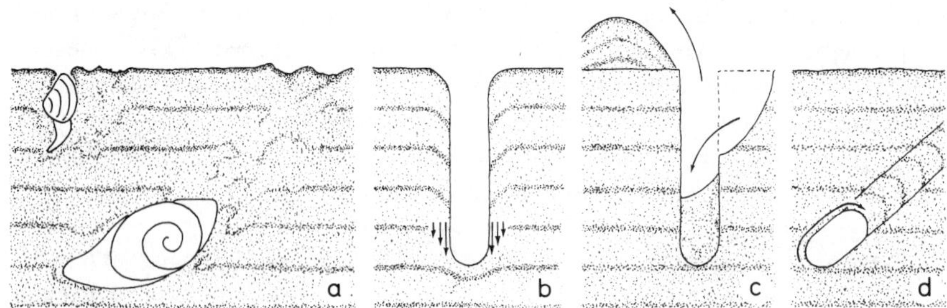

Figure 1.1 Mechanisms by which animals induce mass transport in sediments (a–c modified after Hanor and Marshall 1971). (a) Turbulent or eddy diffusion in soupy sediment. (b) Shear. (c) Upward advection by excavation, followed by downward advection by passive filling. (d) Lateral advection by backfilling, producing an active fill.

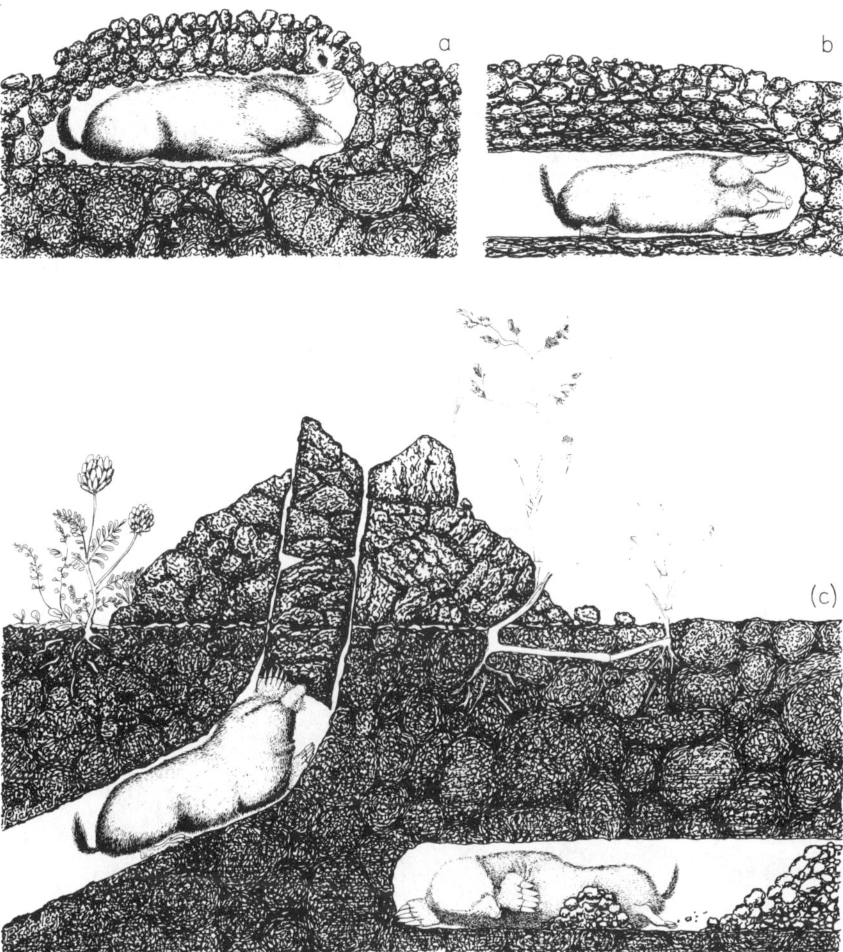

Figure 1.2 Burrowing strategy determined by substrate consistency, as illustrated by the mole. (a) Surficial burrow in uncompacted surface soil produced by intrusion; the roof collapses behind the animal. (b) Compression burrowing at a deeper level leaves an open burrow and no spoils; the fabric of the substrate is plastically deformed. (c) At a still deeper level, the compact soil has to be excavated and spoils are advected to the surface, leaving an open burrow. Modified after Mellanby (1971) and Buch (1980).

substrate. For example, an octopus seeking burrowing prey in the sea floor creates microtectonic disruption of the displaced sediment (Fig. 1.3).

Another group of intrusion structures are panic escape traces where, for example, an animal that has been buried by sudden sedimentation struggles upwards seeking contact with the new surface (section 9.2.9; Figs 3.4e, 3.23c and 4.8e).

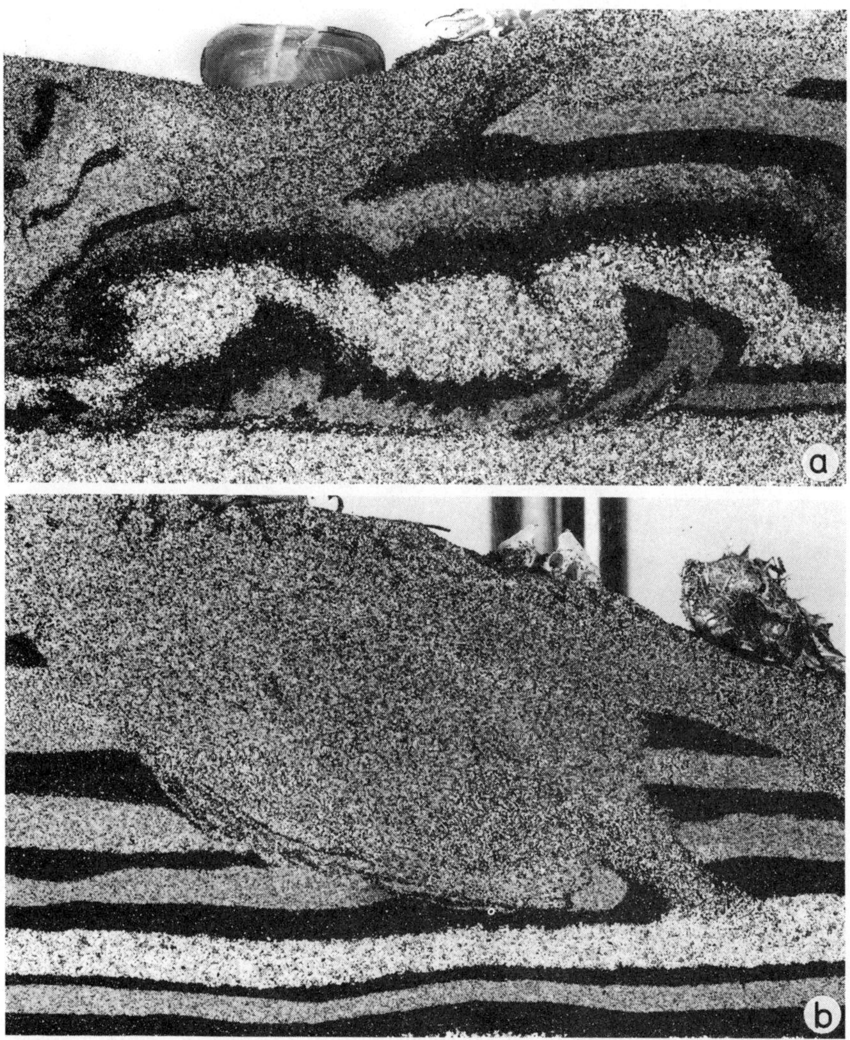

Figure 1.3 Structures produced by an octopus (*Octopus vulgaris*) in laminated sand in a narrow aquarium (viewed from either side). The aquarium was placed in the sea floor, and occupied by the octopus after it had captured and eaten the original inhabitant, the bivalve whose valves now lie on the sediment surface: –4 m, Kefallinia, Greece.

1.2.2 Compression

Compression structures are those where the burrower forces a passage through the substrate by pressing it aside and compacting the sediment (section 8.3.2e). As the animal progresses it leaves behind it an open burrow, the firm, compacted walls of which may be smooth or ornamented by characteristic impressions left by the compacting process (section 7.2).

Moles and amphisbaenids again illustrate this burrowing strategy (Fig. 1.2b). Compressional burrowing is common among invertebrates, including bivalves and many kinds of worms. The open tunnels and compactional aspects of this method considerably change the bulk properties of the sediment. Grains are shifted by processes of eddy diffusion or shear, whereas the open burrows allow **passive filling** by particle **advection** (Fig. 1.1c) from above or **active filling** from other parts of the burrow system (sections 5.2 and 8.3.4).

1.2.3 Excavation

In sediment that is already somewhat compacted, the easiest mode of constructing an open burrow is by excavation. This process involves loosening the sediment ahead and conveying the spoils elsewhere, commonly out of the burrow entrance onto the substrate surface.

Most fossorial mammals in the terrestrial realm excavate their burrows in this way (Fig. 1.2c). In aquatic environments, crustaceans and fish are active excavators. Normally the sediment is conveyed outside the burrow in a basket formed by the limbs or, in the case of fish, in the mouth. A strong irrigation current may be used to waft the grains out of the burrow. Alternately, some or all of the material may be ingested, passed internally along the gut and voided outside the burrow, or deposited elsewhere within it.

Particle advection caused by this process is considerable. The bulk properties of the substrate as such are not greatly altered, however, except by the open burrow systems that are created within it.

1.2.4 Backfill

Perhaps the most sophisticated means of progress through sediment is that of backfill. The animal loosens grains ahead of it, transports these backwards around itself as it moves bodily forwards, and redeposits the grains again behind it (Fig. 1.1d). This procedure involves minimum transport of material, and allows a small burrow to be maintained and moved through the sediment. The diameter of the structure may thus be only slightly greater than that of the animal, allowing a zone of sediment transport around it.

As the grains are manipulated, they become sorted for weight, size, palatability, etc. Some particles are ingested, thereby taking an internal route to the rear of the animal. The backfill normally has a laminated **meniscus structure** that reflects the depositional pattern of the material (Fig. 1.4; section 8.3.4).

It is almost axiomatic that the amount of work involved in transporting a solid through a dense medium will be directly related to the cross-

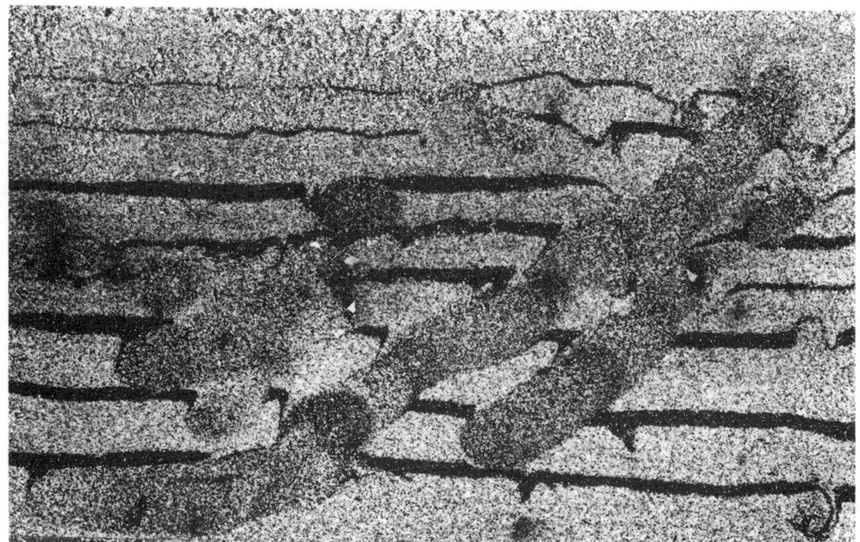

Figure 1.4 Backfill structure produced by a single spatangoid echinoid (*Echinocardium mediterraneum*) in laminated sand in a narrow aquarium placed in the sea floor: –4 m, Kefallinia, Greece.

sectional area of the body. For this reason, most burrowing animals are elongate worms. Backfill burrowing, however, is so efficient that many such burrowers are not elongated. Several spatangoid echinoids are subspherical (Kanazawa 1992).

1.2.5 Spreite

Backfill is such an efficient strategy that major liberties may be taken. Thus, a number of animals have found it advantageous to move broadside-on (sideways) through the substrate instead of axially. The structure resulting from such lateral shifting of a burrow is a **spreite**. We know next to nothing about spreite production in extant animals and our interpretation of these structures is deduced from trace fossil morphology (Seilacher 1953a, 1967a). Some spreiten result from up and down equilibrium movements of a U-tube in response to sea-floor fluctuations (section 9.2.8) although the majority may correspond to endogenic 'strip mining' activities of deposit feeders (section 9.4.5). Lateral movement greatly increases the active surface area of the animal in contact with the sediment and promotes food-gathering activities.

1.2.6 Bioturbation

These sediment-mixing activities lead to an obliteration of existing sediment structures (Fig. 1.5). Mixing generally is considered to reduce the

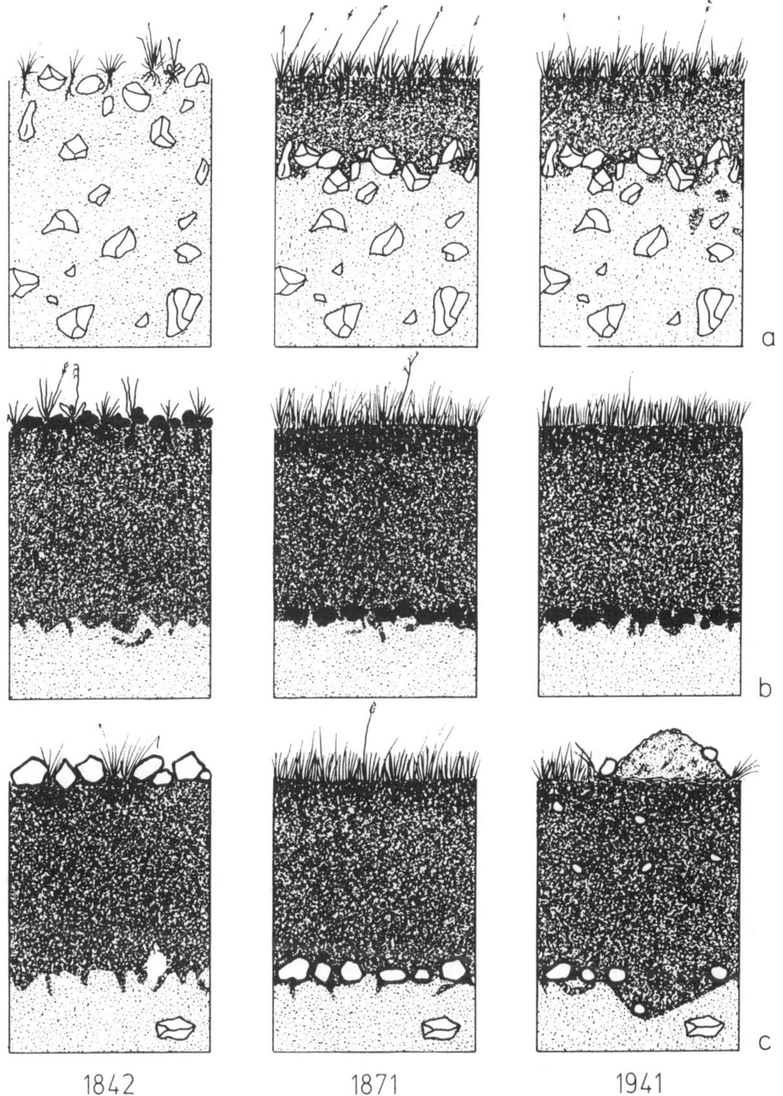

Figure 1.5 Soil profiles illustrating Darwin's (1881) experiment on earthworm bioturbation. (a) When ploughing of a field ceased in 1841, the topsoil contained evenly distributed flint clasts. By 1871, earthworms had advected fine material upwards, and the clasts had sunk to the base of the zone of worm bioturbation, 7.5 cm below the surface. The upper layer was now flint-free loam. Kieth (1942) re-investigated the site and found the fabric unchanged except that **reverse-conveyor** activity, probably by other species of worms, had taken loam down among the flints. (b) By 1871, a layer of cinders placed on the surface of a field in 1842 had accumulated 18 cm down, at the base of the loam. The situation was unchanged in 1941. (c) In a third field, chalk rubble was laid on the surface in 1842. By 1871 the layer had migrated down to 18 cm through earthworm activity. In 1941 a different pattern was found. Moles had entered the field and were advecting chalk clasts to the surface again. Earthworm activity continued to bury these and work them down. As the upper 7.5 cm now contained clasts it was considered that moles must have arrived about 20 years earlier (Kieth 1942). Modified after Jewell (1958).

degree of order of the sediment and tends toward homogeneity. However, some endobenthic activity re-orders or 'unmixes' the sediment by introducing new material and new structures, which may be preservable as discrete trace fossils (section 5.2.2). The term **bioturbation** was introduced by Richter (1952) to cover these processes. Microscale homogenization by meiofauna was called **cryptobioturbation** by Howard and Frey (1975).

1.3 The substrate

The consistency of the sedimentary medium exercises a major control over the structure of the endobenthic community. Substrate consistency determines how animals burrow and why, and thus is a limiting factor on the species present in the community and on the behaviour patterns available to them. Consistency varies spatially, in both lateral and vertical senses. Thus it is a controlling factor in the vertical and areal distribution of species and behaviour patterns within the sediment.

Many mechanical properties of the sediment are implemental in determining its consistency and although they are all intricately interdependent we may consider them separately.

1.3.1 Grainsize

The absolute grainsize and degree of sorting of the substrate are extremely important to the sediment-processing endobenthos. Sands and muds are different worlds for the animals that inhabit them. It is therefore proper that studies on endobenthic species or communities should include granulometric analyses of the burrowing medium. Alexander *et al.* (1993) placed many species of burrowing bivalves onto different seived size-fractions of sediment, and demonstrated several degrees of sensitivity to grainsize in different species.

Nevertheless, grainsize analyses must be treated with caution. The functional grainsize, as experienced by the burrower, may not resemble the physical grainsize. Muds are commonly pelleted and behave as sands. Sands may also be mucus bound and behave quite unlike pure, clean-washed laboratory sands.

Another problem is that functional grainsize must be considered in relation to the absolute size of the burrower. If juveniles and adults of a species share the same sediment, the apparent grainsize of the substrate will be different for each. Their burrowing strategies will have to be modified in order to cope with the 'different' sediments and this will be reflected in the resulting biogenic structures (Fig. 1.6).

1.3.2 Water content

The void ratio and thus water content of a sediment is critical in controlling the strength of the substrate. The initial water content of a deposit

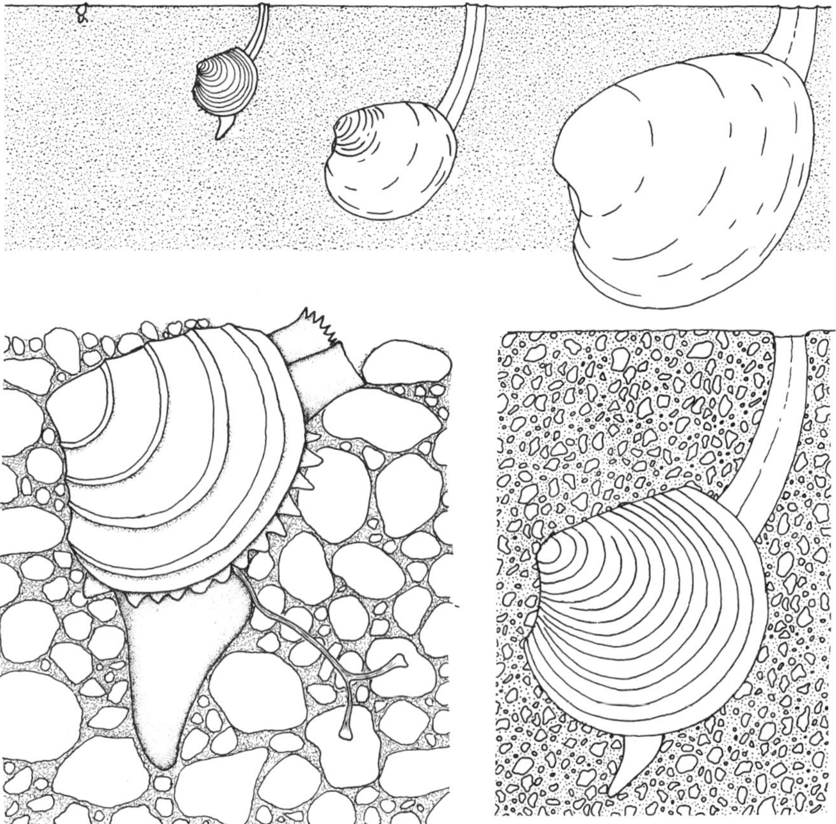

Figure 1.6 Above, four growth stages in *Mercenaria mercenaria*. Below, the first two stages depicted with shell-sizes enlarged to that of the adult phase above. This emphasizes the marked apparent change in substrate grainsize as experienced by the growing bivalve. A different burrowing technique is required in the 'coarse' and 'fine' sediments and a correspondingly different shell ornament is seen. Juvenile *M. mercenaria* may be transported as sand grains (Woodin 1991). Modified after Stanley (1972).

varies considerably, but weight of overburden causes compaction and dewatering within the immediate subsurface so that the vertical profile through the substrate will be zoned with respect to water content.

Mechanical properties of sediment related to water content include **thixotropy** and **dilatancy**. Chapman and Newell (1947) and Chapman (1949) discussed the importance of these properties to burrowing animals. If the strength of the medium (which for the burrower's purpose can be measured as resistance to penetration) decreases with time during the application of force, that sediment exhibits thixotropy. The grain-to-grain microfabric collapses and the sediment becomes fluid. If, on the other hand, the strength of the sediment increases with time as force is applied, then the sediment exhibits dilatancy. As force is applied and distortion of the sediment increases, intergrain friction also increases.

The same sediment may show thixotropic or dilatant properties according to the packing of its grains, its water content, and the mode of application of force. Thus a sediment that behaves thixotropically near the sea floor may show dilatancy a short distance beneath. It will be clear from this that thixotropic properties will favour intrusive and compressive penetration techniques, whereas a dilatant sediment will be conducive to excavation or backfilling.

1.3.3 Shear strength

Because of the complex nature of sediment systems, strength cannot easily be defined as a basic material property; nor is it simple to determine the 'quantity' experimentally. Silva (1974, p. 53) pointed out that 'shear strength parameters of fine-grained sediments are dependent on: (a) void ratio, water content, porosity; (b) electrochemical interparticle forces (i.e. cohesive bonds); (c) constraints on mode of failure and stress systems during experimental tests. It is therefore seen that shear strength of a given sediment is not an invariant property'. Silva did not acknowledge another even more intractable parameter, the mucous binding and organic content of the sediment.

Carney (1981, p. 362) put the problem well: 'the strength of the sediment encountered by the burrower will be determined by the manner in which the animal burrows (i.e. how it places a load on the sediment)'. All in all, a precise statement of shear strength that is ichnologically helpful is not available.

1.3.4 Mucus and the effect of bioturbation

Throughout the above discussion of sediment consistency, a stumbling block has been the variable factor of the organic content. This controlling factor of substrate response to bioturbation has been underplayed or ignored by most workers. Among exceptions to this are Nowell *et al.* (1981), Rhoads and Boyer (1982) and, in freshwater, McCall and Tevesz (1982).

Mucopolysaccharides are a complex and poorly defined group of macromolecules. Different macro- and microorganisms produce a range of chemically different mucus to serve different purposes (e.g. Schäfer 1956, 1962; Pequignat 1970; Myers 1977a). Freely burrowing animals, intruding their way into the topmost sediment, secrete mucus in order to control grains as they pass by the body, thereby reducing the energy expenditure needed for locomotion (Fig. 5.6). Animals establishing burrows in sand drench the surrounding medium with slime in order that the burrow walls may support themselves. Slime is also used by some suspension feeders to trap seston from the irrigation current for food.

These mucus types differ also in their biodegradability and thus in their longevity. Once the grains of a substrate are bridged by such an elastic film the mechanical properties of the sediment are greatly altered.

The argument is a circular one, for it is with the purpose of changing the properties of the substrate that its inhabitants secrete mucoid-binding substances. While the basic condition of the substrate influences which animals can gain access to it as a place to live, the animals that inhabit it in turn influence and modify the properties of the substrate by their life processes (section 5.2). So it is difficult to describe meaningfully the consistency of a bioturbated substrate; standard methods for describing sediments are inadequate for understanding many aspects of organism–sediment relationships (cf. Whitlatch 1981; Jones and Jago 1987).

1.3.5 Terminology of substrate consistency

Because of these problems, only a very simple classification has come into use by trace fossil workers (Ekdale *et al.* 1984a; Ekdale 1985). This may be summarized as follows.

A highly watery sediment is referred to as a **soupground**. This substrate has a fluid consistency; the grains are hardly in contact or are separated by mucoid substances, and animals may 'swim' through the substrate. Nevertheless, mucus- and other organic-walled tubes may be constructed by sedentary endobenthos, in some cases as upward extensions of burrows constructed in more dewatered sediment lower down (Fig. 5.6a).

Perhaps the most widespread aquatic substrates are **softground** and **looseground**, which consist of watery sediment in which the grains are in contact. Softground suggests mud or silty mud, whereas looseground was recently introduced for sands and gravels (Goldring 1995a). These categories include thixotropic sediment and the soft dilatant sediments. The initial establishment of a permanent burrow in softground requires some wall-support mechanism, either through compression alone or together with mucus impregnation. Excavation and backfill burrowing techniques become practicable in looseground and towards the firmer end of the softground range.

Advanced dewatering and compaction produces **firmground**. Such sediments characteristically are exposed at the sea floor after erosion of upper layers to expose stiffer, deeper levels. Relic sediments of this kind are typically reworked by excavation, the compact form of the fabric precluding the use of compressive processes.

These three categories cover increasing consolidation of the substrate, but do not involve cementation. Introduction of cement into pore spaces and the concommitant rigidifying of the substrate creates **hardground**; bioturbational processes cease and are replaced by bioerosional processes.

The suffix -ground implies that the substrate is exposed at the depositional interface. However, within burrowing depth beneath that interface the texture of the substrate may change significantly. Thus, beneath soupground, endobenthos may encounter softground-like and firmground consistencies. In these cases the term is modified as **concealed firmground**, which indicates that the firm substrate is inhabited but is not exposed at the sea floor.

1.4 Tubes and walls

Unconsolidated sediment is difficult to process unless grains are embedded in mucus. (Terrestrial arthropods, unable to secrete mucus, spin threads and webs for the same purpose.) At the interface between a benthic animal and its substrate, mucus is normally secreted, even if the animal is merely passing over the surface. Freely burrowing forms passing through soupground and softground leave trails of mucus to ease their passage (Myers 1977a).

The establishment of a burrow within soup- or softground normally entails soaking the surrounding sediment in mucus (Fig. 5.6). This both supports the wall to prevent collapse, and seals off the permeability to allow effective canalization of the irrigation stream (Schäfer 1956).

In more permanent burrows, inhabited longer, the wall may receive more attention (section 8.3.2). Extra mucus may be applied, and this may

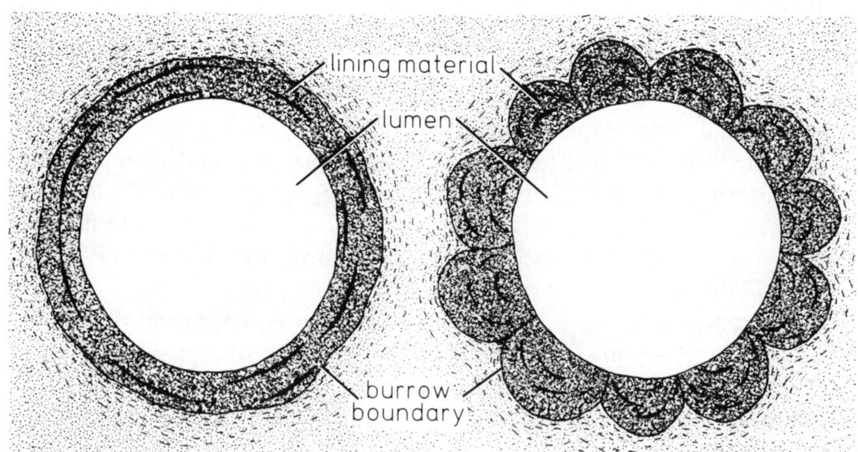

Figure 1.7 The two components of a burrow wall. A lining of clayey material applied by the burrower to the inside of the wall, concentrically laminated (left) and pelleted (right); and a compactional deformation of the substrate immediately external to the burrow boundary.

passively trap fine-grained sediment from the irrigation current and become dusty. The animal may actively incorporate food remains and detritus into the wall; mud pellets, sand grains, skeletal fragments, sea-grass leaves, etc., may be collected, shaped if necessary, and built like brickwork into the wall construction (e.g. Fager 1964; Anderson and Meadows 1978).

The **wall** thus comprises two components; the compacted fabric outside the original **burrow boundary**, and the actively built-up (or passively accumulated) **lining** inside (Fig. 1.7). A burrow having a strongly constructed organic or particulate lining is commonly referred to as a **tube** (Hertweck 1972). A tube may be extended above the floor as a **chimney** (section 3.7).

1.5 Physical induction of flow in burrows

Physical effects are exploited by burrowing animals to induce the flow of fluid (air or water) through their burrows with a minimum expenditure of metabolic energy (Vogel 1978). According to Bernoulli's Principle, the pressure of a steadily moving fluid must decrease whenever its velocity increases so that its total energy remains constant. Thus, a flow will be induced in a burrow having unequal aperture sizes; the larger the aperture the greater the 'viscous entrainment' caused by the passing current (Fig. 1.8a). Raising one aperture as a chimney will increase the speed of current over it and thereby increase viscous entrainment (Fig. 1.8b). If one end of a U-tube is raised on a mound the velocity of the bottom current will also increase as it passes over it and its pressure is accordingly

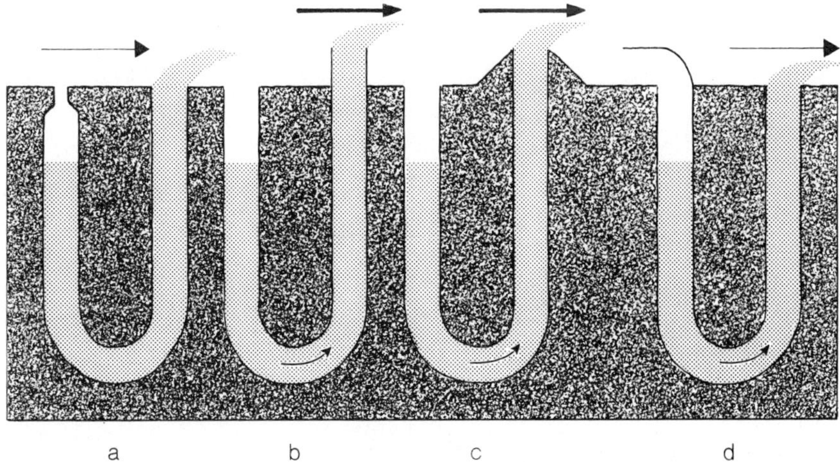

Figure 1.8 Four means of inducing flow through a burrow in a current-swept sea floor. Thickness of arrow indicates current velocity. Inspired by Vogel (1978).

lower (Fig. 1.8c). In all three examples, the direction of the bottom current is not important and can be variable. A fourth type of flow induction may be created with a horizontal tube directed into the static pressure of the oncoming current (Fig. 1.8d). This system of course requires precise alignment with bottom flow.

1.6 Animal–sediment ecology

Many published autecological studies carry little message for the ichnologist or sedimentologist. In some cases, studies have been undertaken by geologists in order to emphasize the bio-sedimentological interactions. However, large gaps exist in our understanding of animal–sediment relationships. The few species that have been studied in detail mostly derive from easily accessible habitats. We know virtually nothing about spreite producers, although their structures dominate much of the fossil record. Moreover, certain groups have been neglected that are small relics today of what must have been powerful bioturbating forces in the geological past. Such groups include the hemichordates, pogonophores and priapulids, and, in particular, the aplacophoran molluscs, of which no species has received ichnological treatment.

Making do with what we have, however, the following chapters treat a number of animals that illustrate most of the known sediment-processing activities that are involved in bioturbation.

2 Sediment stirrers

2.1 Interstitial meiofauna and microfauna

Wherever the sea floor is sufficiently stable and oxygenated for endobenthic colonization, the uppermost few centimetres of sediment team with interstitial microorganisms (e.g. Swedmark 1964; Fenchel 1969; Reise and Ax 1979; Gooday *et al.* 1992; Meadows *et al.* 1994). In sands, these organisms are mostly smaller than the grains they live among, and include ostracodes, nematodes, copepods, turbellarians, archiannelids and juvenile macrobenthos such as polychaetes and molluscs (Fig. 2.1).

The constant activity of these organisms, many of which graze on organic coatings on grains, causes continual dislodgement of particles. Reichelt (1991) observed nematodes establish thread-like mucus-supported burrow systems among grains, which persisted several hours until destroyed by other organisms. Ott (1993) described a chemosymbiont nematode, covered with sulphide-oxidizing bacteria, that travelled constantly between the oxic surface sediment and the anoxic below.

However, a few members of the **meiofauna** such as tanaid crustaceans, produce larger, more long-lasting burrows. Many radiolaria and foraminifera also enter the upper few millimetres of marine muds (Fig. 2.2; Lipps 1983; Gooday 1986; Miller 1988; Gooday *et al.* 1992; Meadows *et al.* 1994) and modify the sediment texture. Agglutinating foraminifera and the lesser-known deep-sea groups Komokiidae and Xenophyophoria manipulate sediment grains and construct their skeletons of them (Tendal 1972; Tendal and Hessler 1977).

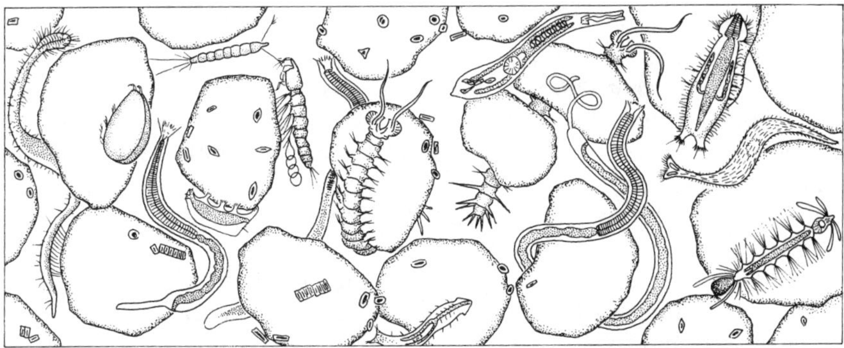

Figure 2.1 A selection of meiofaunal organisms occupying the intergranular space within a marine sand. Diatoms adhere to the grains. Data from Thorson (1968).

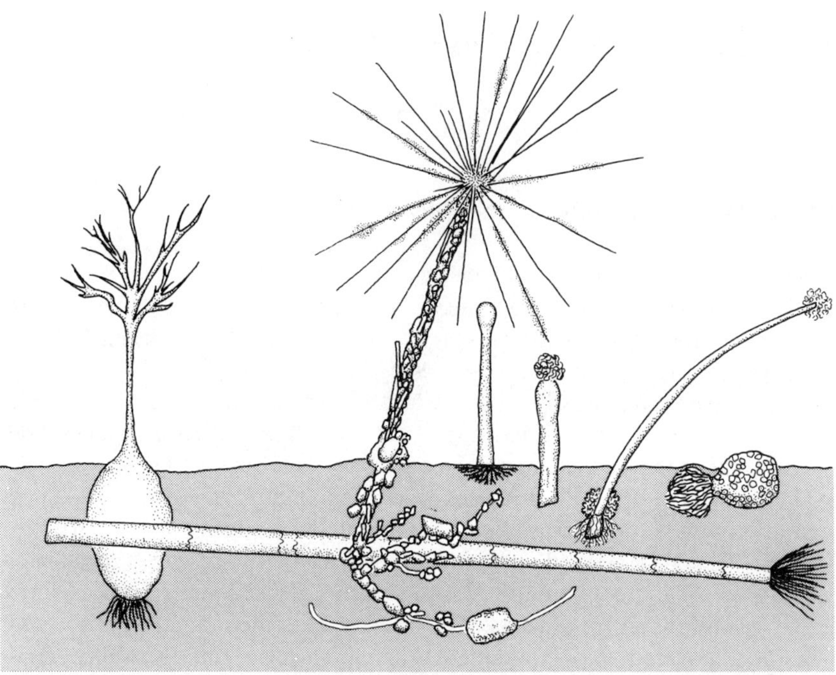

Figure 2.2 Life positions of some benthic foraminifera that disturb muddy sediment, about x2.5. Left to right: *Pelosina arborescens*, *Marsipella arenaria*, *Hyperammina* sp., *Pilulina* sp., *Bathysiphon* sp. and *Saccammina sphaerica*. Lying within the mud, *Bathysiphon filiformis*. Modified after Christensen (1971).

Cullen (1973) described the effect of meiofaunal activity upon **exogenic** traces. In an aquarium containing sea-floor sediments that had been passed through a 1 mm sieve, a hermit crab was allowed to produce its characteristic trackway of footprints before being removed from the tank. Over the following few days the quality of the trackway gradually deteriorated and the structure ultimately disappeared, obliterated by the work of the meiofauna.

The homogenizing effect of the meiofauna is greatest at the depositional interface and diminishes downwards through the uppermost millimetres to perhaps 1.5 cm. Below the redox barrier (**RPD**), the oxidized walls of deeper burrows support meiofaunal communities (Reise 1981; section 5.1.2). Ronan (1977) found that meiofaunal activity obliterated burrows in aquaria within a period of months.

The degrading work of the interstitial fauna is demonstrated by photographs of the deep-sea floor. Deep-sea sediments are generally totally bioturbated; yet, even in hadal depths (Lemche *et al.* 1976), the sea floor is not totally covered with elegant traces. On the contrary, these tend to be nicely spaced, separated by evenly micro-hummocky surfaces. This is because the exogenic traces are constantly being destroyed by meiofaunal activity. In some sea-floor photographs in which trace-making animals are seen leaving fresh trails, the older parts of these traces have already become degraded within the limits of the pictures (Heezen and Hollister 1971, figs 4.20 and 5.1; Hollister *et al.* 1975, fig. 21.18).

Time-lapse photography of the deep-ocean floor reveals gradual obliteration of surface tracks through meiofaunal activity over periods of weeks (Paul 1977). Cullen's (1973) hermit-crab track disappeared in a matter of days; a comparable anomuran crab track at nearly –5000 m on the Pacific Ocean floor took months to disappear (Paul *et al.* 1978).

It should be realized, of course, that anaerobic microbes continue below the base of the oxic zone. They have been reported active to depths of hundreds of metres subsurface (Parkes *et al.* 1993).

2.2 Haustoriid amphipods

There is no abrupt break in size between meiofauna and macrofauna, owing to the presence of growing juveniles. In the centimetre class, too large to be considered true meiofauna, amphipods are important sediment processers in a wide range of aquatic environments from the non-marine to the abyssal.

Nicolaisen and Kanneworff (1969) studied two brackish-water species, *Bathyporeia pilosa* and *B. sarsi.* These amphipods work at slightly different depths within sand, to 5 and 7 cm down, respectively, and move through it more or less continually. Oriented dorsal-side down, the

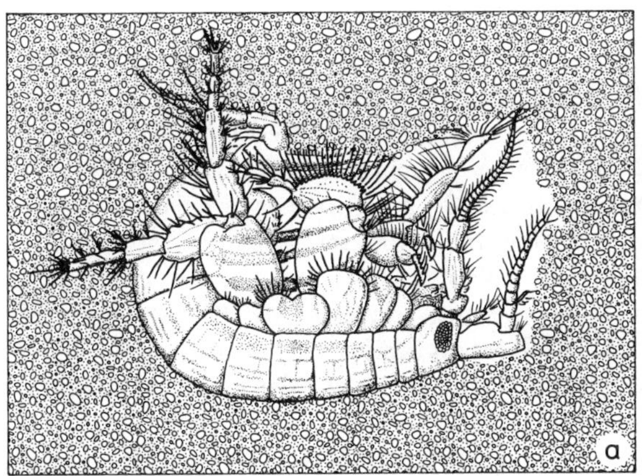

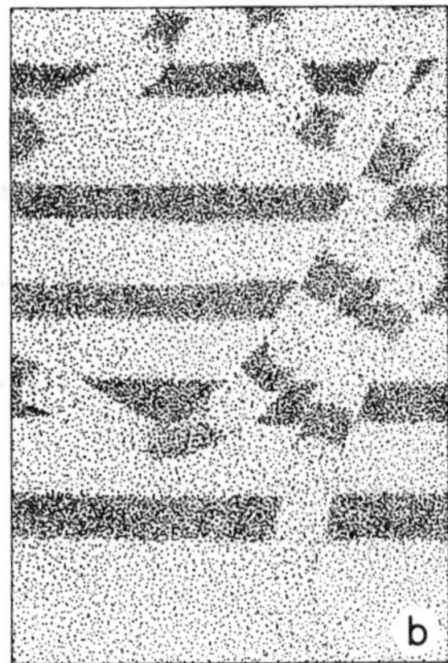

Figure 2.3 (a) The amphipod *Bathyporeia sarsi* manipulating sand in its burrow. The roof and sides are supported by differently specialized limbs. (b) Backfill structures produced in laminated sand by *Haustorius* sp., as seen in a radiograph. Modified after Nicolaisen and Kanneworff (1969) and Howard and Elders (1970).

animals maintain above them a small open space, the burrow, in which to process grains. Little or no mucus seems to be employed, the burrow being supported by limbs (Fig. 2.3a). The remaining limbs manipulate individual grains, passing them from front to rear of the burrow and removing organic coatings for ingestion. In this way, the animal, together with its burrow, migrate forward through the substrate.

Other species, e.g. *Haustorius arenarius*, *Urothoe marina* and species of *Bathyporeia*, burrow dorsal-side upward (Dennell 1933; Watkin 1939, 1940), but the feeding activity is similar.

Howard and Elders (1970) studied the work of seven species of haustoriids in artificially laminated sediment; minute meniscate backfills were produced (Fig. 2.3b), but continued reworking of the sediment ultimately obliterated lamination and produced an apparently homogeneous, cryptobioturbated sediment. The seven species not only showed slightly different sediment depth preferences (section 5.4.2), but also produced slightly different sediment structures.

This is not the only form of activity of sand-burrowing amphipods. Hertweck (1972) described open vertical shafts nearly 20 cm deep constructed by the talitrid, *Talitrus saltator*, in the Mediterranean backshore. The talitrid sand-hopper *Talorchestia deshayesii*, although an epibenthic algal feeder, burrows into sand beaches in a similar way (Reid 1938). In deeper water, *Haploops tubicola* constructs a tough vertical tube (Fig. 5.13d). Corophiid amphipods are treated later (section 3.4.1).

2.3 Intruders in soft substrate

Many animals move freely and more or less continuously within softgrounds. Predatory snails offer good examples of such activity, and naticids in particular have been well studied.

Trueman (1968c) described the burrowing process of three naticid species in sand. The shell of these snails is largely covered by soft parts and the animal employs the **double-anchor technique** of substrate penetration. The propodium, which extends anteriorly and covers the head, is expanded as a **terminal anchor**, gripping the sediment while the posterior is drawn up towards it (Fig. 2.4a). Blood is then transferred posteriorly from the propodium, which consequently deflates, to the mesopodium, expanding this to form the **penetration anchor**. This anchorage point allows the now thin propodium to be intruded into the sediment ahead.

During this process, much mucus enters the surrounding sediment. No true burrow is produced, the respiration chamber being the mantle cavity within the shell, and the sediment closes again behind the animal. A chaotic biodeformational structure is created in the substrate by the compression, entrainment in mucus, transport past the animal and redeposition by eddy diffusion behind. The disturbance structure was shown in

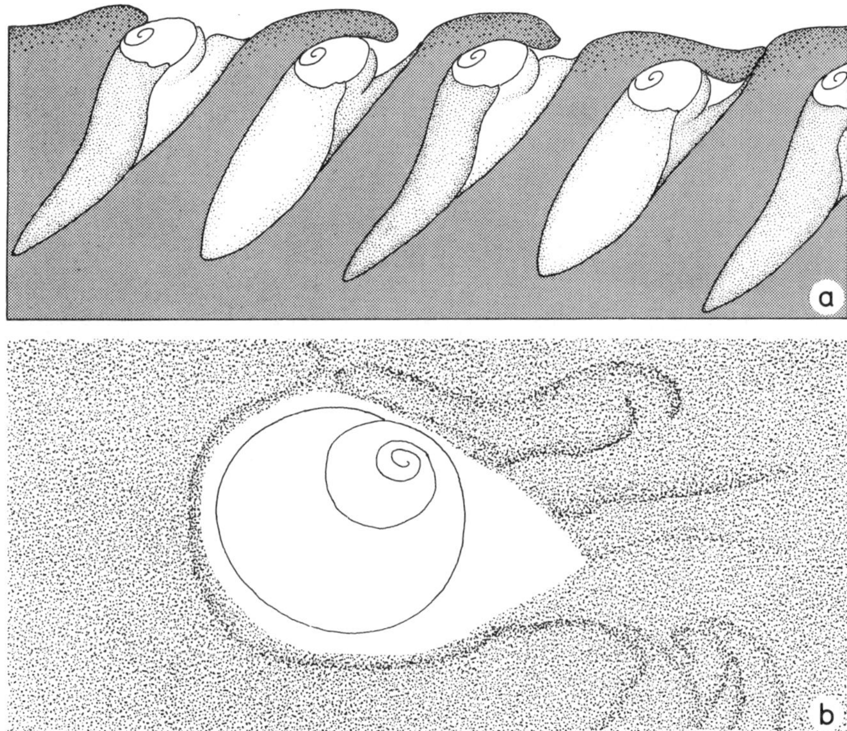

Figure 2.4 Naticid snails. (a) *Polinices josephinus* entering the substrate using the double-anchor technique. Inflated parts pale. (b) Biodeformational structures produced as *P. duplicatus* intrudes its way through the substrate. Inspired by Trueman (1968c) and a radiograph by Frey and Howard (1972). See also Fig. 6.3.

radiographs of naticid and other snails by Howard (1968) and Frey and Howard (1972) (Figs 2.4b and 6.3). Schäfer (1956) compared the trace of the naticid *Lunatia nitida* to the effect of an inert ball moving through the substrate; a disturbed zone was formed, at the boundary of which sedimentary laminae are deformed.

2.4 Swimming through the substrate

The cephalochordates, or lancelets, are extraordinarily efficient endobenthic wanderers that have developed an unusual mobility within their sedimentary medium. Species of *Branchiostoma* modify their behaviour according to the consistency of the substrate (Fig. 2.5). Preference is shown for well-sorted sand. By artificially increasing the sharpness of the grains by fracturing them, or removing organic coatings from the grains,

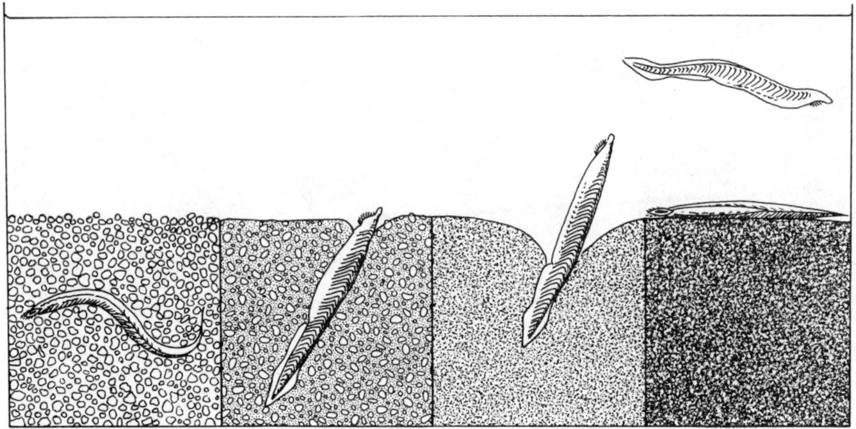

Figure 2.5 Behaviour of lancelets in substrates of different consistency. From left to right: in permeable, clean sand the animals live within the sediment; in fine sand, the mouth must be exposed to take in water, but this can be pumped out of the pharyngeal clefts into the sediment; in muddy sand, the whole pharynx has to be free of the substrate; lancelets cannot enter mud. Modified after Webb and Hill (1958).

sand is rendered unsuitable for lancelet colonization (Webb and Hill 1958).

Schäfer (1956, 1962) confirmed that *B. lanceolatum* passed through the sediment using movements similar to those for swimming in the water above. Whereas the body of the animal is slime-coated, no mucus is imparted to the sediment, which remains loose and collapses behind; eddy diffusion produces minor deformation structures. Hagmeier and Hinrichs (1931) found that in aquaria, a dense population of lancelets created well-ordered inverted graded bedding in a previously homogeneous sediment after six months' occupation. The effect was noticeable even after a week's activity.

Ricketts and Calvin (1962) provided a lively account of the activity of lancelets. 'It is hard to believe that any animal can burrow as quickly ... They hurl themselves head first into the sand ... Seemingly they can burrow through packed sand as rapidly as most fish can swim.'

3 The work of worms (mostly)

Many authors subdivide fossorial animals into two groups: those having a hard skeleton at or near the surface, and those that are entirely or functionally soft (e.g. Trueman and Ansell 1969; Elders 1975). The 'soft' burrowers include many elongated forms which, although it annoys my zoologist friends, I call **worms**.

3.1 Two worms in soft mud

3.1.1 A priapulid worm

Priapulida is a phylum of unsegmented worms that contains only six living genera, and which has a poor body fossil record. Probably having Precambrian origins, today's representatives are likely to be the remnants of a large phylum that was important in the Palaeozoic, having been out-competed since by more efficient burrowers (Van der Land 1970). Five or more burrowing species occurred together in the Middle Cambrian Burgess Shale (Conway Morris 1977).

Priapulids have an anterior introvert or praesoma armed with rows of spines; a main trunk; and a caudal appendage. Burrowing behaviour has been studied in only a single species, *Priapulus caudatus*, although movements observed in a second species, *Halicryptus spinulosus*, suggest similar behaviour (Friedrich and Langeloh 1936). These and four other species are mud-dwelling carnivores (Van der Land 1970).

According to Schäfer (1962), Hammond (1970) and Elder and Hunter (1980), *P. caudatus* progresses slowly through its watery mud substrate using a double-anchor technique (Fig. 3.1).

Owing to the fluidity of the substrate, the technique is near the limits of its usefulness. Thus, as the praesoma is everted and intrudes forward into the mud ahead, the penetration anchor may slip backwards. However, the trunk is held in a curved pose so as to increase friction. The praesoma is then inflated as a terminal anchor, and the contracting trunk is drawn forward on that anchorage. As the praesoma deflates, a peristaltic wave passing forwards along the trunk eases this into the space vacated by the praesoma, and a penetration anchor redevelops at the rear. This burrowing cycle takes a minute and advances the animal 25 per cent of its length.

No burrow is produced, according to these authors, the sediment closing behind the animal (Fig. 3.1f). Other observers, however, have repor-

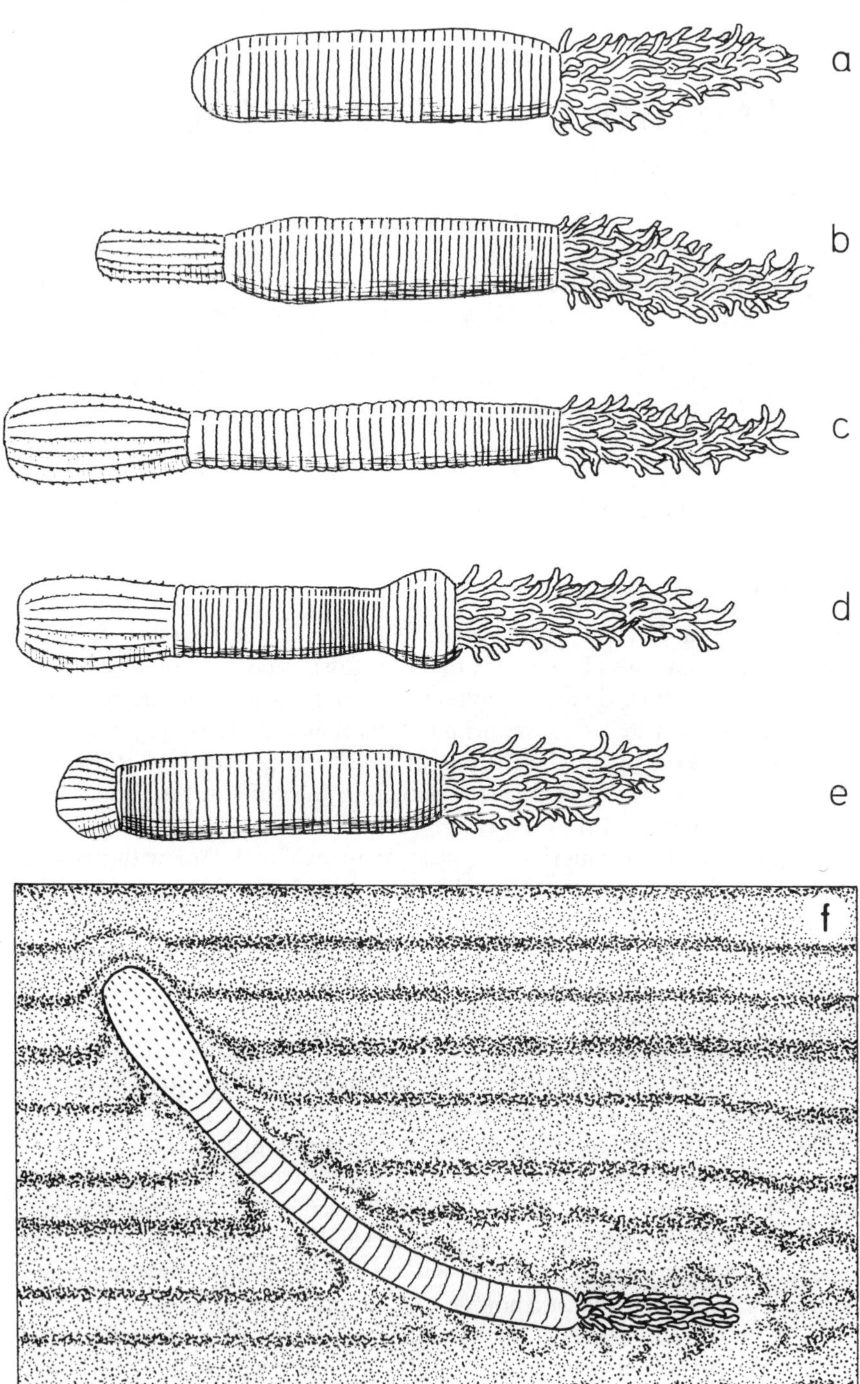

Figure 3.1 *Priapulus caudatus*: (a–e) successive forms taken during the burrowing cycle; (f) biodeformation by *P. caudatus*; the curved pose increases purchase on the penetration anchor. Modified after Schäfer (1962), Hammond (1970) and Elder and Hunter (1980).

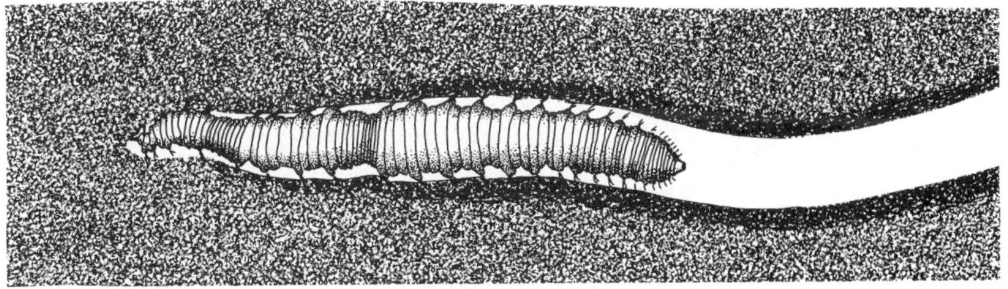

Figure 3.2 *Polyphysia crassa* constructing a burrow in mud. Three peristaltic contractile waves are seen moving forwards. Projection of the fans of setae backwards in the dilated parts of the trunk provide an effective penetration anchor. Modified after Elder (1973).

ted the existence of a burrow, which is irrigated by peristalsis (e.g. Mettam 1969).

3.1.2 A carnivorous polychaete worm

A different approach to semifluid penetration is shown by the carnivorous scalibregmiatid polychaete *Polyphysia crassa.* Elder (1973) revealed that, instead of employing a double-anchor technique like so many worms, this species scrapes the flocculated substrate ahead of it out to the sides, using its prostomium (Fig. 3.2). No digging cycle is observed; the head end of the animal progresses in a continuous, non-cyclical way into the substrate while the rest of the body advances by means of forward peristaltic waves.

An excavation technique of this type is unexpected in soupground, although common in sediments of higher shear strength. Nevertheless, after the prostomium of *P. crassa* has liberated grains or flocculae from their bonding, the parapodia then push the mud laterally and posteriorly, and an open burrow is produced (Fig. 5.13d). Thrusting movements of the tail dilate the burrow and consolidate the walls with mucus (Elder 1973).

Fauchald (1974) suggested that the superficial detritus soupground was the first substrate occupied by the primitive Precambrian polychaetes before improved burrowing efficiency permitted colonization of firmer substrates. Nevertheless, while priapulids may be primitive burrowers, the majority of occupants of this type of watery substrate today are probably advanced burrowers that have modified their efficient burrowing techniques to cope with the semifluid environment.

3.2 Sea anemones and other cnidarians

3.2.1 Actinaria

The sea anemones include a number of species that are able to establish themselves in soft sediments. Thanks to studies by Ansell and Trueman

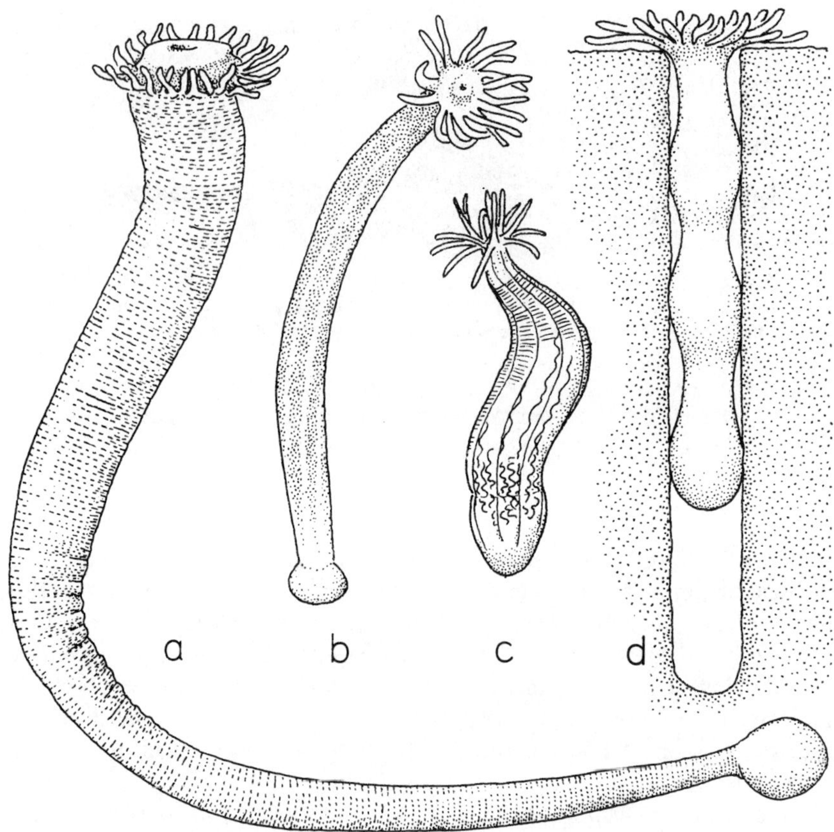

Figure 3.3 Fossorial actinian anemones. (a) The large *Harenactis attenuata*, about half natural size and (b) *Edwardsiella californica* twice natural size, showing the inflated physal bulb. (c) *Edwardsia tricolor* natural size. (d) An actinian in its simple mucus-supported burrow irrigated by peristaltic movements. Modified after Ricketts and Calvin (1962), Morton and Miller (1968) and Sassaman and Mangum (1972).

(1968) on a form having a physa, and Mangum (1970) on one having a pedal disc (the two types of distal end possessed by these anemones), the burrowing activity of these relatively simple animals is well understood.

Anemones burrow using peristaltic waves of contraction of circular muscles, passing from tentacle crown to base. The basal disc is modified as a knob-like physa that can be inflated as a terminal anchor or made pointed for intrusion into the substrate (Fig. 3.3). Using the double-anchor technique, the anemones slowly enter the uppermost sediment. The burrow is minimal, being supported by mucus and the body of the inhabitant; it is little larger than the anemone and is irrigated by peristaltic waves (Fig. 3.3d).

Some endobenthic anemones are by no means feeble burrowers. When irritated, the animal withdraws into its tube and inflates its physa strongly. Ricketts and Calvin (1962, p. 240) described how difficult it is to extract the large *Harenactis attenuata* from the substrate in this contracted condition, its physa gripping tenaciously some 45 cm below the sediment surface (Fig. 3.3a).

3.2.2 Ceriantharia

This small order of anthozoans may have greater significance for ichnology than the actinarians, because they build semi-permanent, walled burrows. Endobenthic cerianthids appear to have two contrasting lifestyles. Most species construct a single vertical shaft and are microcarnivores feeding on passing prey. Some species, on the other hand, construct long, horizontal, branched galleries several centimetres down (Picton and Manuel 1985; Jensen 1992). Let us begin with the shafts.

The burrow is lined with mucus in which are embedded innumerable expended cnidae. These cnidae are of a type uniquely used for tube building (Mariscal *et al.* 1977), their long and matted threads forming a strong walling material. Removed from its burrow, a naked anemone immediately produces a new sheath of cnidal threads around itself as it slowly re-enters the substrate.

In an established burrow, other foreign matter may be incorporated into the tube and the lined burrow is extended downwards some distance below the animal (Fig. 3.4b). If danger threatens from above, the burrower retreats to the bottom of its shaft. Individuals of *Cerianthus membranaceus*, when disturbed in their burrows in the sea floor, are seen to take hold of the aperture of the pliable tube with their tentacles as they retreat and to pull this closed behind them, leaving no visible evidence of their location.

The burrow is permanent in that the animal does not voluntarily leave it or move it. In aquaria, individuals may live for several years without shifting their burrow if undisturbed. However, I have watched a young *C. membranaceus* in an aquarium jump out of its tube when prodded from below by an advancing burrowing *Echinocardium mediterraneum*.

Schäfer (1962) described sediment processing by *C. lloydii* in the North Sea. This species constructs a burrow nearly 1 cm in diameter, and somewhat longer than its body. The structure is lined with cnidae, mucus and sand grains. During periods of gradual accretion of the sea floor, the animal maintains equilibrium with the sediment surface, extending the burrow upwards in pace with deposition (Fig. 3.4c). In this way, long **equilibrium structures** can arise, seemingly quite out of proportion to the size of the animal (section 9.2.8). Remane (1940) recorded such a *Cerianthus* staircase 1 m high.

If an increase in depositional rate should bury the animal, a new behavioural response is seen. Tube building activity ceases; the animal bunches its tentacles and intrudes its way up to the new surface, where it

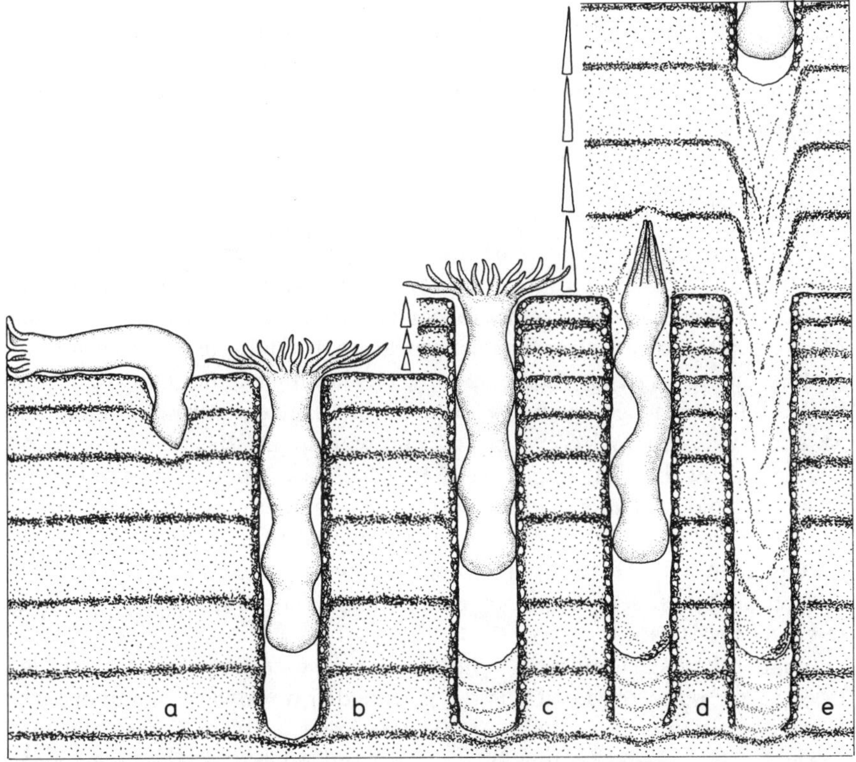

Figure 3.4 Behaviour of *Cerianthus lloydii* under different conditions. (a) Initial entry into the substrate after exhumation: protrusive compression. (b) Established in a **domichnion**, lined with cnidae and sand grains. (c) Retrusive adjustment in response to small increments of sediment, producing an **equilibrichnion** (section 9.2.8). (d–e) Burial by more rapid deposition releases retrusive escape reactions and produces a **fugichnion** (section 9.2.9) up to the new sea floor, where a domichnion is re-instated. Modified after Schäfer (1962).

establishes a new burrow (Fig. 3.4d and e). This **escape trace** should not be confused with the equilibrium structure, which is merely a modification of the normal tube (section 9.2.9).

Frey (1970) found the vertical tubes of *Ceriantheopsis americanus* to measure about 1.5 cm wide by up to 35 cm deep. Minor side branches occurred that were difficult to explain. Possibly they arose during repair and re-establishment of the tube after erosional damage, or represent burrows of asexually budded-off individuals.

Giant species produce impressive burrows. In California, the intertidal *C. aestuari* was described by Ricketts and Calvin (1962) as living 'in a black, parchment-like tube that is covered with muck and lined with slime. More energetic diggers than the writers have taken specimens with tubes six feet long'. The great size of burrow and special walling material mentioned here may indicate the *C. aestuari* belongs to the other group of cerianthids.

Three cerianthids have been described as inhabiting extensive, deeply emplaced, branching burrows, *Cerianthus* sp. in shallow water (Ricketts and Calvin 1962) and two deep-water species: *Arachnanthus sarsi* (Picton and Manuel 1985) and *C. vogti* (Jensen 1992).

C. vogti is the best known. Between 1244 and 2926 m in the Norwegian Sea, the mud bottom contains dense populations of the cerianthid. The animal lives in an extensive, branched, horizontal tube system, probably several metres long, at a depth of about 10 cm, well below the RPD. The tube is walled in two layers. The outer surface consists of the surrounding sediment entangled in a network of a loose, light green felt-like mat of discharged cnidae. Inside this the tube is smooth, leathery and black. The outer diameter of the tube is 2 cm.

Natural seeps of methane and hydrogen sulphide are present in the area (Hovland and Thomsen 1989) and the cerianthid populations are particularly dense in the vicinity of these seeps, together with abundant pogonophores. This led Jensen (1992) to suggest that the burrows acted as 'gas pipelines' in which CH_4 and H_2S might accumulate. These gases could supply sulphide- and methane-oxidizing symbionts in a bacterial garden in the thick walls of the burrow (section 1.1.8).

3.2.3 Sea pens

The pennatulaceans are an order of octocorals that in typical development have a feather-like rachis bearing numerous colonial individuals. The stem-like base is firmly anchored in the sediment and therefore is capable of producing biogenic structures (Fig. 5.13c). Little is known, however, of the activity of these colonies, although some species are fairly mobile. The chief function of the stalk is presumably anchorage. In the American Pacific intertidal species *Stylatula elongata*, the whole colony withdraws into its burrow at low tide. When extended at high tide, it can snap down into the sea floor when disturbed (Rickets and Calvin 1962).

B. Bett (pers. comm. 1994) has observed the deep-sea species *Pennatula aculeata*, when 'nudged' by the camera-frame, to withdraw entirely into the substrate, only to re-emerge again later, apparently none-the-worse for the experience.

It is also known that pennatulids rotate freely in order to 'face the current'. Indeed, the orientation of pennatulids is a useful indicator of current direction in sea-floor photographs (Ohta 1984).

Some authors have suggested pennatulaceans as originators of the trace fossil *Zoophycos*, the stem of the colony serving some deposit-feeding function. Indeed, Bradley (1973, 1980, 1981) extended this suggestion to a whole range of ichnogenera. However, until we understand the behaviour and supposed sediment processing of the pennatulacean stalk more fully, such suggestions remain little better than speculation.

3.3 U-burrows for suspension feeders

Of the numerous animals that follow this lifestyle I choose two contrasting examples as representative: *Chaetopterus variopedatus* and echiuran worms. Polychaete and echiuran worms are not closely related, but their lifestyles show remarkable similarities. Both are well known and have reached the textbooks (e.g. Barnes 1980), but they illustrate ichnological principles so well that I will repeat their stories here.

3.3.1 The chaetopterid worm

Chaetopterus variopedatus was originally described from the Adriatic Sea and it is doubtful if the American material described under this name (Enders 1908; MacGinitie 1939) can be identified with this species (Mary

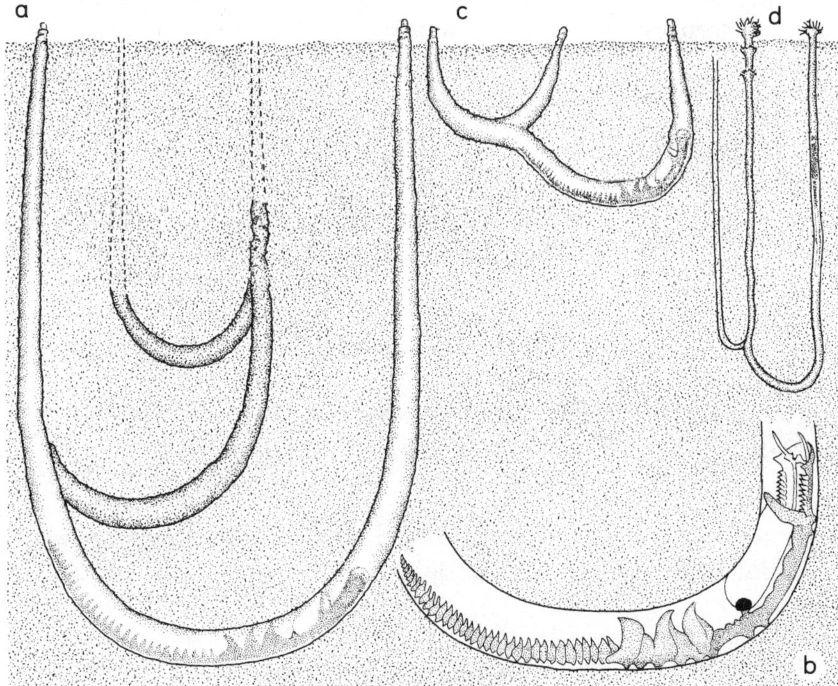

Figure 3.5 (a) *Chaetopterus variopedatus* in its U-tube, with remains of earlier growth stages. From observations in shallow water at Kefallinia, Greece. (b) *C. variopedatus* in feeding position within its tube. The seston net is set and is being continuously rolled up as a bolus (black). (c) *Chaetopterus* sp. in a similar U-tube, showing a similar mode of enlargement. (d) The U-tube of the polychaete *Lanice conchilega*. This tube is extended in a similar 'W' manner. Modified after Enders (1908), Seilacher (1951) and Barnes (1980).

Petersen, pers. comm. 1988), although their habits are apparently similar. Occupying subtidal sandy substrates, *C. variopedatus* constructs a U-tube lined with a strong, parchment-like secretion that incorporates grains of the surrounding sediment. The tube of the adult worm can be over 2 cm in diameter at the base of the U, but tapers up the vertical legs, and narrows further at the apertures, which are raised on short chimneys. The tube in large, subtidal individuals may exceed 1.2 m in length (Fig. 3.5a). Tubes of *Chaetopterus* sp. described by Enders (1908) from the intertidal zone were half that length.

Enders (1908) watched newly metamorphosed larvae construct their first tubes, less than 1 mm in diameter and about 2 cm long. These small tubes are enlarged by the growing worm by cutting a slit in the tube and excavating a wider U-burrow as an extension of the previous one (Fig. 3.5c). The excavated sand is expelled from the opposite end of the first tube. The burrow wall is strengthened and added to during the period of occupancy.

When the new burrow is complete, the old limb is sealed off by a parchment wall. Such abandoned limbs are destroyed by the activities of other endobionts, so that the complete growth history of the burrower is not normally preserved (Fig. 3.5a). This manner of U-tube extension is seen in other tube worms (Figs. 3.5d and 3.6).

C. variopedatus is a suspension feeder, a highly specialized, obligatory one. Behind the head, two large notopodia curve up like horns against the burrow wall. From these, a mucous net is spun that extends poster-

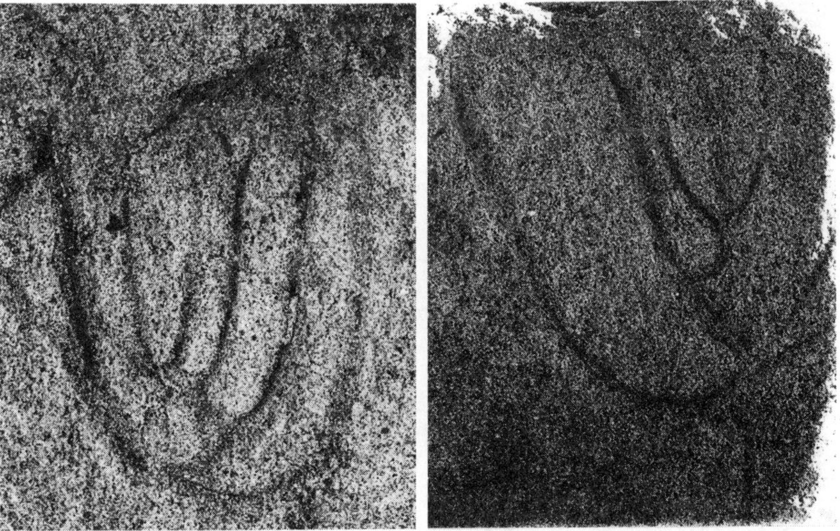

Figure 3.6 U-tubes from the Middle Jurassic Vardekløft Formation of Jameson Land, East Greenland, about half natural size, showing a similar enlargement pattern to that of *Chaetopterus variopedatus* (Fig. 3.5). Droser *et al.* (1994) illustrated similar trace fossils in Cambro–Ordovician sandstone. See also Figs 3.8, 3.14, 4.29, 4.33 and 5.4.

iorly to a food cup (Fig. 3.5b). The irrigation current induced by three fan parapodia further back along the worm has to pass through the net, which catches all suspended material. As it clogs with seston, the net is rolled into a food-ball at the cup while further net material continues to be secreted by the notopodia. When the ball is sufficiently large, net secretion ceases and the bolus is passed forward to the mouth for ingestion (Barnes 1980).

Other genera of the family construct single-aperture vertical tubes, although the feeding style is similar according to Barnes (1964, 1965). Some species of *Mesochaetopterus* extend their tubes over 1 m into the substrate (MacGinitie and MacGinitie 1949).

3.3.2 The fat innkeeper

In spite of their similar habits and burrow shape, it would be difficult to envisage two worms that looked less alike than *Urechis caupo* and *C. variopedatus*. The plump and smooth *U. caupo* is called innkeeper because of the commensal guests that share its burrow. But such hospitality is not unusual and is seen in burrowers of many kinds (section 5.1).

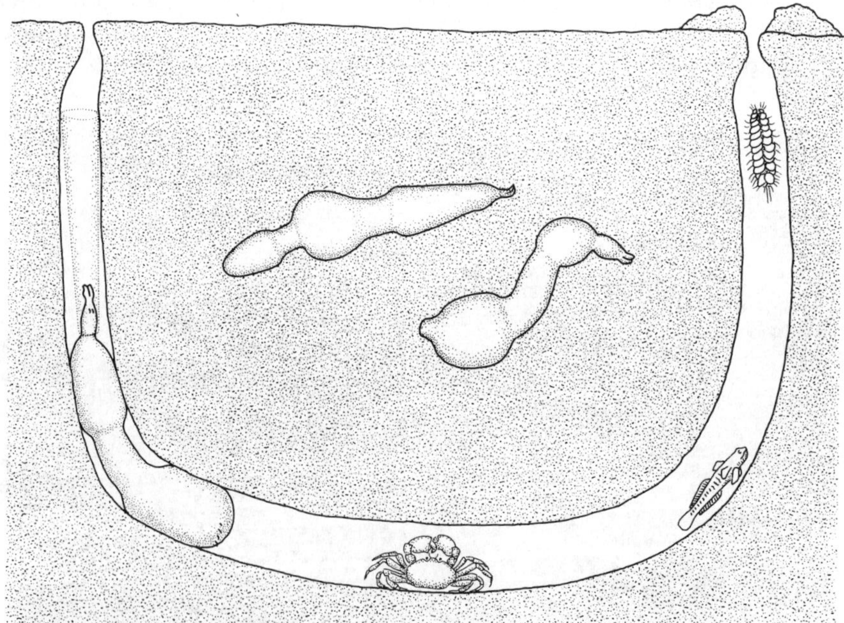

Figure 3.7 The echiuran worm *Urechis caupo* in its U-burrow, passing rings of water peristaltically past its body and thereby filling the slime net with seston. Three commensals that share the burrow are indicated. Sketched between the limbs are two individuals of *U. caupo* showing peristaltic movements. Modified after Fischer and MacGinitie (1928) and Ricketts and Calvin (1962).

Fischer and MacGinitie (1928) described laboratory experiments in which *U. caupo* individuals were observed in glass tubing 'burrows'. When feeding, the animal secretes a conical net of mucus across the lumen of the burrow through which the irrigation current must pass. Peristaltic movement of the worm creates the current, and the delicate net, originally invisible, eventually becomes clogged with seston. Particles as small as 40 Å are trapped (MacGinitie 1945). Then, as in *Chaetopterus*, the net and its contents are eaten.

U. caupo advances through the substrate by intruding the very short proboscis first, loosening the sediment, then crawling forward to produce a terminal anchor. Incidentally, these peristaltic movements are far more efficient in these non-septate worms than in the annelids (Mettam 1969; Schembri and Jaccarini 1977).

A permanent burrow is produced, a large, unlined U having vertical shafts and a horizontal middle section (Fig. 3.7).

Enlargement of the burrow is achieved by excavation, using two setae at the anterior end to quarry, and a ring of tail setae pushing the spoils backwards out of the burrow. *Urechis* has an enlarged cloaca for respiration, and it uses this to blow material out of the burrow. Narrowed burrow apertures increase the velocity of the current as it leaves, thereby improving cleaning efficiency.

Urechis is a very unusual echiuran. Fisher (1946, p. 265) considered that its four or five species, spread around the world, appear to represent 'a very ancient stock, one that may have flowered into many species during Paleozoic times. It belongs to the honorable company of *Lingula* and those other aristocrats sometimes referred to as "living fossils"'.

3.3.3 A less unusual echiuran

One other species of echiuran has been studied in detail, *Echiurus echiurus*, but not while living in glass tubing. *E. echiurus* excavates in more or less the same way as *U. caupo*, except that its larger proboscis is folded back out of the way and the anterior and posterior bristles alone are used. The burrow walls are lined with mucus and compressed by the peristaltic double-anchor movements of the fat worm (Gislén 1940; Schäfer 1962). The burrows of *E. echiurus* in fine muddy sand are U-shaped, as much as 30–50 cm deep, and narrowed at the apertures (Fig. 3.8).

Enlargement follows the pattern shown by *Chaetopterus*, a new leg being excavated in connection with the old U, the abandoned leg filling with sediment (Fig. 3.8a). In boxcores, Reineck *et al.* (1967) demonstrated spreiten associated with *E. echiurus* burrows, both protrusive and retrusive (section 8.4). This is chiefly the case in burrows of adult animals (Fig. 3.8b). The burrows clearly had shifted upwards and downwards, although a reason for this was not immediately obvious.

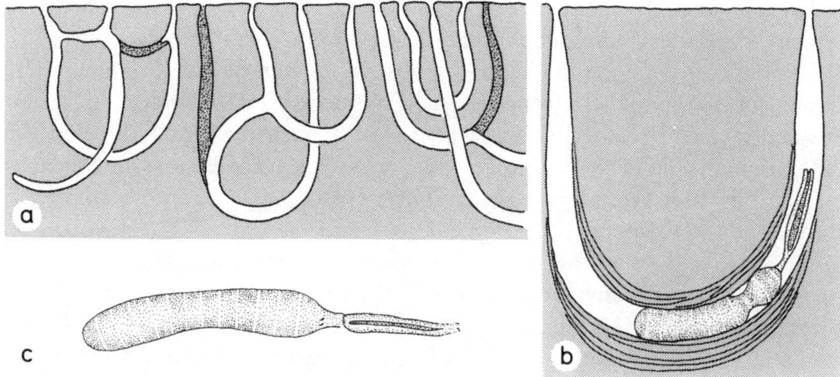

Figure 3.8 *Echiurus echiurus.* (a) Burrows of rapidly growing juveniles showing various means of enlarging the U, and reaching about 12 cm into the substrate. (b) Burrow of an adult worm about 50 cm deep, showing spreite structure. (c) *E. echiurus* at rest. Modified after Gislén (1940) and Reineck *et al.* (1967).

Gislén (1940) observed roof material being scraped off in order to bury faecal pellets in the floor of the burrow, producing a retrusive spreite lamina. Reineck *et al.* (1967) suggested that when small quantities of sediment entered the burrow this was trampled into the floor. If large quantities entered, however, this material was burrowed under and plastered onto the roof, creating a protrusive spreite lamina; and as faecal material does not normally leave the burrow, this must also be pressed into the burrow wall (Elders 1975).

The prehensile proboscis of *E. echiurus* is used in detritus feeding (Gislén 1940; Schäfer 1962; Nyholm and Bornö 1969), licking the sediment surface radially around the burrow. No sediment is taken up, only particles of organic detritus. The proboscis is also used, it seems, for grazing on the walls of the burrow (Reineck *et al.* 1967), a process that results in spreite formation in periods when no extraneous sediment enters the burrow. However, details of this sediment processing are unknown.

3.3.4 Circular arguments on U-shaped burrows

We know a great deal about the feeding technique of *Urechis caupo* but little about the sediment structures it produces. On the other hand, we know little about the feeding habits of *E. echiurus* but a lot about its burrow structures. *U. caupo* was mainly examined while living in a glass tube, where it had no choice but to suspension feed. Its reduced proboscis may indicate that it can feed in no other way. And yet, an individual of *U. caupo* has been seen to collect sediment inefficiently with its proboscis while lying outside its burrow in an aquarium (Fisher 1946).

E. echiurus, on the other hand, has been studied in its opaque substrate where an 'invisible' seston net, if it exists, might not be noticed, but the animals had access to detritus and sediment and fed on those resources. After having efficiently grazed the area around the oral end of its burrow clean of detritus, and sometimes reversing within the burrow and grazing off around the other end too, the animal 'rests' (Gislén 1940).

In a deep-water species, Smith *et al.* (1986) noted the same phenomenon; 120 hour periods of activity were followed by 30 days of quiescence. It might seem that, while new detritus accumulates and grows, the animal waits inactively, out of sight in its burrow. Surely it is more likely that an alternative feeding process is being used. Maybe the burrow walls are being exploited; burial of faeces to produce a spreite lamina would suggest a form of gardening (section 1.1.6). Chuang (1962) reported that another species, *Ochetostoma erythrogrammon*, browsed particles from its burrow walls.

Or the numerous mucus-secreting tubercles may be producing a seston net. We do not know; we have not looked. Yet, the bald facts as we know them tell us that *U. caupo* is a suspension feeder much like *Chaetopterus variopedatus*, whereas *Echiurus echiurus* is a detritus feeder. This contrast in trophic styles would seem unlikely, since the burrows of the two echiurans are so similar.

3.3.5 Spoke burrows, U-burrows and L-burrows

Photography of the abyssal sea floor has revealed very common, eye-catching, radiating structures (e.g. Ewing and Davis 1967, figs 53–9; Hollister *et al.* 1975, fig. 21.15; Young *et al.* 1985; Gaillard 1991, fig. 3A and B). Already by 1970, Häntzschel had compared some of these to trace fossils. Some of the structures are produced by polychaete worms and others may be traces or bodies of Xenophyophoria (Tendal 1972, pl. 16D, 1980).

One group of radiating structures is distinctive, having broad bladed spokes. These were interpreted by Ohta (1984) from photographs as the work of Echiura; in a few cases the echiuran proboscis was extended and visible. With the improvement of technology, we now have *in situ* video film and time-lapse photography of shallow and deep-water echiurans and presumed echiurans (Rice *et al.* 1991).

Jaccarini and Schemri (1977) suggested that echiurans feed with their proboscis in two ways. Some species extend the proboscis onto the sea floor and transport detritus selectively to the mouth by ciliary currents; practically no sedimentary disturbance is caused, although removal of organics produces a visible colour difference in sea-floor photographs. Other species actively scrape surficial sediment into the burrow as they retract the proboscis; this leaves a radiating groove having some preservation potential as a trace fossil. Bett and Rice (1993) reported a third

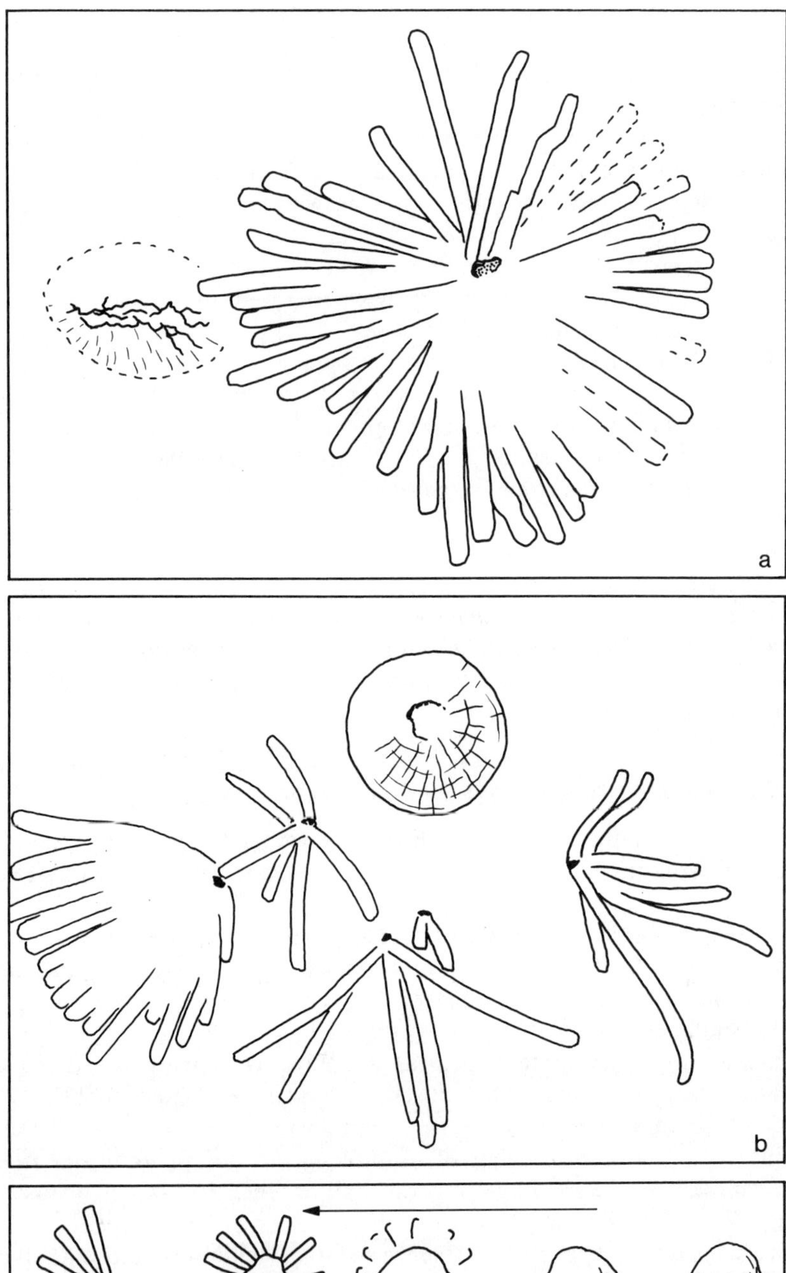

Figure 3.9 Three variations on a deep-sea echiural theme, from sea-floor photographs. (a) 'Spoke burrow' beside a 'gashed mound' (left). The burrow must be L-shaped (Fig. 3.10). (b) A mound having a central pit, associated with five spoked apertures. This appears to be a U-burrow where the feeding shaft is shifted to new detritus areas whereas the tail shaft

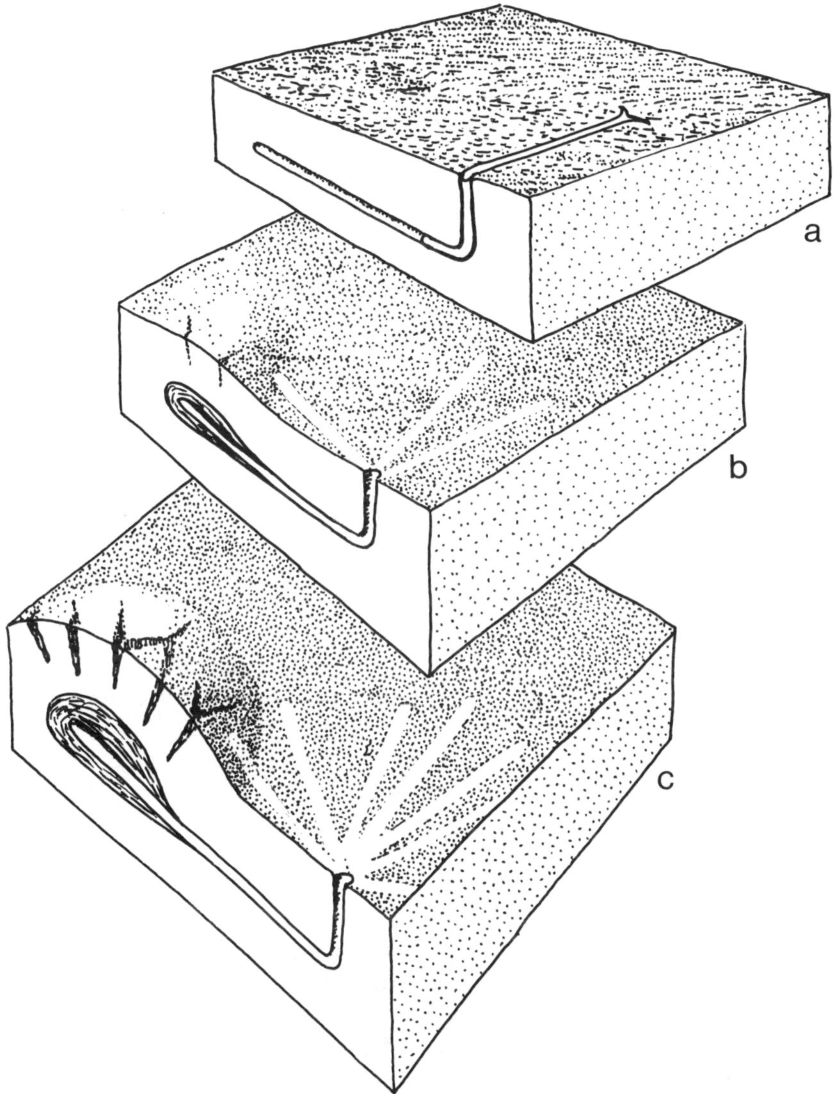

Figure 3.10 Development of a gashed mound and a spoke trace from the activity of an abyssal echiuran worm in an L-burrow. Modified after de Vaugelas (1989). (a) A new burrow, the proboscis extended from the aperture. (b) Activity of the proboscis produces radiating traces. (c) As excrement is retained in the burrow, it is assumed that this is pressed into the wall, creating the gashed mound above as the lining volume increases.

---

remains stationary. (c) Clear example of the W-technique of moving a U-burrow (e.g. Fig. 3.14). Migrating leftwards, successive spoked apertures are abandoned and a mound is built up on them as tail shaft replaces head shaft. (a) and (b) traced from photographs published by Ohta (1984); (c) Brian Bett (pers. comm. 1994).

style, in which the proboscis is extended over the detritus and then swept sideways. This left little trace in the sediment.

In connection with this activity, several indications of L-shaped echiuran burrows have been seen, having only a single aperture (Fig. 3.9). De Vaugelas (1989) suggested a model for an L-shaped burrow for deep-water echiurans (Fig. 3.10). Details of this model were criticized by Bett and Rice (1993), who nevertheless noted that faecal material does not seem to be returned to the surface and may be retained in the burrow.

The existence of deep-sea echiurans having an L-shaped burrow in which they conceal their excrement (reverse conveyors) and feeding in a circular manner at the sea floor, led Kotake (1992) to suggest an echiuran as the marker of *Zoophycos*. As the animal grows in length, the corresponding endobenthic circular movement would describe a spiral. However, *Zoophycos* spreiten are far more regular than any echiuran spoke burrow yet described.

The intertidal species *Prashadus pirotansis* extends its proboscis from the top of a mound, and thus also appears to produce a single-entrance burrow (Hughes and Crisp 1976). Hughes *et al.* (1993) made a detailed description of another, larger shallow-water species, *Maxmuelleria lankesteri*. This species also constructs an L-shaped burrow 1 m deep and 2 m long. The animal scrapes large quantities of surficial sediment into its burrow during active periods. During the following quiescent period, the animal is presumed to be sorting the sediment in its burrow. Soft sediment is then passed out of the aperture and eventually a large mound is developed. Some faecal material is also expelled, but some pellets are incorporated in the burrow wall, as in *Echiurus echiurus* (Kershaw *et al.* 1983; Hughes *et al.* 1993). This, and reingesting the expelled mound sediment, may be interpreted as gardening.

Risk (1973) described a burrow of *Listriolobus pelodes* as having up to four active openings at the sea floor, so we still have much to learn. He also produced convincing evidence for radiating proboscis traces preserved on Silurian sandstone bedding planes. Corresponding burrows were only partially preserved and had probably collapsed because they were not backfilled. Living fossils they may be, but echiurans are certainly very important today as tracemakers and bioturbators. Their trace fossils must be abundant in the geological record if we could only identify them.

3.4 U-burrows for detritus feeders

3.4.1 The tidal flat shrimp

If the burrows of *Echiurus echiurus* leave much uncertainty about their function, those of the amphipod *Corophium volutator* are more clearly interpretable. This animal excavates with skeletonized limbs, and so is not a worm even by my definition; however, this is a suitable place to discuss its work.

The burrows of this small crustacean have long been studied in the North Sea tidal flats. They are basically U-shaped and about 4 cm deep (Linke 1939; Häntzschel 1939; Schäfer 1962; Farrow 1975). Around one of the apertures the amphipod makes radiating feeding traces as it rakes the detritus with its enlarged second antennae. The corollary of this detritus feeding is the appearance of a pile of faecal pellets around the other aperture of the U-burrow.

Sectioning of the burrow, using a special embedding technique and thin sections, revealed the presence of a protrusive spreite between the limbs of the U (Fig. 3.11a; Reineck 1958, pl. 4; Seilacher 1967a, pl. 1a and c).

The presence of the spreite does not necessarily suggest that subsurface deposit feeding is taking place. The amphipods frequently leave their burrows and establish new ones, and the spreite is probably the result of normal burrow construction. By transferring sediment from floor to ceiling at the vertex of the U, this part of the burrow migrates downwards and the limbs lengthen. Yeo and Risk (1981) claimed to have found both retrusive and protrusive spreiten in association with *C. volutator*, in the style of the equilibrium spreiten of the trace fossil *Diplocraterion parallelum*

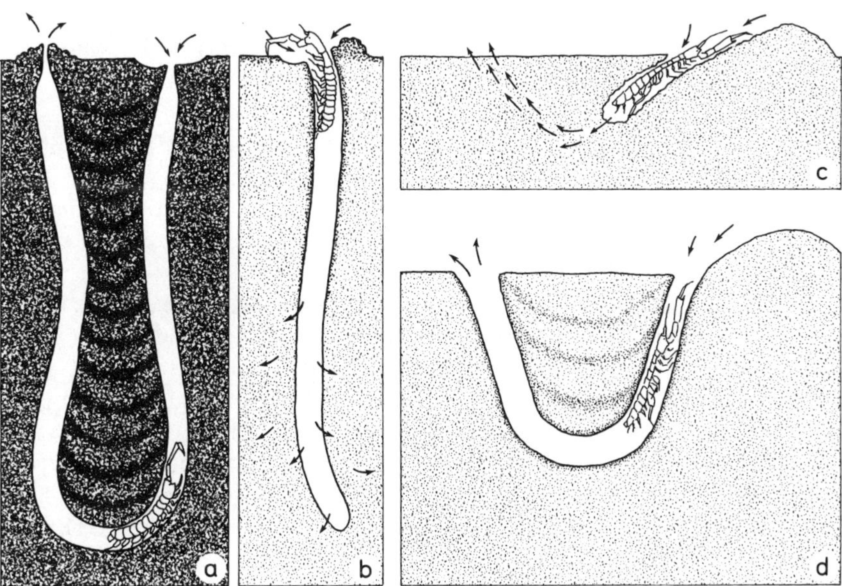

Figure 3.11 (a) U-burrow of *Corophium volutator* in mud and (b) in sand. Arrows indicate water currents. Modified after Seilacher (1953a). (c) A stage in burrow construction by *C. arenarium*. Passage of respiration water through the sand weakens the substrate and eases construction of a second shaft. This U-burrow is then gradually deepened. (d). Data from Ingle (1966).

(Fig. 8.11). However, it is uncertain whether these authors actually observed retrusive and 'yoyo' spreiten, or merely assumed their existence.

Rather high up a dry tidal flat, Mortensen (1900) noted both apertures capped by a plug of excrement, maybe helping to maintain interior humidity. By lifting the plugs, the amphipod fed at either end of the U-burrow (cf. Hertweck 1970a, fig. 71).

In sand, *C. volutator* shows radically different behaviour (Fig. 3.11b). In the more permeable substrate it is not necessary to maintain a full U-burrow; a single shaft is sufficient, the irrigation water passing out into the porosity of the sand. Only detritus feeding would now seem likely, the exploitation grooves and pellet piles developing side-by-side at the single aperture. The pellets are removed and the detritus layer refreshened at each flood tide.

Schäfer (1962) noted that shafts of *C. volutator* may be branched. Possibly this may represent the breeding shafts reported by Thamdrup (1935) in which juveniles make their excursions from the burrow of the adult.

Another species, *C. arenarium*, is restricted to sand, where it constructs a mucus-lined U-burrow (Fig. 3.11c and d). This species thus would appear to be a suspension feeder.

3.4.2 Life of the lugworm

In its most usual lifestyle, *Arenicola marina* is a detritus feeder, although it also swallows much sediment. As first described by Thamdrup (1935), the worm constructs a J-shaped, mucus-supported burrow in loose sand (Fig. 3.12a). Wells (1948) pointed out that during eversion of the pharynx, the spiny advancing buccal mass actually excavated the sediment on a small scale. Nevertheless, following this initial penetration phase, a double-anchor compressional burrowing technique is used, during which sediment is not swallowed (Wells 1944).

The basic burrow was described in detail by Wells (1945). At the distal end of the burrow, ingestion of sand produces a conical **head shaft** of collapsing sediment. The ingested sediment is extruded at the **tail shaft**. Collapse of the sand in the head shaft is assisted by the worm, which moves up into the column and, expanding its terminal anchor, draws the sediment downwards (Wells 1945), veritably 'dragging its anchor'.

This activity produces a pit at the sea floor and detritus is funnelled down the rapidly flowing core of the shaft (Rijken 1979, pl. 1). The pit acts as a trap in which further detritus accumulates (cf. Lampitt 1985).

Water is pumped through the burrow in the reverse direction to sediment transport. Entering past the faecal castings, it leaves again along the path of least resistance, the head shaft of loosely packed sand. Indeed, the upward flow helps to keep the head-shaft sediment fluid. This current oxygenates the sand around the burrow.

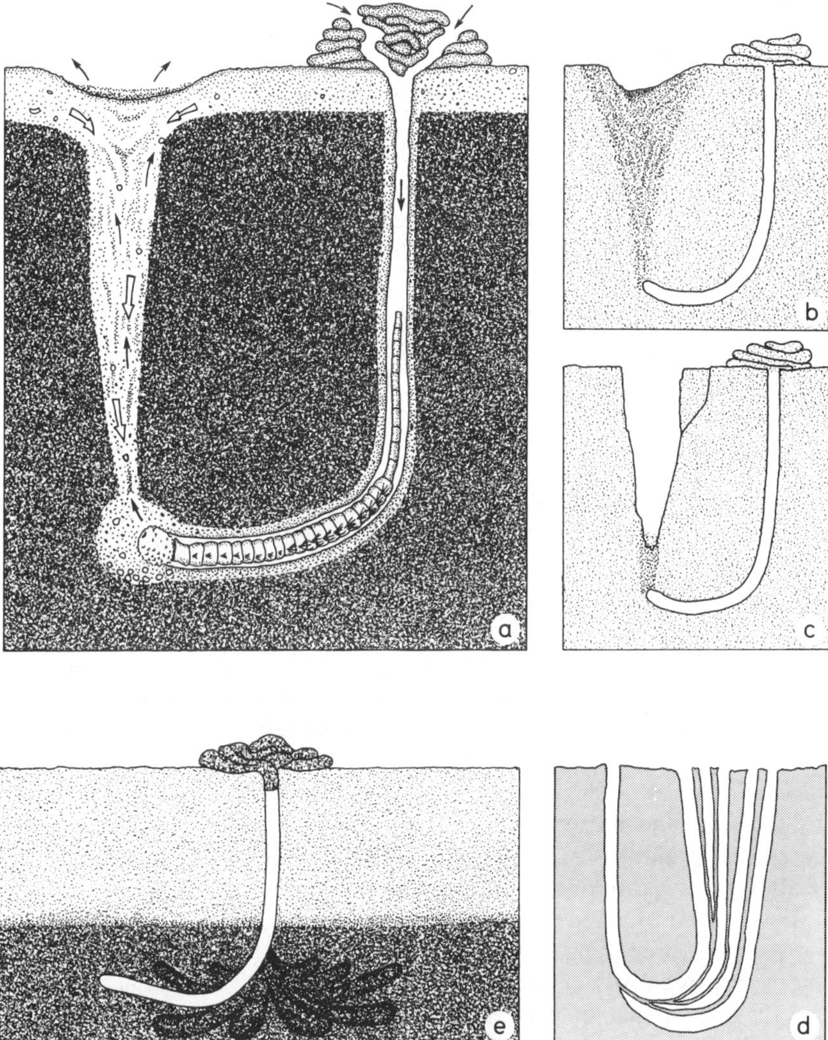

Figure 3.12 The work of *Arenicola marina*. (a) The standard system whereby sediment is conveyed down the head shaft (left), selectively ingested at its base, and deposited as castings above the tail shaft. Coarser grains accumulate below the head. Oxygenated sediment pale. White arrows show sediment movement, black arrows water currents. The thorny buccal parts are shown everted as a terminal anchor. (b) Burrow variants occurring in loose and (c) firm sediment. (d) Unusual variant where *A. marina* succeeds in exploiting clayey sediment. (e) Cumulative structure produced where the worm, deposit feeding in an organic-rich substrate, moved the burrow radially around a stationary tail shaft. Modified after Schäfer (1962), Rijken (1979) and other sources mentioned in the text.

Krüger (1959) pointed out that seston borne on this irrigation current would be trapped by the head-shaft sediment and eaten by the worm, implying a suspension-feeding function for the burrow. Head-shaft sand was found to be especially rich in detrital organic particles (Jacobsen 1967).

Hylleberg (1975) showed that the Pacific lugworms *Abarenicola pacifica* and *Ab. vagabunda* could not sustain themselves on filtering in that way. However, he observed higher concentrations of organic matter in the faecal castings than in the sediment the worms were feeding on and suggested gardening as an explanation. The increased surface area of the pit above the head shaft, together with the stream of oxygenated water through it, would create a suitable milieu for culturing microorganisms. Thus, we glimpse yet another aspect of the lugworm's varied diet. However, the raised organic matter content in the castings could alternatively be produced by rapid colonization of these by microorganisms (Longbottom 1970).

There is a marked increase in activity of meiofauna (Reise and Ax 1979; Reise 1981) and microbes (Reichardt 1987) within the head shaft and oxygenated burrow walls. Although the worm does not eat these animals, it ingests the bacteria on which they thrive. A similar richness in meiofauna is associated with the burrows of many animals (section 5.1.2).

This standard model of arenicolid feeding does not represent the only activity available. Schäfer (1962) and Rijken (1979) demonstrated that *A. marina* could also thrive in much finer-grained and firmer sediments where the sinking head shaft will not function (Fig. 3.12b–d). The sediment structures arising from these activities differ from those produced by the soft sand burrow. Rijken (1979) also observed *A. marina* making successive probes in organic-rich sediment, instead of producing a head shaft. This is a case of true deposit feeding (Fig. 3.12e).

Few authors have considered the long-term activity of the worm and its sediment processing. Wells (1945) and Rijken (1979) pointed out that the worm frequently shifts the position of its head shaft while maintaining the same tail shaft (Fig. 3.12e). *Abarenicola pacifica* shifts its position every three days (Woodin 1991). Rotating in this way around the tail axis will produce a radiating pattern that is familiar in many trace fossils (section 6.3).

In vertical section, prolonged bioturbation by lugworms produces a characteristic texture (Fig. 3.13; Reineck 1958, pl. 3, fig. 8), which has been recognized in early Holocene sands (Hansen 1977).

The unmixing effect that these worms have on sediment produces graded bedding (van Straaten 1952, 1956; Trewin and Welsh 1976). Grains of all sizes are drawn down the head shaft, but the largest are too bulky to be eaten. These accumulate as a layer at the head level of the worms, as the population moves through the sediment (Fig. 3.13). Smaller grains adhere better than larger ones to the mucus on the proboscis (Baumfalk

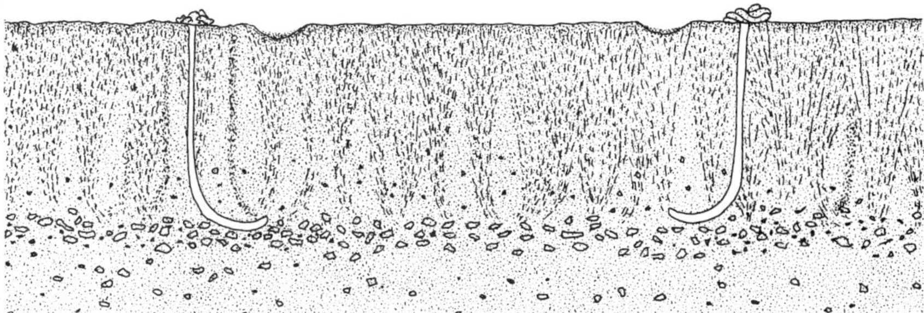

Figure 3.13 Graded bed fabric produced by bioturbation dominated by *Arenicola marina* activity. Based on Schäfer (1962).

1979) so that faecal strings ejected at the sea floor are slightly enriched in fines. Similar restructuring of sediments by *Ab. pacifica*, causing resuspension of clay particles, has been described by Swinbanks (1981a).

The rapid reworking of sediment by lugworms has always attracted attention (Davidson 1891). Swinbanks (1981a) computed that *Ab. pacifica* in British Columbia reworked the topmost 10 cm of sediment entirely within 100 days. Working at a similar rate, *A. marina* populations in the Dutch tidal flats rework to a depth of 33 cm, the summer rate being of one order of magnitude greater than winter values (Cadée 1976).

3.4.3 The funnel U-burrow

Many animals construct U-burrows that have an entrance shaped as a funnel. Powell (1977) termed this mode of life **funnel feeding**. Kudenov (1978) made a detailed study of a funnel-feeding maldanid worm, which is recommended reading.

Some endobenthic holothurians are funnel feeders. Myers (1977a) and Powell (1977) demonstrated this type of sediment processing in the slender apodian *Leptosynapta tenuis* of the north Atlantic coast of North America. *L. tenuis* loosens the sediment ahead of it with its tentacles and moves forward using the double-anchor technique (Clark 1964; Hunter and Elder 1967). It then thrusts its body against the burrow wall to compact it and apply mucus. The holothurian swallows the sediment as it burrows, so we have here a complex mixture of techniques involving true excavation as well as compression.

A U-burrow is established (Fig. 3.14) having a funnel at the head end and a pile of excrement at the other. Every day or two the burrow is shifted in a manner comparable to that of *Chaetopterus variopedatus* (and quite unlike *Arenicola marina*). Burrowing out from the funnel shaft, a new U is produced. The old funnel shaft now becomes the new tail shaft and spoils from construction of the new burrow are backfilled (via the

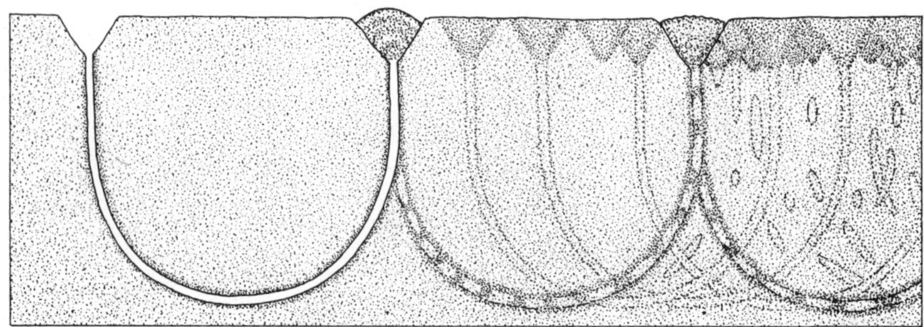

Figure 3.14 The funnel U-burrow of *Leptosynapta tenuis*, the slender holothurian itself not shown. Shifting of the burrow causes rapid bioturbation of the uppermost 3 m at the level of the faecal-filled funnels, but the U-burrows produce much less disturbance, deep down in the anoxic zone. Data from Myers (1977a).

gut) into the abandoned burrow lumen. A new funnel develops around the tentacle crown as superficial sediment is ingested, and the previous funnel now becomes filled with faecal material.

Unlike *A. marina*, irrigation water enters the funnel and is pumped out through the faecal castings. Powell (1977) found a raised organic carbon level in the castings relative to the surrounding surface sediment. *L. tenuis* occasionally reverses its position in its burrow and reingests its faecal castings, but too infrequently to represent a normal contribution to the diet. The high organic content of the castings may be due to accelerated bacterial growth in the irrigation current, but I prefer to attribute it to the raised nutritional value of the detritus that accumulates in the funnel traps where the worm feeds. In any case, in time, the sediment-processing holothurian population will create an uppermost layer in the substrate composed of maturing excremental material for later reconsumption (Fig. 3.14).

Myers (1977a) reported that, in aquaria, forming and filling funnels completely bioturbates the uppermost 3 cm of sediment. The holothurian does not ingest the finest or coarsest particles in the sediment available to it, yet biogenic graded bedding was not reported.

Rhoads (1974) reported *Leptosynapta* sp. as feeding differently, in a head-down vertical position. Powell (1977) found some individuals reach out of the burrow to scrape up detritus radially around the opening; and also to feed from the wall lining. Thus, as in the lugworms, these detritus/deposit feeders adjust their sediment processing according to the prevailing environmental conditions.

3.4.4 A pedate holothurian

Not all holothurians occupying U-burrows display funnel feeding. *Thyone briareus*, for example, is a suspension feeder (Pearse 1908). This is a

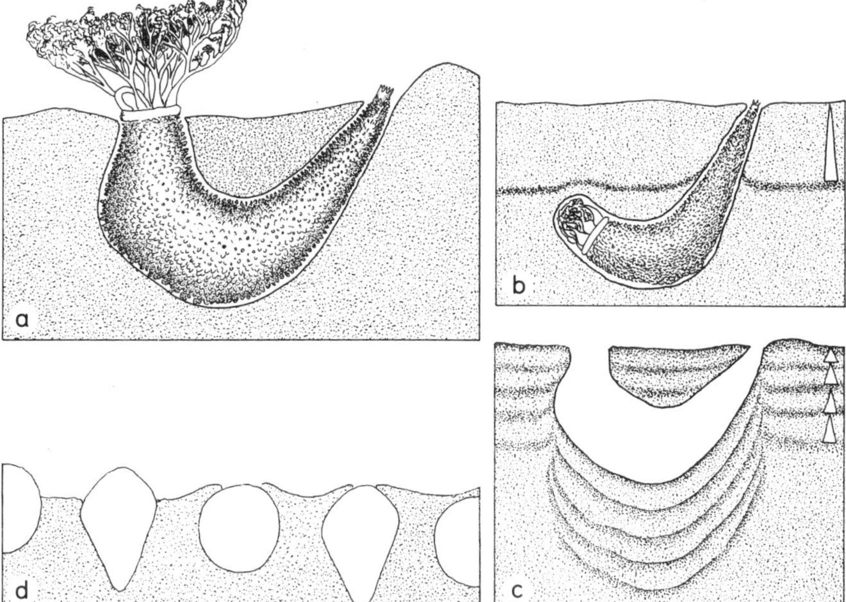

Figure 3.15 The holothurian *Thyone briareus*. (a) An animal in its U-shaped burrow. (b) Following rapid accretion of the sea floor, the animal first regains contact with seawater at the posterior end for respiration. (c) Small increments of sediment cause the animal to produce a retrusive equilibrium spreite. (d) Cross-sectional form of an animal entering the substrate broadside-on. Data from Pearse (1908) and radiagraphs by Howard (1968).

pedate holothurian and prefers to attach itself to buried hard objects with its podia and draw itself down, but it can also enter softground without such aid. It does this broadside-on, very slowly, deforming its body cross-section as it does so (Fig. 3.15d).

T. briareus can tolerate sudden burial beneath up to 15 cm of sand and burrow up to the new surface (Fig. 3.15c). The animal burrows up exactly as it did on the way down, producing a retrusive spreite, which may be regarded as an equilibrium structure rather than an escape trace (section 9.2.8).

3.4.5 Some enteropneusts

Funnel-feeder U-burrowers are found in a wide range of unrelated worms (e.g. Fig. 5.13d). Many enteropneusts of the genera *Balanoglossus* and *Saccoglossus*, for example, are burrowers. These sluggish animals glide within their burrows by means of cilia and peristaltic contractions of the proboscis. Sediment is ingested unselectively (Rao 1954; Suchanek and Colin 1986), and this process creates a funnel at one end of the U-burrow and a pile of castings at the other.

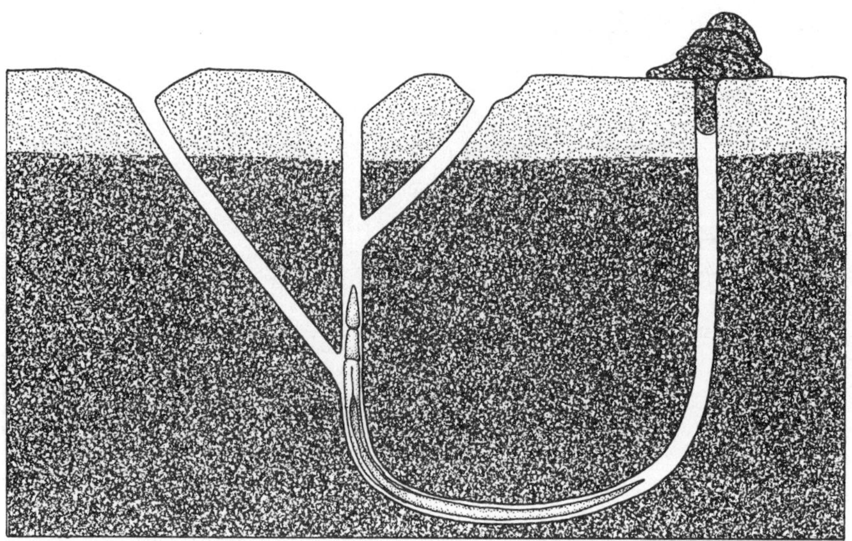

Figure 3.16 Burrow system of the enteropneust *Balanoglossus clavigerus* penetrating 60 cm into the black, anoxic substrate. Modified after Stiasny (1910).

According to Stiasny (1910), *Balanoglossus clavigerus* in the Mediterranean Sea lives well beneath the RPD, and black (anoxic) faecal castings are suggestive of deep deposit feeding. However, the animal also produces funnels (Fig. 3.16) and was said frequently to stretch out of its burrow to exploit detritus. Morton and Miller (1968) mentioned *Balanoglossus* burrows that had a spiral head shaft, and the entire burrow of *Saccoglossus inhacensis* is tightly spiralled (Van der Horst 1934). Some enteropneusts are large: *B. gigas* attains a length of 1.5 m and constructs burrows 3 m long in which it funnel feeds on detritus (Powell 1977; Barnes 1980).

The most detailed study that has been made of a burrowing enteropneust is that of *B. aurantiacus* by Duncan (1987). This west Atlantic coastal species produces a slender U-burrow 5–6 mm in diameter, having an oval cross-section, and thinly lined with mucus (Fig. 3.17). The one aperture is narrowed, and the other opens on a mound of excrement, so the ideal conditions exist for passive ventilation (section 1.5). The plane of the U lies perpendicular to the sea floor. At right-angles to this plane, a second aperture is opened, close to the narrow aperture, and this develops as a funnel. Thus, the U-burrow has two entrances, the one functioning for water circulation and the other for feeding. Only a single funnel is produced at a time.

The animal is extremely extensible, and when fully attenuated it fills only half the diameter of the burrow. This allows the worm to reverse its position in the burrow. It always reverses before establishing a new

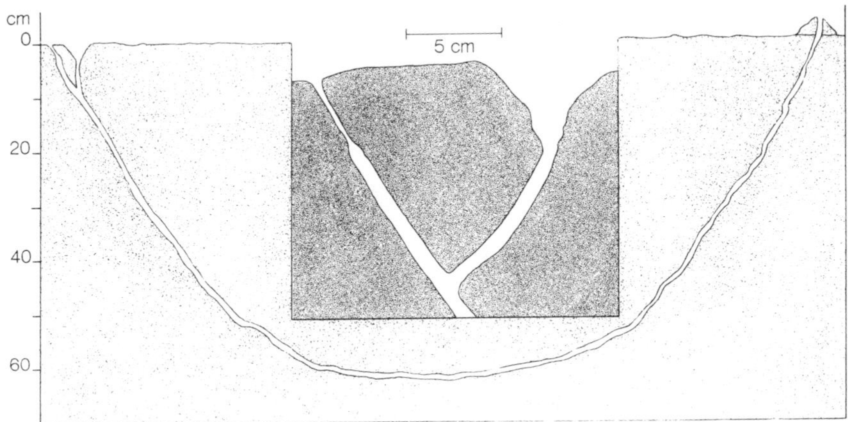

Figure 3.17 The U-burrow of *Balanoglossus aurantiacus*. The twin apertures at left are enlarged in the inset. Note that the funnel side-branch is not actually in the plane of the U as shown here. Modified after Duncan (1987).

burrow and, before doing this, defaecates in the funnel. The animal is remarkably mobile and establishes a new home every one or two days. Duncan (1987) never saw the worm extend outside its burrow and it did not eat its excrement. Animals placed on the sediment surface in aquaria burrowed in in a matter of hours. Evidently this animal is a key bioturbator (sections 5.2.1 and 6.4).

Duncan emphasized that, from the little we know of other species of enteropneusts, the burrow of *B. aurantiacus* is apparently unique, and that the many burrowing species show a wide variety of behaviour (e.g. Brambell and Cole 1939; Brambell and Goodhart 1941; Knight-Jones 1953; Burdon-Jones 1951, 1956).

Romano-Wetzel (1989) and Jensen (1992) described a quite different lifestyle for a deep-sea enteropneust *Stereobalanus canadensis*. This acorn worm lives colonially in branching networks at several levels down to 10 cm beneath the sea floor. The distinct levels are interconnected by diagonal or vertical shafts, and faecal pellets accumulate in the deepest network. Above these 'nests' a mound develops, presumably of advected sediment from burrow excavation. The burrows are circular in cross-section, 6 mm in diameter, and are morphologically close to the trace fossil *Thalassinoides* (Fig. 3.18).

3.4.6 *Rings of pits around a mound*

Among the plethora of distinctive structures captured on film of the abyssal sea floor, a particularly widespread and common form is a mound surrounded by a ring of pits (e.g. Ewing and Davis 1967, figs 77 and 78; Hollister *et al.* 1975, figs 21.A and 6.B; Lampitt 1985; Gaillard

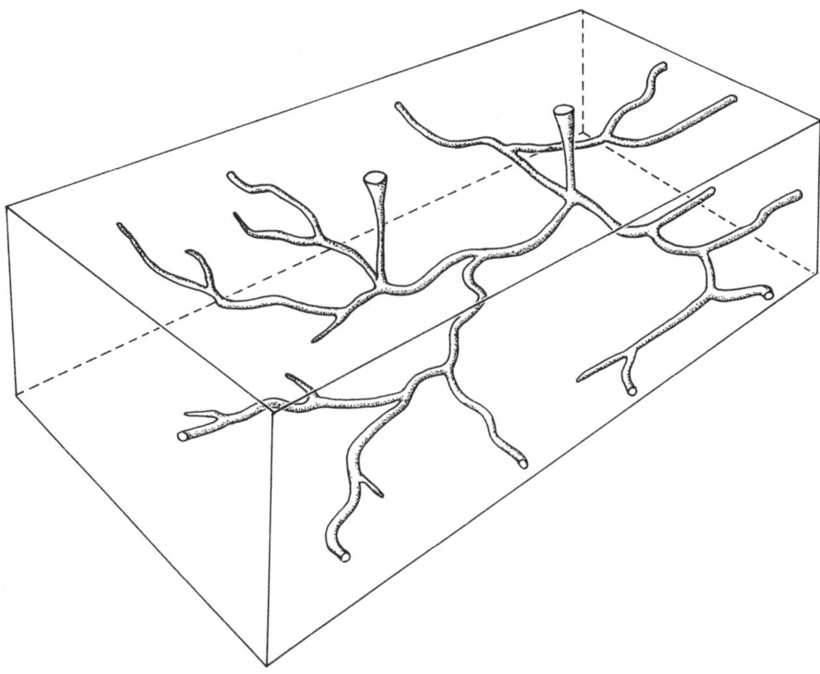

Figure 3.18 The burrow system of *Stereobalanus canadensis* as seen in the uppermost 12 cm of a box core. Based on Romero-Wetzel (1989).

1991, fig. 5D, E, G and H; Fig. 3.19). But the tracemaker does not show itself and remained incognito until Mauviel *et al.* (1987) happened to capture trace and tracemaker in a box core. Parts of an enteropneust were collected, genus indeterminate.

The box core sample revealed the pits to be slightly ellipsoidal funnels

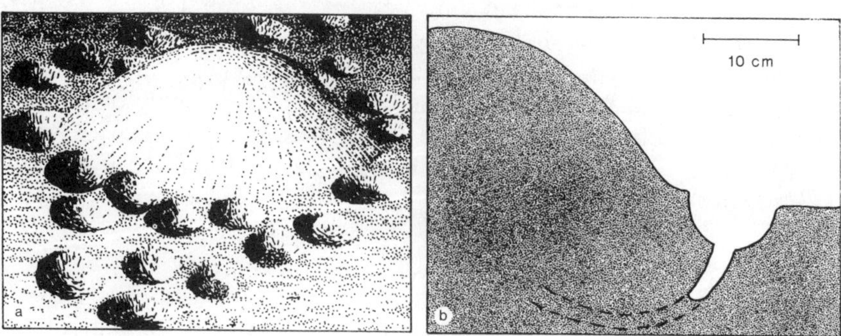

Figure 3.19 Pits around a mound. (a) Well developed example from a sea-floor photograph. Many such structures have just a single ring of pits around. (b) Cross-section of mound and a pit, from a box core. Based on Mauviel *et al.* (1987).

10–13 cm in longest diameter. At a depth of 5 cm these narrowed to a burrow of 2 cm diameter which could be traced downwards for 12 cm (Fig. 3.19). Further connection with the mound was not demonstrated. However, the most reasonable interpretation would be a U-burrow in which the head shaft was frequently shifted around a stationary tail shaft, in much the same way as described for *Arenicola marina* (Fig. 3.12e). A central pit on the mound is rarely detectable.

3.4.7 Poisonous worms

A variety of animal species secrete highly toxic halogenated compounds (all brominated). These species include a few corals, a terebellid polychaete, an arenicolid polychaete and a few phoronid worms (Woodin *et al.* 1987). In contrast, almost all investigated enteropneusts contain large quantities of the toxins. In particular, the well-developed lining smells of these compounds (Brambell and Cole 1939; Knight-Jones 1953; Ashworth and Cormier 1967; King 1986). The chemistry varies, surprisingly within species, community by community, but bromophenols are particularly important.

It is hard to understand how the animal survives the effect of its own secretions, in view of the toxicity of such compounds. After all, the related chlorinated phenols are well known for their potency as disinfectants. Possibly its main purpose is to deter predators. One very obvious result of the toxins in the burrow wall and burrower is the reduction or complete inhibition of microbial and meiofaunal colonization of the burrow.

King (1986) outlined the consequences of this. Inhibition of aerobic activity can decrease biological oxygen uptake, resulting in a larger zone of chemical sulphide and ferrous iron oxidation. This in turn would decrease the influx of toxic H_2S from reduced sediment outside the burrow. In addition, oxidation of ferrous iron would result in the deposition of iron oxyhydroxides that provide a better barrier to sulphide influx (i.e. use a toxin to fight a toxin!). Indeed the burrow wall of *Saccoglossus kowalewskii* is lined with a 2 mm thick deposit of iron oxyhydroxides. The diagenetic stage is set.

King (1986) also pointed out that decreased microbial metabolism would increase longevity of lining mucus. It is clear that a decreased requirement for replacing mucus could result in greater allocation of energy for growth and reproduction.

Jensen *et al.* (1992) were struck by the manner in which somewhat similar burrow systems produced by an enteropneust and an echiuran in the same sediment had contrasting microbial and meiofaunal colonization. The echiuran burrow walls contained a diverse community of nematodes, foraminifera and bacteria, whereas the toxic walls of the enteropneust were devoid of guests.

3.5 Deposit-feeder conveyors

Many benthic animals spend their lives orientated vertically or obliquely in the substrate, head downward. Ingesting the sediment at depth, they excrete at the sea floor and thereby cause significant upward transport of particles. Rhoads (1974) coined the term 'conveyor-belt' for such vertical advection activity. Following Thayer (1983), I shall shorten this to **conveyor**.

3.5.1 A fat holothurian conveyor

The molpadid holothurians lack podia. One species, *Molpadia oolitica*, was studied by Rhoads and Young (1971) in Cape Cod Bay, Massachusetts, where it occurs in muds at water depths in excess of –22 m. The animal feeds some 20 cm below the sea floor, selectively ingesting sedi-

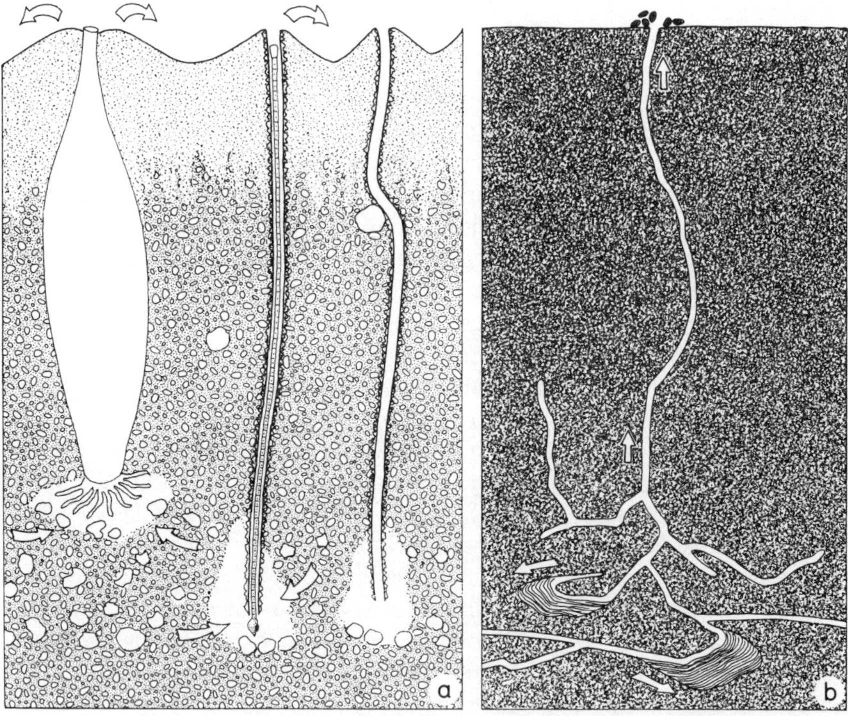

Figure 3.20 Conveyor deposit-feeder systems. (a) *Molpadia oolitica* feeding 20 cm below the sea floor together with two individuals of *Clymenella torquata* in their tubes of sand bound with mucus. Fine particles only are ingested and their advection upwards (arrows) causes the entire substrate to subside downwards, producing a graded bed and leaving a residual layer of coarse grains at head level. (b) *Heteromastus filiformis* (burrow diameter exaggerated) showing two directions of particle advection (arrows). Modified after Schäfer (1962) and Rhoads (1967, 1974).

ment collected on its tentacles. Excrement is ejected onto the sea floor where it accumulates as conical mounds. Feeding results in the accumulation of coarser, non-ingested sediment at the lower end of the animal (Fig. 3.20a). How the animal shifts its position to a new location, which must be necessary at intervals, is not recorded.

The sedimentological significance of *M. oolitica* activity is considerable. Not only does conveyor processing on this scale by a substantial population rapidly mix the top 20 cm of sediment, but a coarse-grained residual layer develops at the feeding level. Ecologically, the activity is also significant. The faecal mounds impart a steep topography to the sea floor, which influences the distribution of other benthic species (section 5.1.2).

Another holothurian, *Trochodota* sp. in New Zealand waters, appears to behave similarly to *M. oolitica*.

3.5.2 A slender polychaete conveyor

The capitellid worm *Heteromastus filiformis* has been studied in the North Sea by Linke (1939), Schäfer (1962) and Cadée (1979). Inhabiting muddy sediments, this thin worm constructs a shaft to some 20–30 cm below the surface. This is well lined with mucus and is semi-permanent, being used for passing excrement to the sea floor and for respiration. From the base of this shaft, the worm makes oblique to horizontal excursions out into the sediment for feeding (Fig. 3.20b). The galleries so produced are not carefully constructed, and may close again after the worm leaves them (Schäfer 1962), although much of the system is visible in X-radiographs (Howard and Frey 1975, fig. 50). Faecal matter from this deposit-feeding activity is conveyed up to the sediment surface as oval pellets which accumulate as a pile (Linke 1939, fig. 67). These pellets are well consolidated, resisting both physical and bacterial breakdown (Cadée 1979), and may be reworked without losing their identity.

The pellets are black, as the animal works anoxic sediments well below the RPD. Indeed, *H. filiformis* is adapted for life at such depths beneath the sediment surface, its haemoglobin having an extremely high affinity to oxygen (Pals and Pauptit 1979). The worm respires with its tail in the oxygenated water at the top of the shaft (Reise 1981). *H. filiformis* is non-selective in its feeding, its pellets containing all grainsizes except those too large for ingestion. Thus, a residue of coarser grains is produced at feeding depth (Cadée 1979).

H. filiformis causes less particle advection than does *Molpadia oolitica*. However, it fetches the material from a greater depth than the holothurian and thereby mixes sediments and waters of very different chemical and physical composition. The small feeding structures produced within the anoxic sediment should have a reasonable chance of preservation as trace fossils (section 6.2).

Schäfer (1962) noticed another deposit-feeding process in *H. filiformis* that opens up quite new ichnological perspectives. Although almost invisible in the anoxic mud, a lucky fracture of the substrate revealed that the worm had produced a spreite (Fig. 3.20b). The lateral shift of the burrow is presumably also a feeding process, in which case the excrement is redeposited at depth instead of being transported to the surface. The biogenically-packed spreite must have higher preservation potential as a trace fossil than the unfilled feeding burrows (section 6.3).

The presence of these almost invisible spreiten in association with *H. filiformis* burrows means that this worm is not exclusively a conveyor feeder; it also can backfill. We do not know how important this mode of feeding is in this species, or whether or not other capitellids feed in this way.

Maldanid polychaetes have similar lifestyles to capitellids (Fig. 3.21). Mangum (1964) described five sympatric species whose coexistence was possible owing to slight differences in grainsize preferences and depth of feeding. Rhoads (1967) and Featherstone and Risk (1977) made more

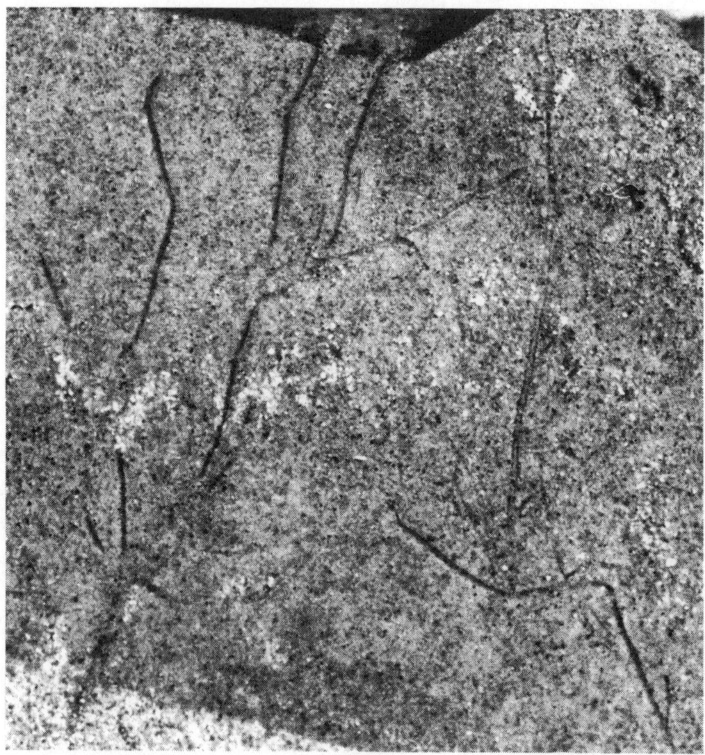

Figure 3.21 *Maldanus* sp. from Kefallinia, Greece, viewed in its burrow through an aquarium wall. The mucus-bound tube is not visible, but extends as a short chimney at the sea floor. x2.

detailed studies of one of these species, *Clymenella torquata* (Fig. 3.20a), which produces a graded bed through its feeding activities (Rhoads and Stanley 1965).

Fisher *et al.* (1980) demonstrated vertical mixing of lake sediments by head-down tubificid oligochaete worms: a non-marine example of conveyor activity.

3.5.3 *Pectinariidae, mobile tube-worms*

The pectinariid polychaetes are well known for the delicate conical tubes of sand grains that they carry about with them. The activity of these worms widely affects substrate consistency and fabric. The investigated species (Fauchald and Jumars 1979) have similar habits, although details vary according to environment.

Pectinariid worms live in fine sand to mud, head down, orientated steeply oblique to nearly vertical in the sediment. The posterior tip of the tube is normally exposed at the surface and feeding activity takes place 5–6 cm below. Sediment is excavated by comb-like bristles arranged in two fans on the head. Tentacles then actively seek food particles in the loosened sediment. Rejected material is passed up the tube and ejected at the surface as pseudofaeces. In this way, a cavity is produced around the head and gradually enlarged. Tentacles glide over the surface of this cavity, pulling at and penetrating the sediment.

If the sediment is relatively nutritious, the animal may remain stationary for long periods, producing one or more vertical shafts from a large cave (Fig. 3.22b). Watson (1927) and Wilcke (1952) described how, in *Lagis koreni*, pumping action of the worm produces sudden water currents that enlarge the cavity. They also saw tentacles extend out of the shafts to exploit detritus from the sediment surface. A mixture of detritus and deposit feeding is implied.

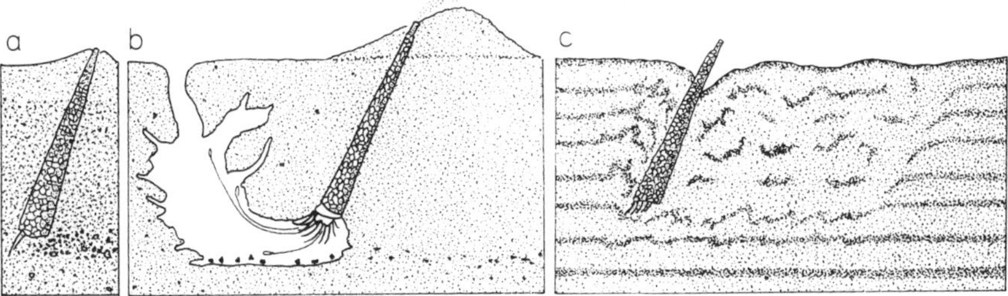

Figure 3.22 Sediment manipulation by pectinid worms. (a) Production of graded bed by selection for the finer fraction during feeding by a population of *Cistenides gouldii*. (b) Feeding *Lagis koreni* produces a cavity and shaft. (c) *L. koreni* wandering through the sediment produces a narrow biodeformational structure. Modified after Wilcke (1952), Schäfer (1962) and Rhoads (1967).

Where the organic content of the substrate is lower, the worm shifts its position frequently, and its conveyor activity produces a small cavity each time it pauses to feed. No shaft is developed. Deposit feeding within the constantly collapsing cavity is highly selective, and only material that is less than 1 mm is ejected at the sediment surface. Coarser grains remain at feeding level (Fig. 3.22a).

In time, a population of pectinariids, reworking the topmost 5–6 cm by selective vertical particle advection, produces graded bedding, as indicated by Gordon (1966) and Rhoads (1967) for *Cistenides gouldii,* and by Ronan (1977) for *Pectinaria californiensis.* Nichols (1974), on the other hand, found that continuous activity in deeper water by *P. californiensis* homogenized the uppermost 5 cm.

The near-vertical tube is moved broadside through the sediment by

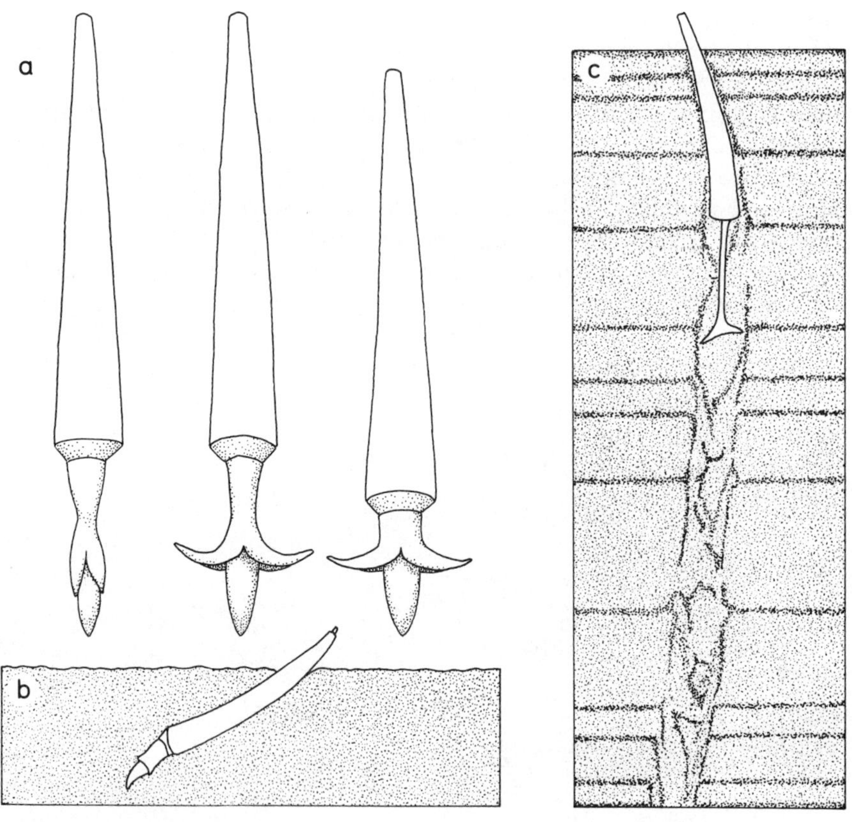

Figure 3.23 Scaphopods. (a) Part of the burrowing cycle in *Dentalium entalis*. The tapering shell resists backslip and acts as a penetration anchor. (b) *D. entalis* in living position. (c) After rapid burial, the foot may nevertheless thrust the shell upwards through the sediment to produce an escape structure. Modified after Morton (1959), Schäfer (1962) and Trueman (1968b).

means of digging activity at the lower end. Swivelling movements maintain the upper end in its constant position in the sediment, a pencil-thin disturbance zone being produced by the wandering of the upright tube (Fig. 3.22c; Schäfer 1962).

Other animals, not necessarily closely related, have similar lifestyles to pectinariids. Schäfer (1962) emphasized the similarity between pectinariids and scaphopod molluscs. However, scaphopods use their foot as a terminal anchor, which will cause compression. Dinamani (1964), on the other hand, drew attention to a liquification of the sediment by foot activity while feeding. We need to know more about scaphopod burrowing.

Another difference is the aptitude that at least some scaphopods show for escaping burial (Fig. 3.23c), a reaction that does not have its equivalent in the Pectinariidae.

3.5.4 *Reverse-conveyor activity*

Systematic vertical advection of particles also occurs in a downward direction. Quantities involved are small in comparison with the upward transport of material just described, but the ichnological and diagenetic aspects of this process are highly significant.

Many organisms that feed at the surface deposit their excrement at depth within the sediment. Myers (1977a) reported such an inverted- or reverse-conveyor strategy in two polychaetes, the terebellid *Polycirrus eximius* and the cirratulid *Tharyx acutus*, both of which feed on surface detritus by means of a spread of tentacles. The spionids *Scolecolepis*

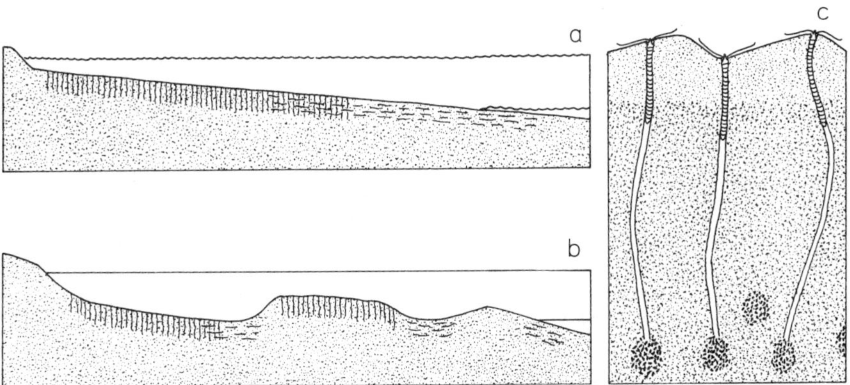

Figure 3.24 Two beach profiles showing the distribution of two polychaete worms, *Scolecolepis squamata* (vertical symbol) and *Paraonis fulgens* (horizontal; section 4.6). Mean high and low water levels are indicated. (a) A flat foreshore 70 m wide showing simple distribution. (b) A topographically more complicated foreshore 200 m wide. (c) The reverse-conveyor worm *S. squamata*. The pale upper zone, 5–10 cm thick, represents the sediment that is reworked with each fair-weather tide. Modified after Wohlenberg (1939) and Röder

squammata and *Pygospio elegans* do likewise, depositing richly organic faecal pellets at depth within clean beach sand (Fig. 3.24c; Reise 1981). A deep-sea bivalve, *Abra longicallus*, feeds on detritus but deposits faecal pellets in excavations deeper within the sea floor (Fig. 4.2), possibly as a means of culturing bacteria as a food source (gardening).

Another important site of downward transport of surface material is the lining of burrow walls using detritus. The terebellid worm described in the next section is an example of this activity. Thick tubes are constructed by this worm to depths of 30 cm using material collected at the sediment surface. The walls are continually being repaired, enlarged and added to so that large quantities of surface material become incorporated within the sediment by this process. Featherstone and Risk (1977) found that up to 50 per cent of the sediment of a tidal flat was composed of burrow walls.

3.6 A thick-walled U-tube

Many terebellid polychaetes are carnivores. Wilson (1980) described a species, *Eupolymnia heterobranchia*, which inhabits a U-burrow and catches larvae of other worms on its spread of tentacles. The capture of even such large prey as the amphipod *Corophium* sp. by the coordinated effort of many tentacles was observed.

However, Aller and Yingst (1978) described a detritus-feeding terebellid, *Amphitrite ornata*, which also lives in a U-burrow, sharing it with a scale worm and a crab. One or both ends of the burrow is normally surrounded by a pile of sand accumulated as a result of feeding (Fig. 3.25).

The head bears numerous highly mobile, ciliated tentacles that radiate

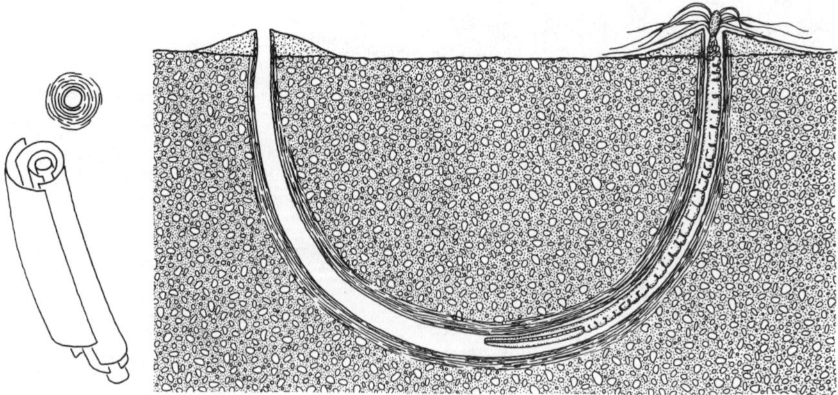

Figure 3.25 The U-burrow of *Amphitrite ornata*, showing (left) the laminated mud wall in cross-section and as an exploded constructional diagram. Data adapted from Rhoads (1967) and Aller and Yingst (1978).

from one aperture of the burrow. Coarse sand and detritus are conveyed via the tentacles to the mouth. The worm defaecates an unpelleted stream of sediment at the same end as it feeds. Reversing its position within the burrow allows the worm to feed at either aperture. Thus the limbs of the U function as single shaft burrows, each used alternately, and the sole requirement for the U shape is for irrigation; peristaltic movements maintain a constant current.

In addition to its tentacular feeding activity, the worm also crops the walls of its tube with its mouth, possibly feeding on bacteria. Thus, also this worm may indulge in a little gardening. The burrow of *A. ornata* is thickly walled with sediment containing a larger proportion of fine material than the ambient sediment. The burrow lumen may be about 0.5 cm in diameter and the wall 1–1.5 cm thick. The wall consists of several concentric layers, each lined on the inner surface by an organic sheet about 5 m thick. Only the innermost layer is a strictly continuous cylinder. The outer layers are split longitudinally and wrap partially around the next layer inward (Fig. 3.25). The material of the wall layers is built up of small bricks of fusiform shape in millimetre size.

Aller and Yingst (1978, p. 232) suggested that 'only the innermost tube and its lining act as the primary dwelling structure. The outer layers result either from slight lateral or vertical movements of the burrow [in the nature of a spreite] or perhaps represent the walls of older, smaller diameter tubes which have been split by the animal after a period of growth, much like moulting'. The animal, at least when adult, cannot make a new burrow if removed from its tube. Thus, it would be expected that some sort of growth vector be represented in the structure of the burrow.

At each growth phase, then, all the earlier tube walls are compressed further into the surrounding sediment by an expansion of the burrow lumen and the application of yet another lining layer. The sediment immediately surrounding the burrow, therefore, must be significantly more compacted than the ambient substrate (cf. Fig. 8.2). Knight-Jones (1953) noticed similar longitudinal splitting of the outer parts of the thick lining of an enteropneust burrow and came to the same conclusion, that the splitting was due to the growth of the occupant.

3.7 Chimney-building worms

Several species of polychaete worms extend a burrow lining above the sea floor as a chimney. Others produce an armoured chimney structure at the top of a weakly lined burrow. Raising the aperture on an armoured pipe conveys some protection against predators and sedimentary burial, raises the aperture to an above-bottom suspension-feeding tier, and has some hydraulic burrow-irrigating advantages (section 1.5).

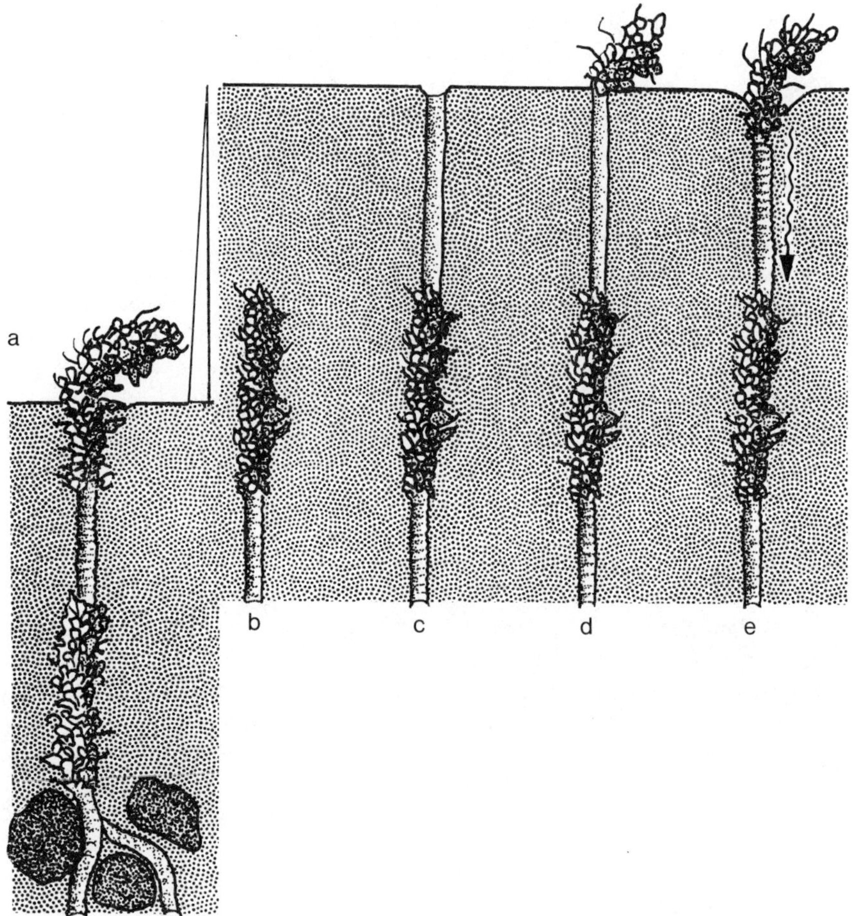

Figure 3.26 Reaction of *Diopatra cuprea* to mass sediment accumulation. (a) Bottom current direction is perpendicular to the page, the chimney bent at right-angles to this. (b) Depositional event: the tube is straightened and (c) an unarmoured portion is constructed to the new surface. (d) A new armoured chimney is built on the sea floor and (e) dragged into the sediment. Modified after Myers (1972).

Chimneys may be built atop single shafts, as in *Diopatra cuprea* and *Owenia fusiformis* (Fager 1964), or in other species as an extension of a leg of a U-burrow, as in *Lanice conchilega* (Fig. 3.5d). Function and structure of the chimney varies in different species, but there is room here for only a single account, and the western North Atlantic polychaete *Diopatra cuprea* is selected. Myers (1970, 1972) has provided good descriptions for the neoichnologist. For equivalent details of *L. conchilega*, see Seilacher (1951) and Ziegelmeier (1952, 1969).

D. cuprea constructs a tube having two forms (Fig. 3.26). Within the sediment a mucus is secreted that hardens on contact with sea water, adhering to a monolayer of grains. Having a diameter of 5–8 mm, this shaft may extend 1 m down. In contrast, a short, inverted J-shaped, reinforced tube is built at the sea floor. This tube is constructed at the sediment surface, and then drawn downwards by the worm so that it projects upwards as a chimney but is firmly anchored. Extending out of its tube, the worm selects shell fragments, pebbles, etc., preferably having tabular shapes, and drags these to the tube margin where they are cemented in place. An imbricate arrangement of grains results (Fig. 3.26).

Deeper parts of the burrow lining may be broken down by microbial activity, but the worm can renew the lining. If the burrow is constructed among pebbles, this renewal may result in branching of the tube.

The J-shaped chimney catches drifting plant fragments and other items which the worm collects as food. Myers (1972) found the aperture to be directed perpendicularly to the bottom current. The worms sometimes occur in linear groups, the line of individuals lying perpendicular to the current and the J-bend of the chimney orientated along the line (Frey and Howard 1969, pl. 3, fig. 3).

If buried under more than 5 cm of sediment, *D. cuprea* straightens the J-tube while the new sediment is still fluid, and then climbs to the new surface, secreting unreinforced lining. At the new surface a new reinforced tube is constructed and dragged downwards (Fig. 3.26).

Gradual erosion of the sea floor causes *D. cuprea* to drag down its reinforced chimney. This causes wrinkling and destruction of the unreinforced tube, but that is easily renewed. If the reinforced tube is pulled down to a buried length of reinforced tube, then the worm changes behaviour, and trims off the top of the chimney. Should erosion expose the whole tube and worm on the sea floor, the worm trims off the tube ends so as to reduce it to the length of the doubled up worm. It then creeps, with its tube, to a suitable place to burrow, and drags the tube down after it. Other tube worms show similar behavioural responses to allow them to cope with complex bottom sedimentation patterns (Fig. 3.27).

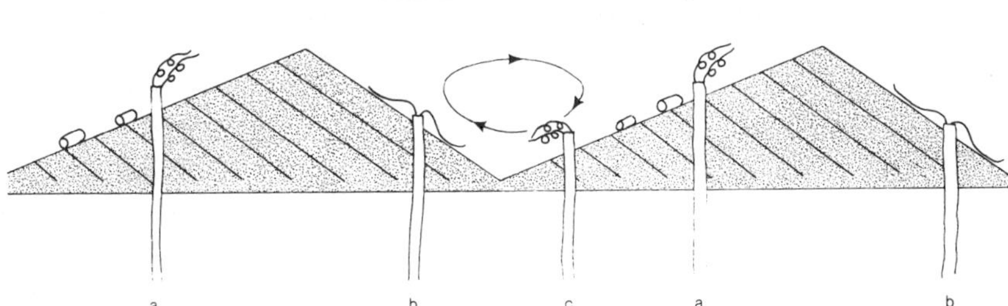

Figure 3.27 Behaviour of the tube-building polychaete *Pseudopolydora kempi* according to the microenvironments created by migrating current ripples. (a) On the stoss side, the worm trims back its tube as it is exposed by erosion, and suspension feeds. (b) The worm builds up the tube in phase with deposition, and detritus feeds. (c) Suspension feeding in the back current in the interripple trough. Flume performance. Modified after Nowell *et al.* (1989).

Growth in diameter in *D. cuprea* is accomplished in a similar way to *Amphitrite ornata* (section 3.6), by slitting the tube. New, concentric layers of mucus are then applied.

Of course, having two distinct types of tube, the *D. cuprea* burrow has been compared with two different trace fossils. Some authors have considered the unreinforced tube a close counterpart of the ichnogenus *Skolithos* (e.g. Skoog *et al.* 1994), whereas Kern (1978) named a shell-armoured tube from the Eocene *Diopatrichnus*.

3.8 Unwhole worms, the Pogonophora

The Pogonophora is a small phylum (i.e. containing few species) that is known predominantly from the deep sea. Pogonophores are very slender animals that, like other 'unwhole animals' entirely lacking a gut, subsist on chemosymbiosis with bacteria, based on reduced compounds, especially sulphides (Reid 1989).

There are two groups: the large vestimentiferan pogonophores, which occur at hydrothermal vents and cold seeps, and are attached to hard substrates; and the small perviate pogonophores, which are widely distributed in reducing sediments in all oceans. All the perviate forms are endobenthic. They occur patchily, but locally they are dominant species of the community, so they should occupy an important place in this book. Unfortunately, however, next to nothing is known of their habits that would be of use to the ichnologist.

The animal lives within an organic tube more or less vertically in the sediment, at least at the upper end; the lower is usually left behind in the sea floor when the boxcorer is raised (Dando and Southward 1986; Fig. 3.28).

In some species, the tube is known to take an irregular course downwards, showing turns and oblique runs. A depth of penetration of 1.5 m is recorded (Ivanov 1960). Some individuals extend to the surface and can be seen in close-up photos of the sea floor (Dando and Southward 1986, fig. 5). Others do not protrude, and their tentacle(s) must be able to take up oxygen from the uppermost sediment through the tube wall. These are truly animals that have learnt to live close to anoxia!

All pogonophores are completely dependent on chemolithotrophic bacteria. It appears that a whole phylum has undergone radiation on a nutritional basis provided by bacterial symbionts, enabling the invasion of normal reducing sediments ranging from restricted environments in shallow seas (Southward *et al.* 1986) to the abyssal plain (Southward 1979). If only we knew something of the bioturbating activity (if any) of this group of worms we might be able to add another dimension to these unfossilizable but fascinating animals. We might also find a tracemaker for some of the deep-tier low-oxygen trace fossils such as *Chondrites* and

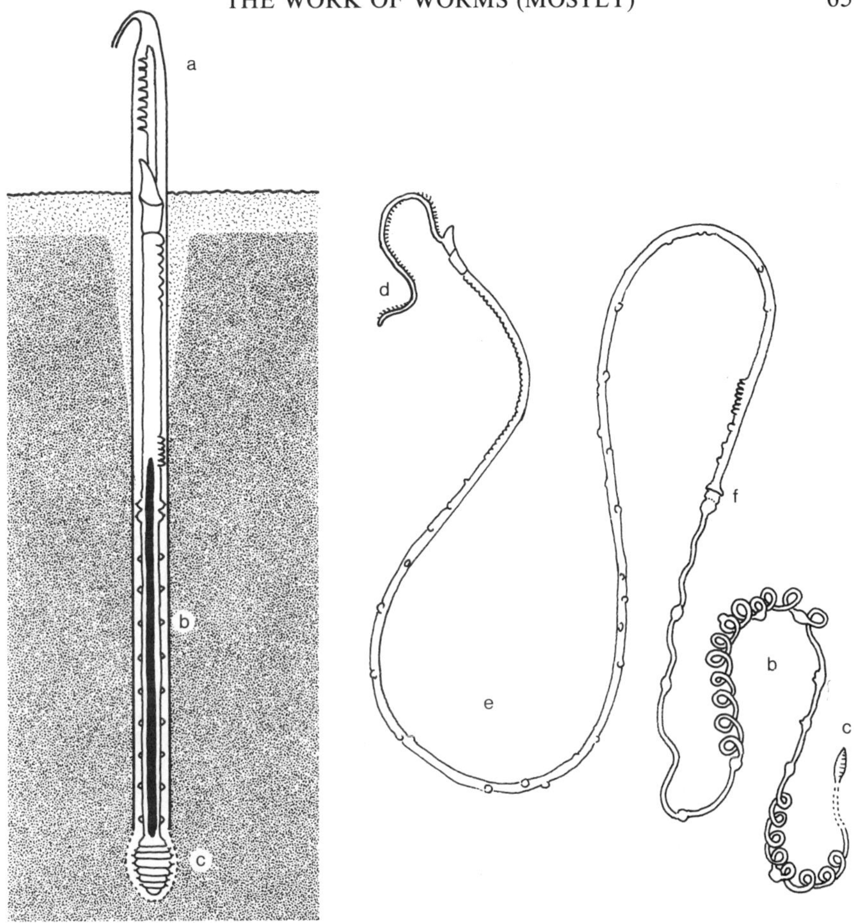

Figure 3.28 A pogonophore represented diagramatically within its tube in position within the sediment (left), and a sketch of an individual removed from its tube. (a) The top of the tube projects above the sediment and is folded over to exclude particles entering. (b) Post-annular region containing bacterial symbionts (black). (c) Opisthosoma, a 'digging organ'. (d) Tentacle (some species have many). (e) Anterior part. (f) Girdles. Modified after Southward *et al.* (1986).

Zoophycos. Very deep slender trace fossils such as some *Trichichnus* (section 10.6.3) and *Bathichnus paramoudrae* (Bromley *et al.* 1975b; Bromley *et al.* 1975a) are possible candidates, but this paucity does not correlate with the abundance of the Pogonophora today. Here is another big gap in our knowledge.

4 Some celebrated burrowers

There are yet four groups of sediment processors that deserve attention, far more than I can give them here. Among these are some that are so behaviourally diverse that even our patchy knowledge of them supports a large literature. I have chosen a few species that carry an ichnological message.

4.1 Bivalves

Burrowing is so much a normal way of life for this molluscan class, and was probably its primitive lifestyle, that it is reasonable to consider as divergent the bivalves that do *not* burrow. Most of the characteristics of the class represent modifications that enabled the early Palaeozoic bivalves to become soft-bottom endobionts. Through the Phanerzoic, bivalves have entered an enormous array of intimate relationships with their sedimentary substrates. Some species of *Donax* burrow so swiftly in sand that they can follow the surfline up and down the beach with the rising and falling of the tide (Trueman 1971). Another extreme form is *Panopea generosa*, a giant that lives well over 1 m below the surface of the sediment (Barnes 1980).

The reader should consult Stanley (1970), the many works of Yonge (e.g. 1939, 1949) and Yonge and Thompson (1976) for a glimpse of this diversity of burrowing behaviour. Ansell and Trueman (1967) and Trueman (e.g. 1975) have studied the details of the bivalve burrowing process. The following accounts of a few species have an ichnological emphasis.

The most characteristic trophic style of the bivalves is suspension feeding, but recently chemosymbiosis has also been shown to be important among bivalves (Reid 1990). Species having these feeding styles tend to live quietly; food comes to them, so there is no need to move unless disturbed. Some bivalves are deposit feeders, however, and these species produce far more disturbance to their substrate, shifting their position and manipulating grains continually.

4.1.1 A deposit feeder

The nuculoid protobranch *Yoldia limatula* has been studied as a sediment processor by Rhoads (1963), Stanley (1970), Aller (1978) and Bender and Davis (1984). The species inhabits shallow marine clay–silt substrates.

Y. limatula is highly mobile and burrows shallowly, the posterior end

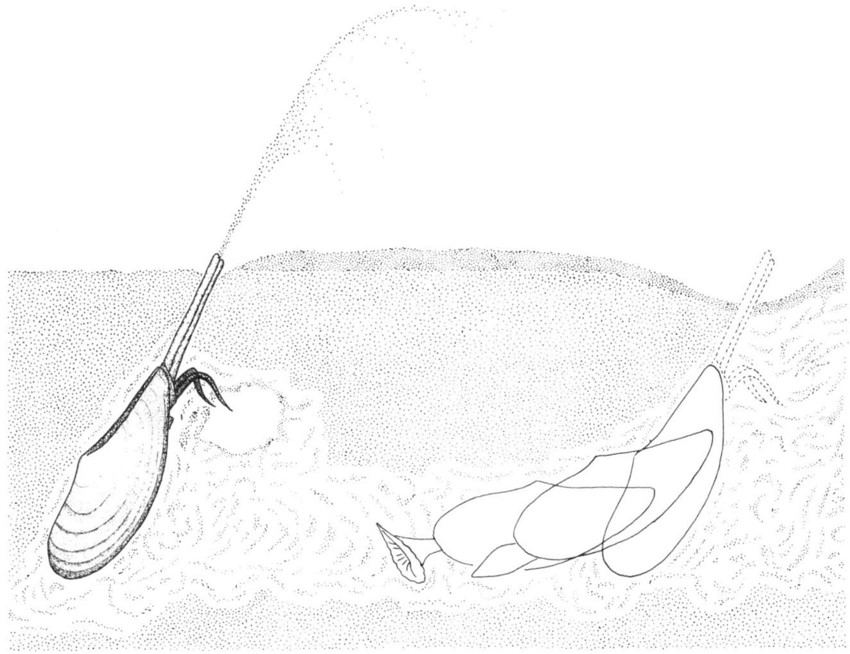

Figure 4.1 *Yoldia limatula*, deposit feeding and shifting from right to left to a new feeding site. Modified after Rhoads (1963, 1974) and Aller and Cochran (1976).

of the shell buried about 1–8 cm beneath the sea floor. Movement through the upper sediment layer is achieved by means of the double-anchor technique, the shell orientated about parallel with the sea floor. The foot has a split termination and when folded together forms a sharp blade with which to penetrate the sediment. As the foot so moves forward, the shell is opened and acts as a penetration anchor. When the foot is fully extended, the two flaps at the end flare open and grip the sediment as a terminal anchor. The shell is now closed and drawn up to the foot (Fig. 4.1).

The nuculoid protobranchs are detritus and deposit feeders that collect sediment for ingestion by means of large, contractile palpal tentacles. Some species sample the surface detritus with these palps, but *Y. limatula* works subsurface material.

In feeding position, the siphon tips are exposed above the soupy sediment. The palps explore the substrate and pass sediment by cilia to the mouth. In the mantle cavity, this material is sorted and much is rejected as inedible. These ejecta leave via the exhalent siphon as a cloud of loose sediment that may settle out nearby. The edible fraction passes the gut and emerges as compact faecal pellets. These are also ejected through the exhalent siphon. X-radiography of animals feeding in aquaria revealed

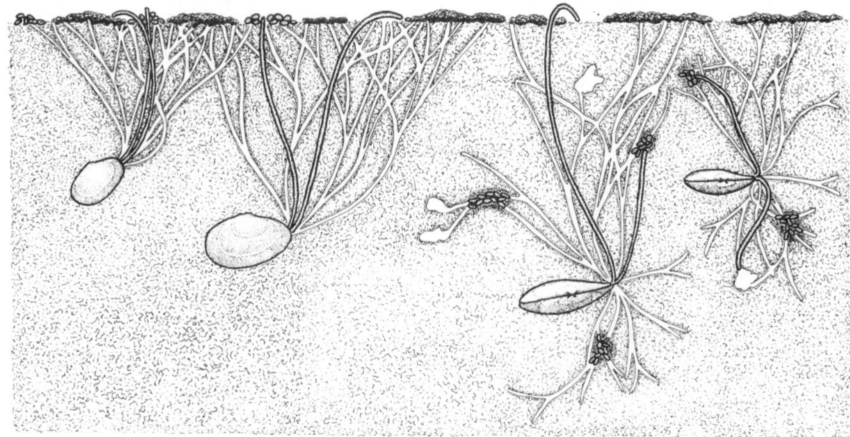

Figure 4.2 The deep-sea tellinid bivalves *Abra nitida* (left) and *A. longicallus*, feeding. Zones of siphonal activity show as networks of abandoned canals. *A. nitida* feeds on surface detritus and places both pseudofaeces and faeces on the sea floor (normal behaviour for tellinids). *A. longicallus*, however, deposit feeds below the surface, and advects pseudofaecal material up to the surface, whereas faecal pellets are deposited at feeding level. This was interpreted by Allen (1983) as gardening activity. Modified after Wikander (1980).

cavities within the sediment caused by the feeding process (Stanley 1970). Bender and Davis (1984) observed a sculpture of distinct grooves on the walls of these cavities. The cavities may become even larger than the animal itself; they can interconnect and collapse to form depressions.

Reworking of sediment by *Y. limatula* thus comprises several processes (Fig. 4.1). Periodic repositioning of the animal causes eddy diffusion to a depth of 3–8 cm, whereas feeding processes convey material from below the surface to above it. Some of this material is bound in compact faecal pellets, while the forceful ejection of pseudofaecal sediment up to 10 cm into the water column causes resuspension of fines.

Bender and Davies (1984) found that expelled solids range up to 200 times the body weight daily; taking the inactive winter months into account, it was calculated that a 14 mm long individual will resuspend 440 g sediment yearly. Thus, feeding results in a fine sediment suspension, an unstable pelletized sediment surface and a coarse-grained subsurface.

Some activities of detritus-feeding veneroid heterodont genera such as *Tellina*, *Macoma* and *Scrobicularia* (Figs 5.12 and 5.13) probably produce similar effects (Schäfer 1962). However, these bivalves feed on surface detritus, either conveying it back to the surface as pellets or burying it at deeper levels (Fig. 4.2). Intertidal *Scrobicularia plana* feeds in different ways according to the tide. At high tide it alternately feeds on anoxic sediment below surface and suspension feeds; at low tide, when predaceous fish and crab activity ceases, the bivalve extends its inhalent siphon out onto the sediment surface and detritus feeds (Hughes 1969).

4.1.2 A jet-propelled suspension feeder

The Tellinacea are rapidly burrowing bivalves most of which are detritus and deposit feeders. Like those mentioned in the previous paragraph, they have separate siphons, the inhalent one typically, as in the Tellinidae, serving as a probing device for sucking in detritus and sediment (Fig. 4.2). In one family, the Solecurtidae, at least some species have turned to suspension feeding, but have retained the separated siphons and the ability to burrow rapidly.

The behaviour and sediment structures produced by *Solecurtus strigilatus* were studied by Dworschak (1987b) and Bromley and Asgaard (1990) in the Mediterranean Sea. *S. strigilatus* prefers a medium sand substrate in the shallow subtidal, where it lives in a J-shaped burrow. The animal is a cylindrical 'worm' that is superficially unlike a bivalve in appearance. The shell is too small to enclose the whole animal; the thin valves bear a divaricate sculpture that obliquely cuts growth lines (Fig. 4.3; Seilacher 1972). A powerful foot extends anteriorly. Posterior to the valves the

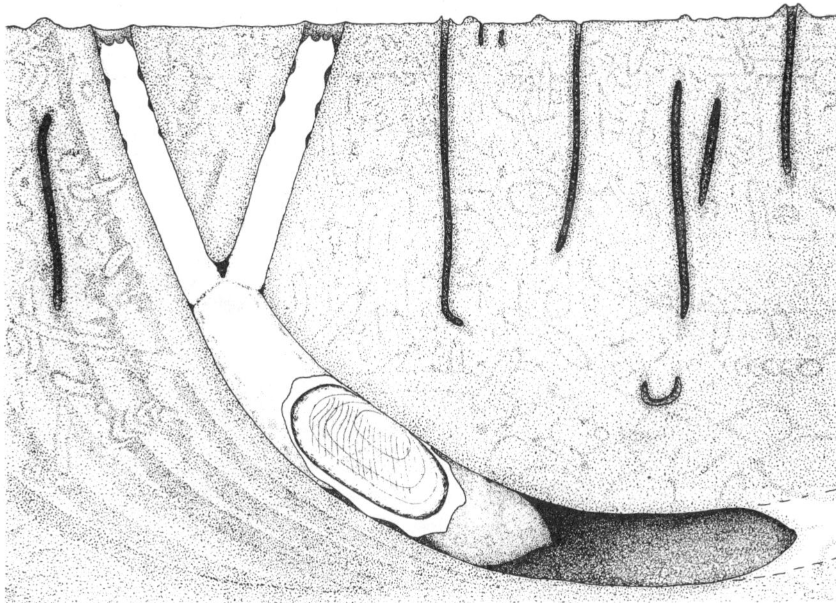

Figure 4.3 *Solecurtus strigilatus* in feeding position. Periodic shifting of the burrow leaves a spreite-like structure. The upper parts of this structure are rapidly obliterated by shallower bioturbation (Fig. 4.6a). Depth of burrow 30 cm. Broken lines at lower right indicate the path taken during escape reactions.

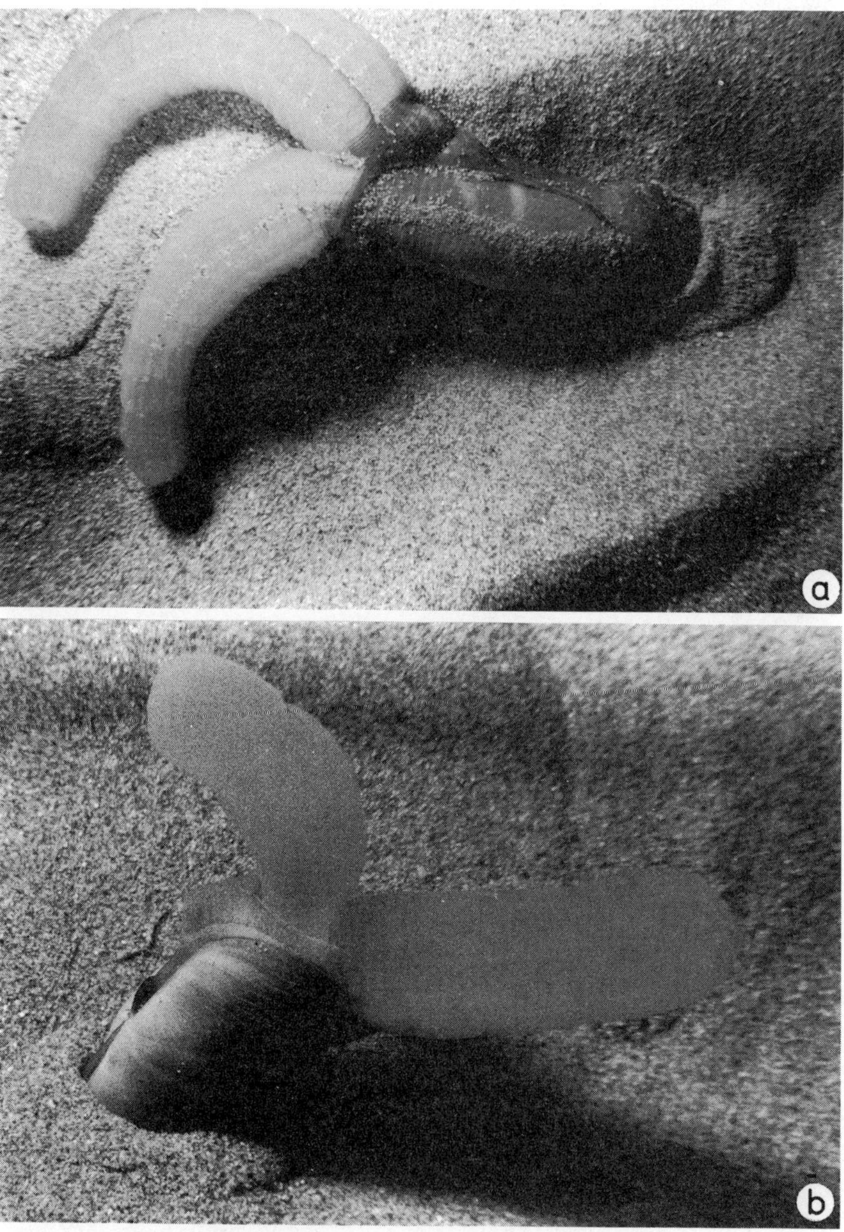

Figure 4.4 Exumed *Solecurtus strigilatus* burrowing. (a) Intruding the foot into the sediment. (b) A few minutes later the animal has drawn itself upright. Half natural size: –2 m, Kefallinia, Greece.

mantle cavity is enlarged as a long, muscular sack, the posterior extension of the mantle cavity (PM). This houses the gills and can relax to a length of more than twice that of the valves, or be contracted right into the shell (Fig. 4.4). At the posterior end of the PM is a pair of large, contractile siphons.

The burrower has effective escape reactions; attempts at digging the animal out causes it to excavate deeper or horizontally into the substrate and to shed the ends of its siphons. The slowly pulsating siphon ends thus liberated into the water act as an effective distraction for predator and curious scientist alike (cf. browsing predation, section 1.1.9).

An exhumed individual intrudes its foot into the substrate and, obtaining purchase with its terminal anchor, normally can raise its long body vertically within 2 min (Fig. 4.4). Within a further 2 min the animal has disappeared into the sand. Once buried, the double-anchor digging cycle begins, or 'push-and-pull action' as Seilacher and Seilacher (1994) called it, comprising five phases (Fig. 4.5):

1. The siphons and PM fill with water. The shell muscles are relaxed and the foot is contracted.
2. The siphons close and contract, forcing water into the distended PM. The foot extends as a folded knife blade.

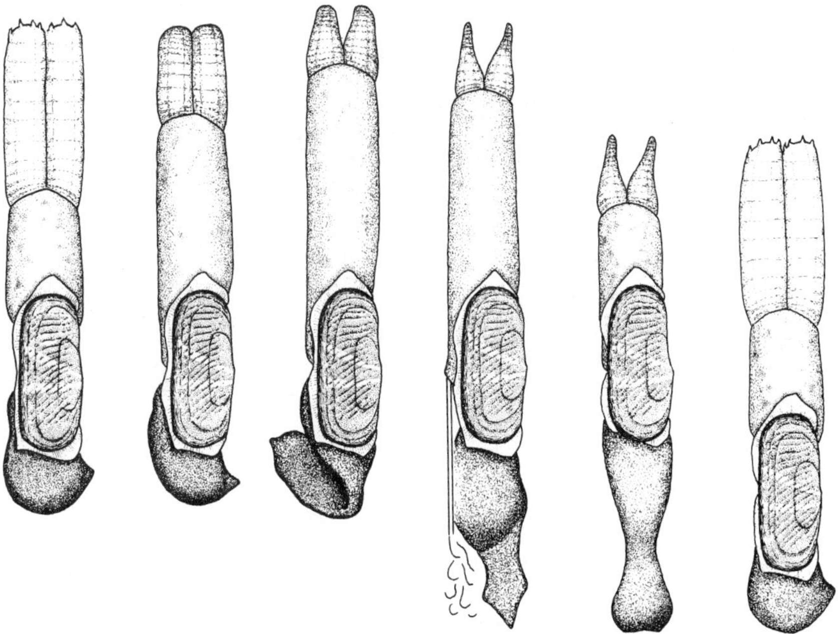

Figure 4.5 The digging cycle in *Solecurtus strigilatus*, modified after Bromley and Asgaard (1990). See text.

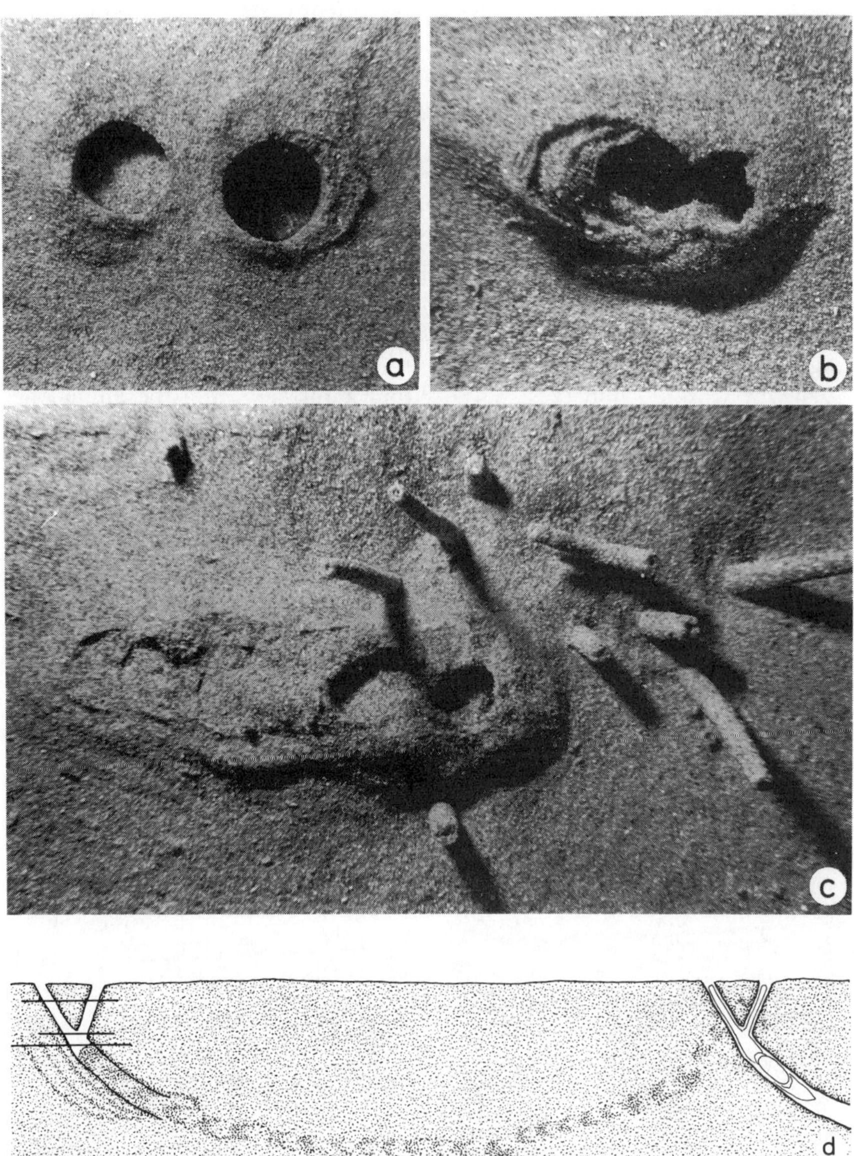

Figure 4.6 Burrows of *Solecurtus strigilatus* sectioned horizontally by carefully wafting the sand away at the sea floor. The positions of the sections are indicated in (d). (a) Mucus-soaked walls of the siphonal shafts resist the scouring action. Backfill from the escaping animal has blocked the lumen. (b) A short spreite is visible. (c) A long spreite soaked with mucus. The tubes are of the polychaete *Lanice* sp.: –2 m, Kefallinia, Greece. (d) Taking these photographs caused the animal to flee laterally, backfilling its domichnion and eventually establishing a new one some 2 m away.

3. Contraction of the PM forces a jet of water anteriorly out of an orifice near the base of the foot. The anterior shell muscles contract, and the hydrostatic pressure of the PM forces the posterior part of the shell against the burrow walls. The ornament of the shell inhibits backward slipping and anchors the shell.
4. The foot whips out along the water jet and plunges into the sediment liquefied by the jet; it curls and swells at its tip to initiate a terminal anchor.
5. The posterior shell muscles contract, releasing the penetration anchor, and the foot draws the shell down into the still liquefied sediment beneath it. This watery sand flows back past the now narrow shell and PM as the body of the mussel displaces it, and backfills the burrow behind. The shell muscles then relax again to re-establish the penetration anchor.
6. The contracted siphons open once more and admit water into the short PM, and expansion of both regions is resumed. When escaping, this cycle is repeated in a 10 sec rhythm.

The burrow walls (Fig. 4.6a–c) are copiously soaked with mucus within 48 hours of establishing a new burrow. In older burrows that have been inhabited for longer, a spreite reveals a rhythmic dorsalward migration of the burrow. This equilibrium structure, which may reflect repeated disturbance by passing epibenthic animals that irritate the siphon tips, resembles trace fossils of the ichnogenus *Teichichnus*. Apart from mucus impregnation, the burrow walls and spreite show no special lining material.

Escape is achieved, following the arc of the burrow, the animal returning to the vicinity of the sea floor about 1–2 m from the original burrow apertures (Fig. 4.6d), where the animal rapidly constructs a new burrow.

4.1.3 Equilibrium and escape in bivalves

The burrowing technique in bivalves is designed to achieve anteriorward or ventralward penetration of the substrate. 'Backwards' burrowing is a quite different matter and in fact few species attempt it. Experiments indicate that few bivalves are able to wriggle upward through even a thin layer of sediment suddenly deposited on top of them. Those that can, not surprisingly, tend to be species found in nearshore, sandy settings (Figs 4.7 and 4.8; Armstrong 1965; Woodin and Martinelli 1991). A detailed study of 25 species' escape potential by Kranz (1974) is recommended reading.

Fully grown *Mya arenaria* are rather immobile. Juveniles, however, can re-burrow, equilibrate and even escape deep burial, the last being accomplished by inverting and digging upwards foot-first (Fig. 4.8). Razor

Figure 4.7 Equilibrium structures produced by an upward migrating population of *Hiatella arctica* in an accreting sand situation. Pleistocene, Lønstrup Klint, Denmark. (a) Several individuals have died during their upward journey and are preserved as body fossils in life position. Traces about 2 cm wide, maximum retrusion observed at 30 cm. Growth during upward movement is not recorded by the trace. (b) Close-up of the meniscus backfill. Photos and data courtesy of Peter Johannesen.

shells, on the other hand, are too long to invert; they just push up (Kranz 1974).

S. strigilatus' most successful predator may be *Octopus vulgaris* (Fig. 1.3), which can rapidly intrude a tentacle into the burrow. Other incipient predators include the starfish *Astropecten* spp., most species of which feed on endobenthic bivalves (Fig. 4.9). The starfish burrow slowly and adult *S. strigilatus* can escape *Astropecten* spp. by rapidly burrowing. Christensen (1970) described how *Tellina fabula* escapes attack from *Astropecten irregularis* by burrowing laterally; once buried, the starfish cannot move laterally. *Cultellus* sp. withdraws rapidly deeper to beyond the reach of the starfish. A similar reaction is seen in *Mercenaria mercenaria* escaping from *Asterias forbesi* (Doering 1981).

Mya arenaria can neither escape laterally nor rapidly downwards; however, the great depth at which this species lives places adults within a refuge that is beyond the reach of most predators.

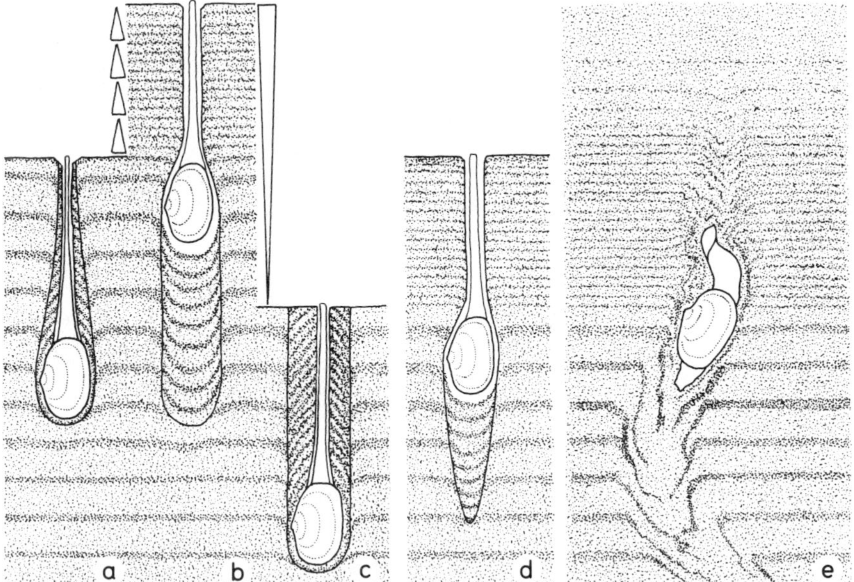

Figure 4.8 Structures produced by juvenile *Mya arenarea*. (a) In a non-depositional sea floor, growth of the bivalve causes a protrusive equilibrium trace. (b) Slow, even deposition causes the bivalve to re-adjust its position repeatedly relative to the accreting sea floor, producing a retrusive equilibrium trace. (c) Response to gradual scour is to dig continually into the substrate, resulting in a protrusive equilibrium structure. (d) Extremely slow, even sedimentation, nevertheless a little faster than the growth rate of the animal, forces the bivalve to creep gradually upwards. The retrusive equilibrium structure reflects the growth vector. (e) A sudden depositional event releases escape reaction in the buried bivalve, which intrudes foot-first upward, creating a spiral escape structure. (a–d) Modified after Reineck (1958, 1970) and (e) after Schäfer (1962).

4.1.4 Bivalve chemosymbionts

Several bivalve groups practise endosymbiosis with sulphuricant microbes. Two of these are well documented and produce distinctive burrows.

(a) Lucinacea

This superfamily comprises three families. All species of two of these, Lucinidae and Thyasiridae, have symbiosis with autolithotrophic bacteria (Dando and Southward 1986; Dando *et al.* 1986; Reid and Brand 1986; Reid 1989). This has brought about considerable modification in the bivalves' anatomy and behaviour. Most species are deep burrowers.

The lucinaceans have only a posterior, exhalent siphon. The inhalent current arrives through an anterior mucous-lined tube, which is constructed by the foot (Allen 1958). Thus, a deep, U-shaped burrow is cre-

Figure 4.9 The starfish *Astropecten* sp. from shallow water, Kefallinia, Greece, seen by an X-radiograph. The stomach is full of newly-injested burrowing bivalves. The starfish cannot move laterally in the sediment. It must hunt its prey from the surface and burrow down for each bivalve. In this way the predator is responsible for considerable bioturbation.

ated. Dando *et al.* (1986) suggested that HS^- could enter the inhalent stream through the wall of the anterior tube, below RPD, and thereby provide the fuel needed by the bacteria, which are lodged in the host's gills.

Dando and Southward (1986) noticed in two species of *Thyasira* that the worm-like foot probed the sediment beneath the shell for about 15 times the shell's diameter, producing ramifying, root-like systems (Fig. 4.10). Similar, but simpler, systems of canals are also produced by lucinids. These canals conform perfectly to the function of sulphide wells (Seilacher 1990).

The Lucinacea is an ancient group of bivalves, and Liljedahl (1992) has proposed that the earliest recognized species, *Ilionia prisca* from the Silurian, also practised chemosymbiosis.

(b) Solemyacea

This small group of species is also ancient. Species of *Solemya* produce a U-burrow and likewise extend this downwards with a deep shaft (Fig. 4.11). These bivalves possess no siphons, but move freely about the U-burrow. Their gills are richly laden with bacteria and the gut is greatly reduced or, in some species, missing altogether. *Solemya* spp. occur in sea-grass beds and in polluted harbour basins where oxygenation is poor.

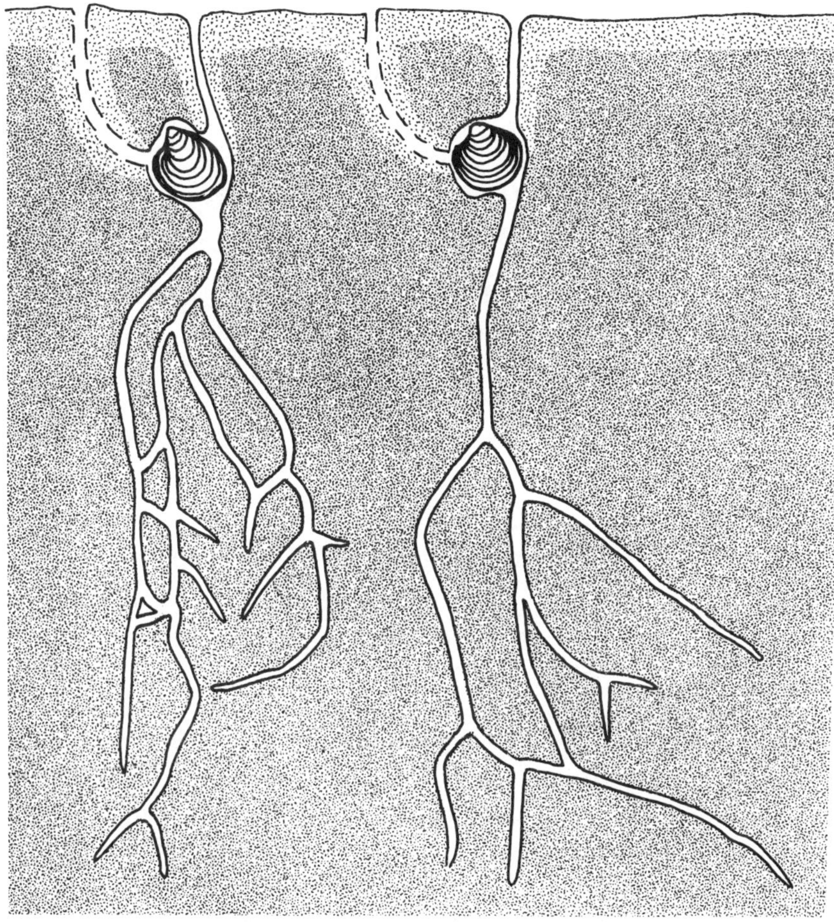

Figure 4.10 Burrow systems of two individuals of *Thyasira flexuosa*. The very extensible foot builds the inhalent tube (broken line) to the surface, and a sulphide well to about 15 cm down. The well canal is shifted from time to time, producing a plexus. Modified after Dando and Southward (1986).

Chemosymbiosis has been demonstrated in modern species by Felbeck (1983), Felbeck *et al.* (1984) and Reid (1989). The shaft is envisaged as a sulphide well from which pore water is pumped from levels well beneath the sea floor.

The cross-section of the U-burrow is elliptical, the longest diameter lying in the plane of the U. On the assumption that this feature is restricted to burrows of Solemyacea, Seilacher (1990) named Y-shaped elliptical trace fossils *Solemyatuba curvatus*. He also considered the elliptical, shaftless U-burrow *Arenicolites curvatus* of the Devonian (Goldring 1964) as the work of a solemyacean and has identified similar structures in the Ordovician.

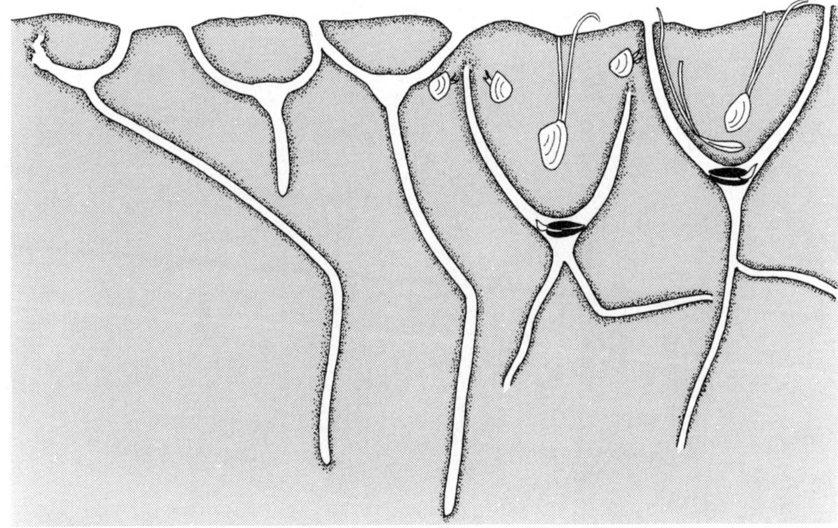

Figure 4.11 Y-burrows of the sulphide chemosymbiont bivalve *Solemya velum*. The animal spends much time in the chamber in the U-form, where the walls are mucus laden. The apertures are frequently closed, partly owing to the interference of the upper-tier bivalve species (tellinids and nuculoids are shown). *S. velum* is capable of withstanding low oxygen levels during these times. The lower shaft, descending deep into anoxic regions, invites comparison with some crustacean burrows (Figs 4.24 and 4.28) and is interpreted as a sulphide well. *S. velum* shifts position every few days, as indicated by the two burrows in contact. Data from Frey (1968), Stanley (1970) and Levinton and Bambach (1975).

Campbell (1992) pointed out that species of *Solemya* also thrive in the proximity of petroleum seeps in methane-rich surroundings. The burrows of these *Solemya* are lined with authigenic carbonates, which should improve their potential for preservation as trace fossils. Campbell (1992) illustrated Miocene examples.

4.2 Two heart urchins of the same genus

The Spatangoida is the order of irregular echinoids best equipped for an endobenthic life (Fig. 4.12). Development of fascioles has provided an efficient means of irrigating a burrow, and specializations of spines and podia (tube feet) have opened up possibilities for exploiting many lifestyles and sediment types. Since the late Cretaceous the Spatangoida has been a very successful group.

Although the biology of many spatangoid species is well known, few studies have been made of the animals' behaviour within their substrate. A breakthrough was made by Nichols (1959), who studied *Spatangus purpureus* and *Echinocardium cordatum* in aquaria and in the field. The studies by Chesher (1963, 1968, 1969) on the burrowing activity of *Moira atropos*, *Brissopsis alta* and *Meoma ventricosa* (Fig. 4.13) are most useful

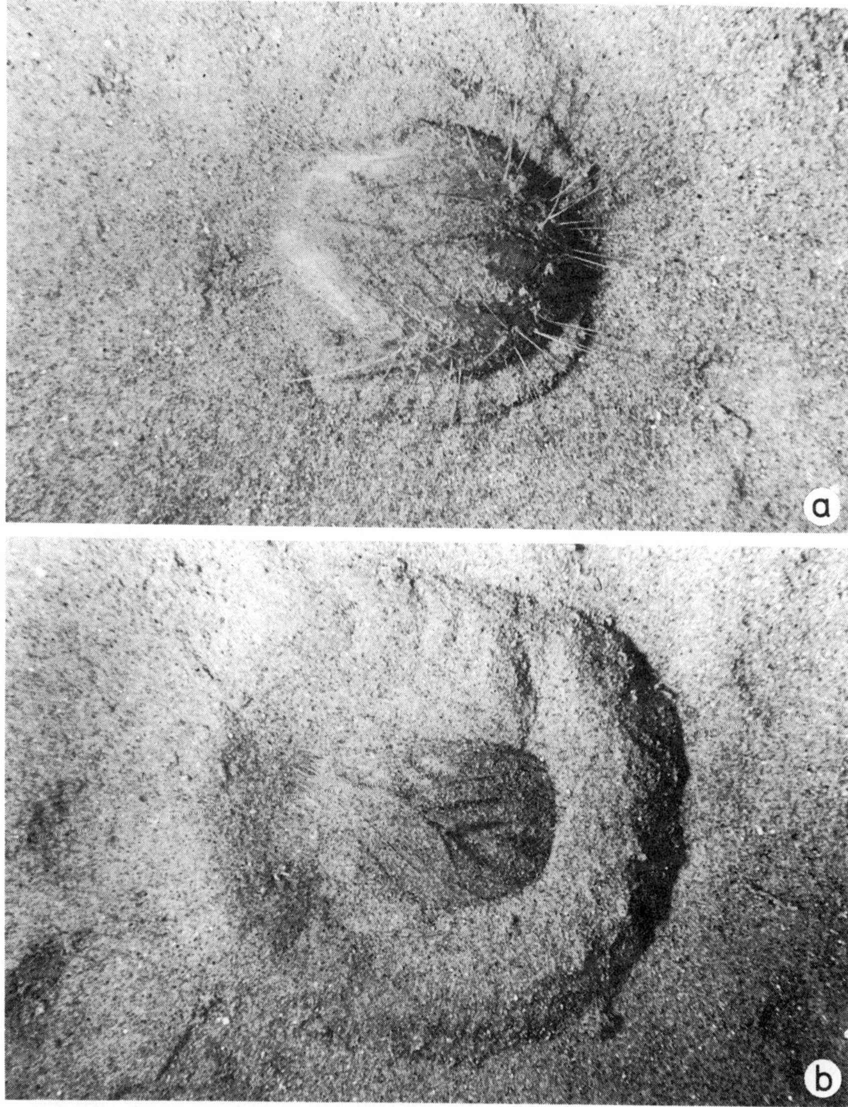

Figure 4.12 Re-entry into the substrate by a spatangoid echinoid. (a) When exhumed, *Lovenia elongata* spreads a fearsome array of barbed spines for protection. (b) As it excavates its way downwards, these are folded down. The animal is covered in less than two minutes and continues down a further 1 cm: –2 m, Gulf of Elat.

for the ichnologist, and the careful investigations by Kanazawa (1991, 1992) have revealed unexpected variety in detail of the spine movements and particle manipulation in different spatangoids.

The species that is best known is *E. cordatum*. Ulla Asgaard, Margit Jensen and I have studied the small individuals of mud-dwelling

Figure 4.13 Two views of the slow-moving giant spatangoid *Meoma ventricosa*, placed on the sea floor 15 minutes earlier. It has nearly covered itself. Note the mucus-laden carpet of sediment passing up over the spines; also, the bioturbated texture of the sediment surface, dominated here by this species: –12 m, Six Men's Bay, Barbados.

communities in the Øresund, between Denmark and Sweden. By way of contrast, we have also examined in the field and laboratory the behaviour of *E. mediterraneum* from sandy softground at Kephallinia, Greece.

4.2.1 An almost anoxic heart urchin

Echinocardium cordatum is surprisingly cosmopolitan as regards the type of substrate. In clean sand in intertidal settings individuals reach 4 cm long and burrow 15–20 cm deep. In deeper water and muddy substrates the urchins do not grow so large and burrow no deeper than 3–5 cm (Nichols 1959; Buchanan 1966).

Small individuals, working in mud at 4 cm below the surface, move forward at a few centimetres per hour (at 9°C). They progress in steps of 6–24 mm, separated by long pauses. At each pause a slender shaft is constructed. About one shaft is produced a day, although disturbed animals may show spurts of three shafts an hour.

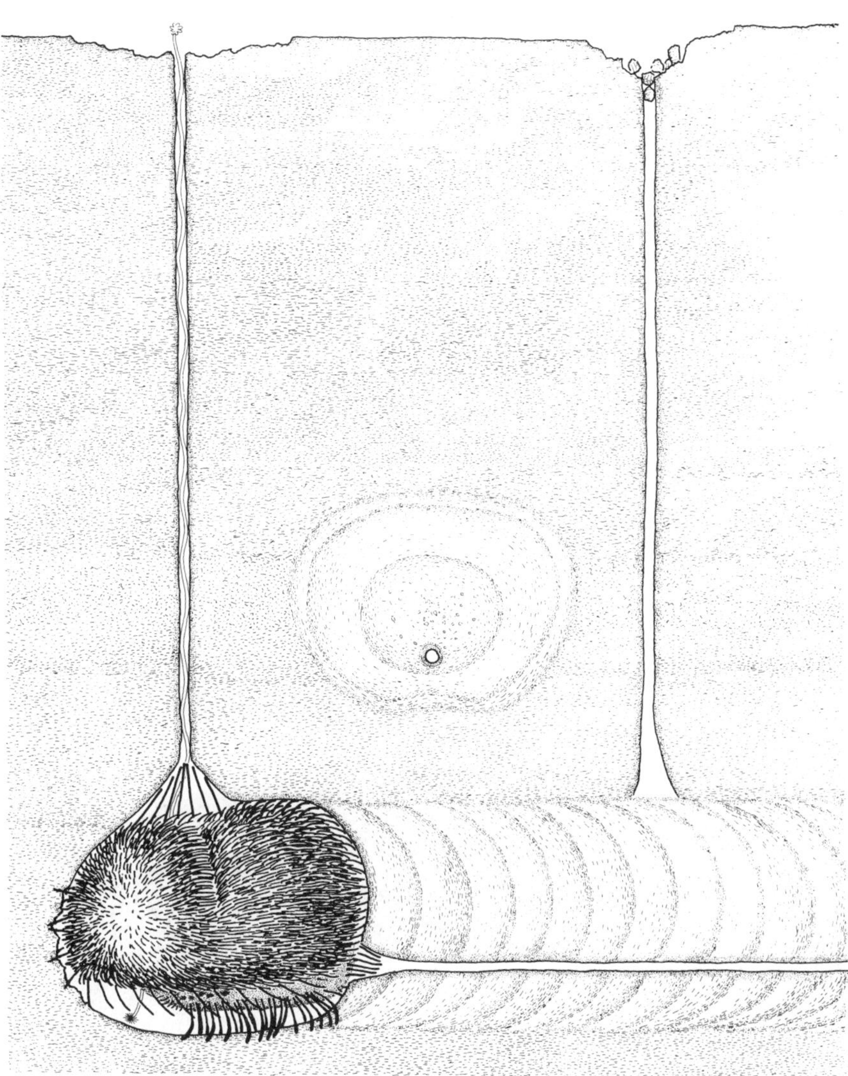

Figure 4.14 *Echinocardium cordatum* in its burrow. Behind it an abandoned shaft remains open in the muddy substrate. Above, a cross-section of the backfill and subcentral drain. About natural size.

E. cordatum constructs a well-defined burrow, comprising three parts. Around the animal is a **burrow chamber** bounded by a mucous envelope that is supported on the curved tips of the spines that cover most of the urchin. The mucous sheath also lines the **shaft**, and a similar, horizontal canal that is constructed as a **drain** behind the animal (Fig. 4.14).

All the spines are covered thinly by mucus-secreting epithelium, so their slimy tips unite with the burrow-wall envelope. Frontally, long spines penetrate the mucous wall and break up the substrate ahead of the animal. On the underside, the tips of propulsive, spade-like spines protrude and dig into the substrate. The only other parts of the animal to protrude outside the confines of the burrow are two types of podia. Food-gathering podia stretch up the shaft and out onto the sea floor to gather detritus; and sensory podia press their pointed ends through the envelope to explore the substrate outside. The remaining podia work within the burrow.

As the animal advances, sediment is entrained in the mucus that covers the transporter spines and the burrow boundary comes to consist of a 'traction carpet' of mucous-bound sediment passing continuously backwards. The elastic carpet is redeposited as a meniscus-backfill lamina behind the animal (Figs 4.14 and 4.15).

Podia excavate a drain behind the animal. It is mucus coated and impermeable, and serves to carry used respiration water away. The drain is inhabited by meiofauna and bacteria, and by the time the urchin has burrowed onward some tens of centimetres the mucus breaks down, allowing escape of the water into the porosity of the sediment.

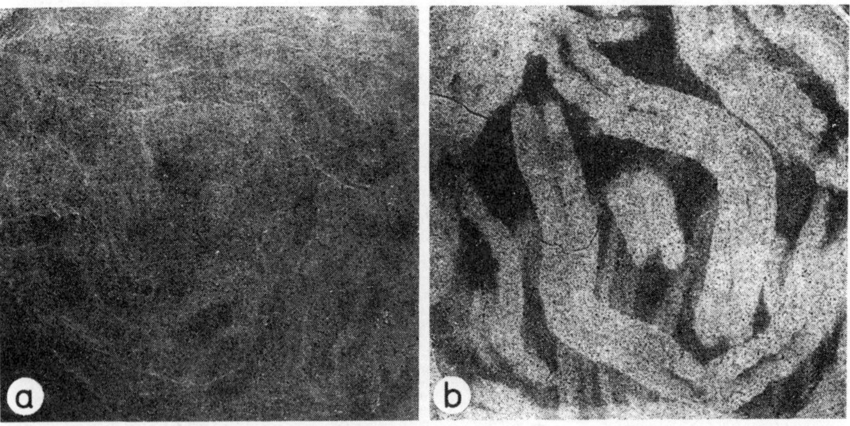

Figure 4.15 Backfill of *Echinocardium cordatum*, seen through the glass floors of aquaria. (a) Unmixing: in a poorly sorted sediment, pale and dark grains are separated in the backfill. (b) Mixing: in a laminated sand, the same activity nevertheless reveals mixing; a light-coloured lamina is advected down into a dark one beneath. Half natural size.

The water flow created by the fascioles is slow but sure. Water enters the shaft aperture, passes over the respiratory podia and leaves via the drain.

Four to six feeding podia stretch up the shaft one at a time and grasp food particles on the sea floor. These are dropped onto the frontal ambulacrum, entrained in mucus and passed along the ambulacral groove, where enzymes are secreted and digestion is initiated, before reaching the mouth on the underside (Pequignat 1970).

Quantities of anoxic sediment are also ingested from the level of burrowing and the gut is always full of sand. At the posterior end of the intestine there is a sizable caecum containing sulphide-oxidizing bacteria (De Ridder and Jangoux 1993). This has led Bromley *et al.* (1995) to suggest that this deeply burrowing species obtains part of its nutrition from symbiosis with these bacteria. Mistakes in the laboratory revealed that *E. cordatum* has an extraordinary resistance to periods of anoxia. The animal is well suited to life within anoxic sediment although, in the long run, both urchin and bacterial symbionts are dependent on the trickle of oxygenated water drawn down from the sea floor.

During surface feeding and tube building, sediment collapses down the shaft and is incorporated into the backfill (Nichols 1959), as is the excrement. Since, in this way, more material is deposited behind the animal than is excavated ahead of it, the backfill must have a slightly higher degree of compaction than the surrounding sediment.

The importance of detritus feeding in this species is well known (Buchanan 1966; Thorson 1968; Pequignat 1970; De Ridder *et al.* 1985). Indeed, the sunken frontal ambulacrum in this and many other spatangoid species would seem to have developed for detritus feeding (Chesher 1963). *E. cordatum* is a deeply burrowing spatangoid, ingesting nutritionally poor, sulphide-rich sediment, probably for chemosymbiosis. It boosts this diet with organic-rich detritus which, during digestion, will produce further sulphides for the bacterial symbionts that are housed in the posterior intestinal caecum.

As a result, the activity of this species causes minor downward advection of sediment. Its passage through the sediment increases its content of combustible organic matter from the excrement and copious mucus incorporated in the backfill.

4.2.2 A very oxic heart urchin

E. mediterraneum superficially resembles *E. cordatum*, though rather less so when deprived of its spines. *E. mediterraneum* is then seen to be not heart-shaped, lacking the frontal groove, and having a circular outline and flat base. Almost nothing has been published of its life habits. The species is common in well-washed to silty, medium-grained sand in shallow water in the eastern Mediterranean.

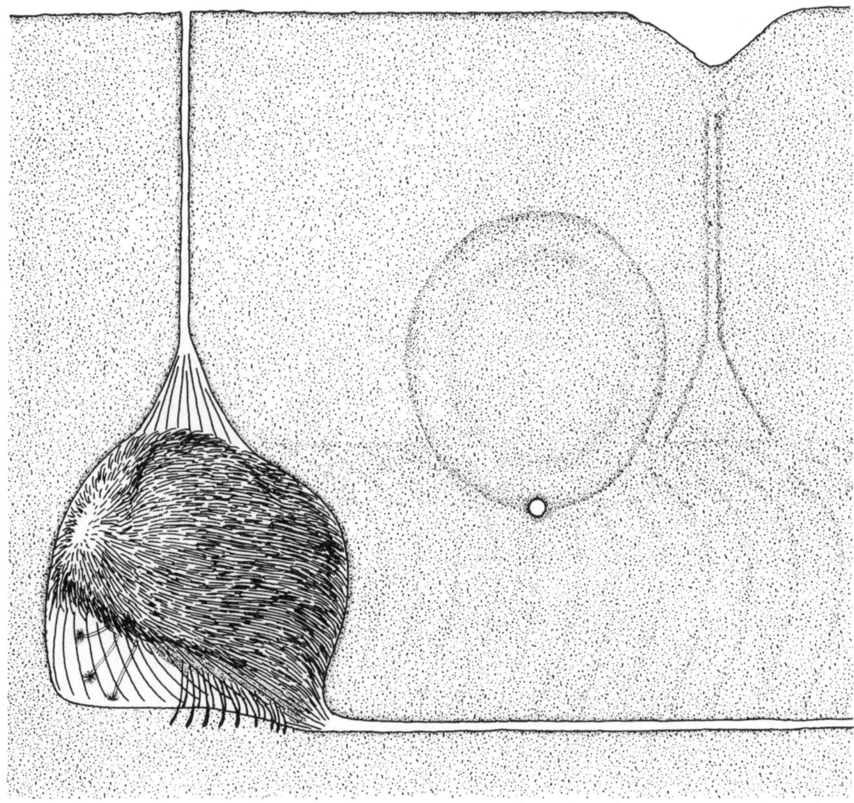

Figure 4.16 *Echinocardium mediterraneum* in life position, deposit feeding within its burrow. The abandoned shaft has collapsed and filled with sand. A cross-section of the backfill shows the low position of the drain.

The animals live at 2–5 cm below the sea floor. Maximum forward progress is considerably faster at their summer temperature than the Øresund *E. cordatum* at theirs: some 10 cm per hour at 15°C. However, in aquaria, stabilized individuals normally progress slowly, at about 1–2 cm per hour, steadily moving forward.

The position of *E. mediterraneum* in its substrate (Fig. 4.16) is quite different from that of *E. cordatum*. The frontal end of the animal is raised so that the body axis lies at about 30° to the horizontal. At intervals a thin shaft is constructed to the surface as in *E. cordatum* but, once completed, no detritus-collecting podia are seen at the aperture.

Around the frontal margin of the sharp ambitus is a row of long, curved spines that define the front of an open space beneath the flat basal surface of the sea urchin. Within this suboral chamber, the large oral podia actively process the sediment (Fig. 4.16). Grains are constantly picked up and dropped or stuffed into the mouth as the animal gradually moves forward through the sediment.

In other details, burrowing resembles that of *E. cordatum*. The backfill contains a single mucus-walled drain that cuts through the meniscus structure and remains open far behind the animal.

E. mediterraneum appears to be purely a deposit feeder. It causes much horizontal transport of grains and only very insignificant vertical advection of particles during shaft construction. As in *E. cordatum*, a characteristic mucus-bound backfill structure is produced (Fig. 1.4).

4.3 Anomuran crustacean burrowers

The burrows of anomuran Crustacea are celebrated for ichnologists, mainly because of constructional parallels drawn between them and the ichnogenera *Ophiomorpha* and *Thalassinoides*, which are so abundant in the rock record (Fig. 8.12).

Since the beginning of the Mesozoic, the anomurans have developed many different trophic strategies within detritus, deposit and suspension feeding, and gardening, in various combinations. The thalassinidean genera *Callianassa*, *Callichirus* and *Upogebia* are the most important today among the fossorial anomurians (Fig. 4.17).

As the burrow morphologies of an increasing number of species are described, a pattern is emerging that could also be recognizable among trace fossils. In the following, a few examples of the characteristic trophic groups and corresponding burrow types are treated.

4.3.1 Callichirus major

A large callianassid species, *Callichirus major*, has invaded the medium-energy sand beaches of the Atlantic coasts of southeast USA. This species was placed in *Callianassa* until the revision of genus *Callichirus* by Manning and Felder (1986). Its behavioural and morphological adaptations to loose substrate and physically stressful environments have rendered this species perhaps the most talked about of all invertebrate tracemakers, owing largely to the similarity of its burrows to the trace fossil *Ophiomorpha nodosa*, as demonstrated by Weimer and Hoyt (1964).

The burrows of *Callichirus major* extend so deeply into the sediment that the lower reaches are largely inaccessible. Epoxy casts of those regions simply cannot be retrieved intact (Frey 1975). According to Pryor (1975) the burrow consists of two elements: a horizontal maze of mud-walled tunnels that are 2–4 m deep and a series of shafts connecting the maze with the surface. In the higher levels of the beach the shafts may reach 5 m deep (Frey *et al.* 1978). No detailed illustration has been published of the deeply hidden maze (Fig. 4.18). However, Pryor (1975) was able to pump 100 individuals out of 10 surface apertures, which he took

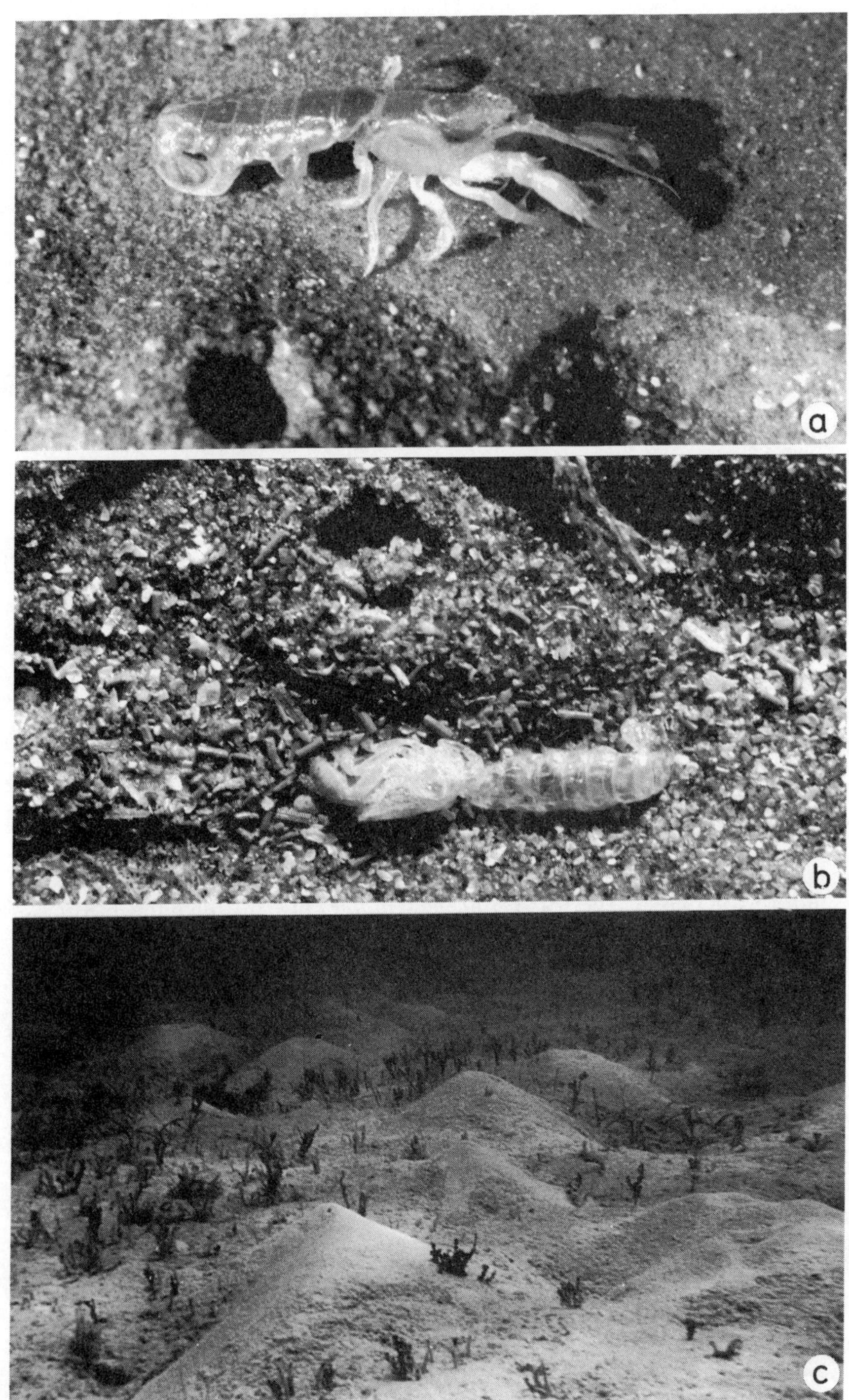
a
b
c

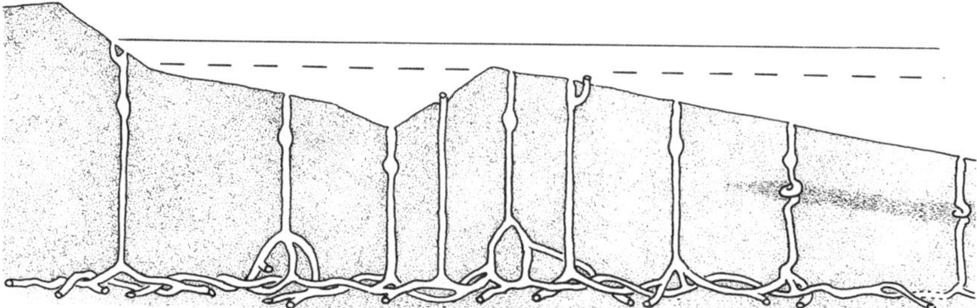

Figure 4.18 Diagrammatic beach profile 3 or 4 m high, showing the burrow system of a population of *Callichirus major*. Burrow thickness exaggerated. Mean high and low tide levels are indicated. Modified after Frey *et al.* (1978).

to indicate that the animals live communally within a continuous burrow system.

Earlier authors considered *Callichirus major* a deposit feeder (Pohl 1946), but the faecal pellets are composed of fine material that is hardly represented in the clean sand substrate and has clearly been derived from suspension. Pryor stated that, while feeding, the animals occupy the upper part of the shafts and set up an inward flow by rapid beating of their pleopods. The water passes through a net of hairs on the maxillipeds; seston is sifted out and passed to the mouth. After 15–30 min, flow is reversed for some minutes, during which time faecal pellets are discharged from the burrow entrance. There is then a pause of a few minutes, possibly during which time animals may move into feeding position.

Callichirus major fits its burrow diameter tightly (Fig. 4.19b) and the narrowed apertures accelerate the current and ensure effective removal of the faecel pellets. The narrow opening also protects the burrow from invasion by predators or sediment; normally the shrimps never leave their secure labyrinth. The turbulent environment produces a need for repair and alteration of the apertures (Fig. 4.20), but the deeper parts of the shafts are relatively permanent and may become overgrown by bryozoans (Pohl 1946).

Figure 4.17 Sediment processing by callianassids. (a) An unhappy crustacean, *Upogebia* sp. (3 cm long), exposed to sunshine and air outside its burrow entrance. Foreshore at Puerto Peñasco, Mexico. (b) The work of the same species. Three types of grains have been advected upwards: around the burrow aperture (top centre) a conical pile of sand excavated during burrow construction 40 cm beneath the surface; cylindrical faecal pellets of mud derived from suspension feeding; and the shed skin of the burrower. The original substrate surface is seen in the foreground. (c) Mounds produced by deposit-feeding *Glypturus acanthochirus* (see Fig. 4.25c): –4 m, Great Bahama Bank, pelleted mud facies.

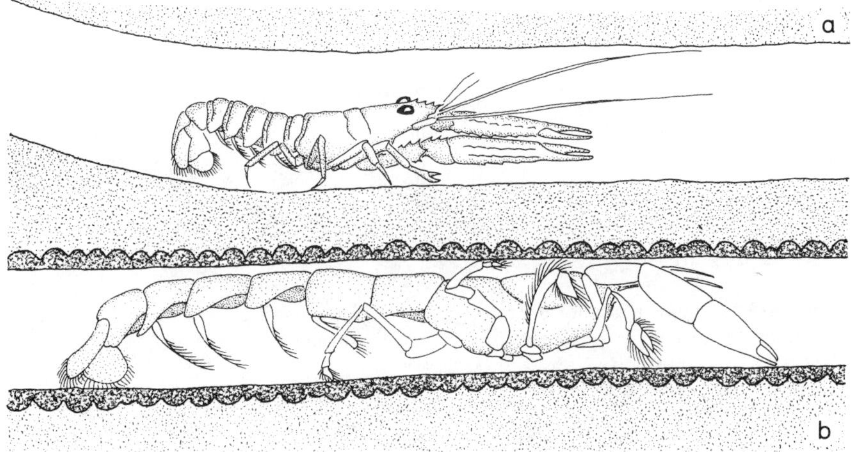

Figure 4.19 Burrows of the lobster *Nephrops norvegicus* (a) and *Callichirus major* (b) drawn at the same diameter, to show the relative degree to which their occupants fill the lumen. Modified after Bromley and Asgaard (1972b).

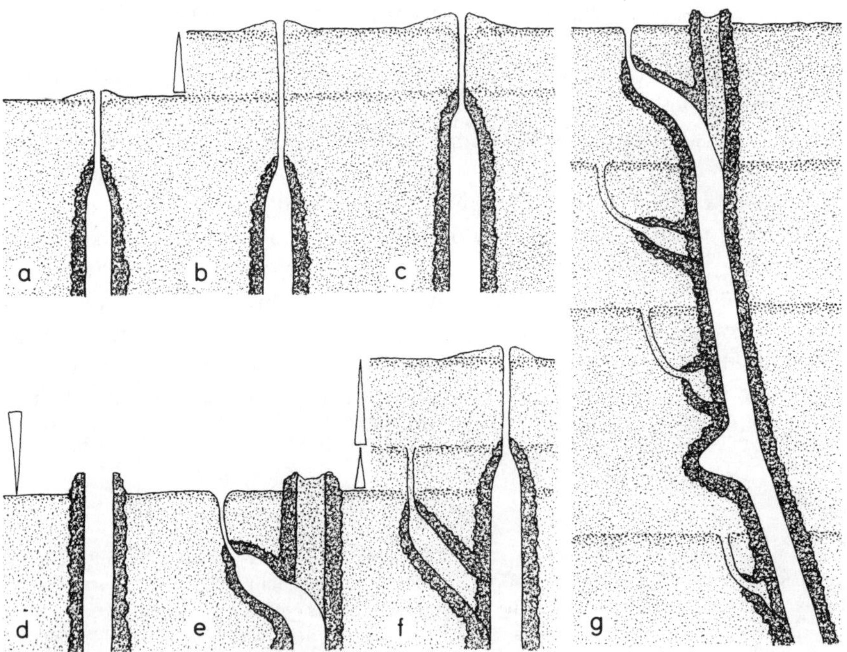

Figure 4.20 (a–f) Response of *Callichirus major* to erosional damage and burial of its burrow apertures. (g) An example of *Ophiomorpha nodosa* from the Pleistocene of Florida. Modified after Howard (1978).

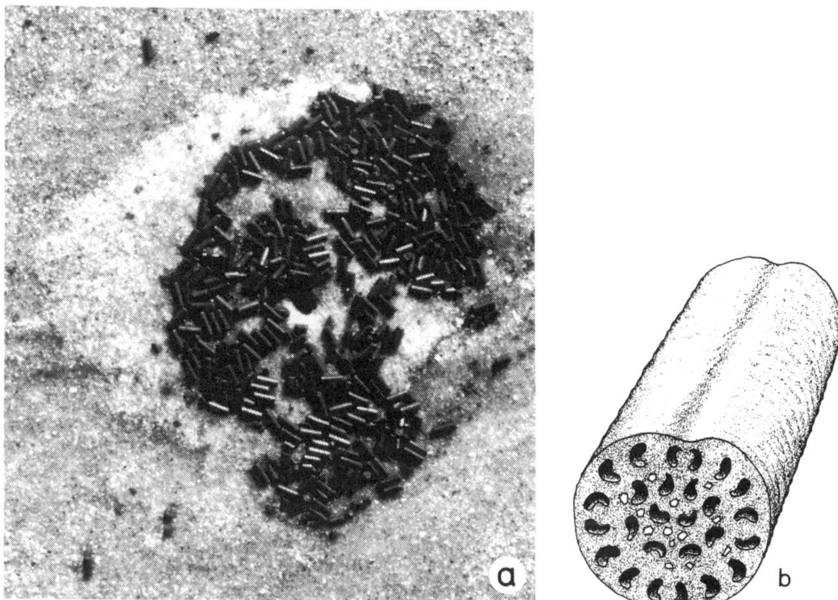

Figure 4.21 Faecal pellets of *Callichirus major*. (a) A narrow burrow aperture on a Georgia beach. Three sediments are represented: the general surface of the foreshore; a water-washed cone of pale sand advected from perhaps several metres below the surface during burrow enlargement; and dark faecal pellets from suspension feeding. (b) A faecal pellet about 1 mm in diameter, modified after Pryor (1975).

The faecal pellets are compact and do not disintegrate readily. As is usual among anomuran crustaceans (Moore 1939), the pellets are rod-shaped and contain a number of canals running their length, 26 in this species (Fig. 4.21). Copious quantities of faecal pellets are produced. Pryor recorded 40 pellets per minute leaving individual apertures during excurrent intervals. As they consist of fine sediment that is normally in suspension mode in the beach environment, the deposition of these sand-size pellets constitutes a significant biodepositional input of mud into otherwise clean sands. Reworking of the pellets can accumulate the mud in extensive beds many centimetres thick (Pryor 1975).

Another increment of mud contributed biogenically to the sand by *Callichirus major* is in the form of the wall material of the burrow. In order to stabilize the loose sand, sandy mud bound with gelatinous mucus is applied to the walls by the animals in the form of round pellets. These project into the surrounding sand as hemispherical warts on the outside of the lining layer, but inside they are smoothed flat. The shafts receive the thickest layers, in places as thick as the diameter of the burrow lumen (Pohl 1946). This mud does not appear to derive from excrement as the rigid faecal pellets would be visible if present. Since they are not available in the substrate sediment, the fine constituents of the wall material also must be caught from suspension.

4.3.2 Boxworks for deposit feeding

Unlike *Callichirus major*, most of the fossorial species of *Callianassa* and *Callichirus* appear to have a basically deposit-feeding mode of life. Even though but a small number of species has been studied, a pattern among their sediment-processing activities is appearing. It may be premature to attempt to classify burrow types, but herein I shall risk making such a preliminary gesture. It would seem that three recurring plans of deposit-feeder architecture are recognizable: mazes and boxworks, spirals, and dendritic structures.

Callianassa californiensis is perhaps the best known constructor of maze systems; the species occurs mainly in muddy sand in tidal flats of the northeast Pacific. However, many other species seem to show similar behaviour, such as *Callichirus islagrande*, preferring sand, and *Callianassa jamaicense louisianensis*, preferring mud (Phillips 1971).

MacGinitie (1934) described the mode of burrowing in *Callianassa californiensis*, the animal carrying excavated sediment in a basket of hairs on its limbs. The diameter of the burrow is minimal and turning chambers are constructed where the animal may somersault and reverse its direction. The animal works continually; when not actively burrowing it is cleaning itself. The excavated material is sorted for food; indegestible material exceeds digestible and is heaped up, ultimately to be carried to an exit and wafted outside. The animal never leaves its burrow.

Communication between neighbouring burrows is avoided, breakthroughs being blocked and burrowed around. Abandoned passages collapse and spoils are carried up to the sea floor, producing conical mounds around exits. This material buries the detritus and, with the continuation of this conveyor activity by the whole population, the organic-rich surface material is progressively worked down to feeding levels. By this time a nice microbial culture has matured; bacteria are probably the main food (MacGinitie 1932). In this way the uppermost 70 cm of the sediment may be reworked by a dense population in less than a year, producing graded bedding (MacGinitie 1934; Warme 1967).

The South African species *Callianassa kraussi* produces similar effects (Branch and Pringle 1987). Its bioturbating activity caused a marked increase in quantities of bacteria in the sediment, and living diatoms were buried to a depth of at least 40 cm.

Feeding on buried organic-rich sediment brings *C. californiensis* into intimate contact with reducing pore waters; R. Thompson and Pritchard (1969) found this species to have an unusual ability to survive under anoxic conditions.

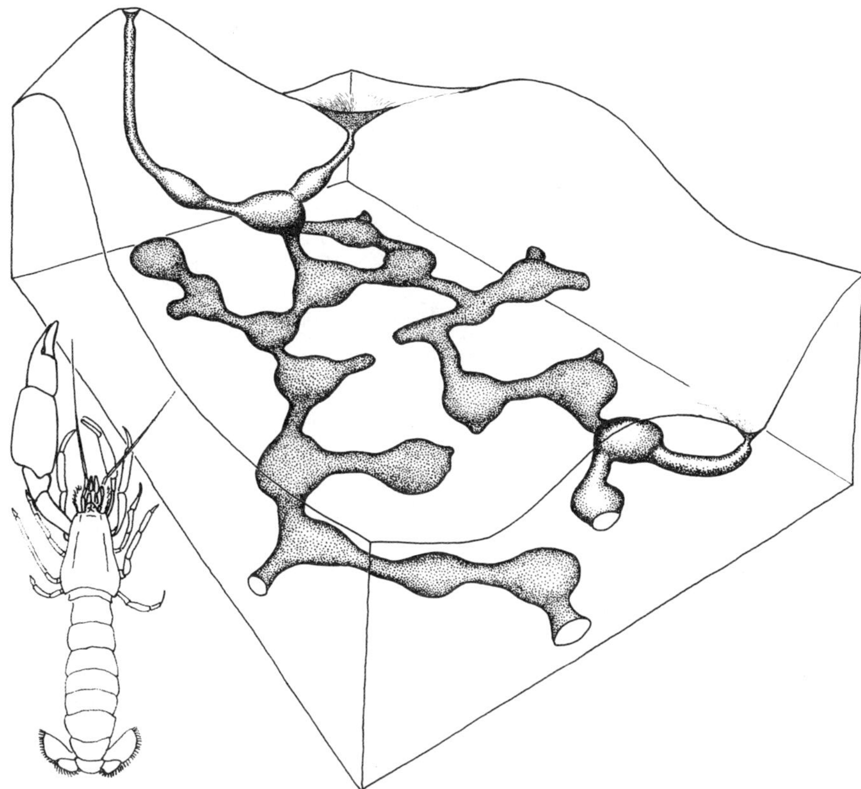

Figure 4.22 Burrow maze of *Callianassa californiensis*. The deeper of the two tiers is stippled darker. Based on Swinbanks and Murray (1981).

Swinbanks and Murray (1981) and Swinbanks and Luternauer (1987) made resin casts of several burrows (Fig. 4.22).

The apertures are constricted and lead via oblique shafts to a generally horizontal gallery. This is much branched dichotomously, having turning chambers at nodes. There is wall lining only at the apertural necks. Indeed, L. Thompson and Pritchard (1969) found the burrows so weakly constructed that they readily collapsed. But this collapse of the feeding galleries is an integral feature of the deposit processing; only the respiration canals should preferably show some stability.

This picture of a purely deposit-feeding system is disturbed by a report by Powell (1974) that the stomach contents of *Callianassa californiensis* may include significant quantities of plankton. This would imply dietary supplement from suspension feeding during high tide.

Several species of *Callianassa* behave similarly to *Callianassa californiensis* (Fig. 4.23). In the upper shoreface of Georgia, burrows of *Callianassa biformis* in sand are straight and little branched, having strong

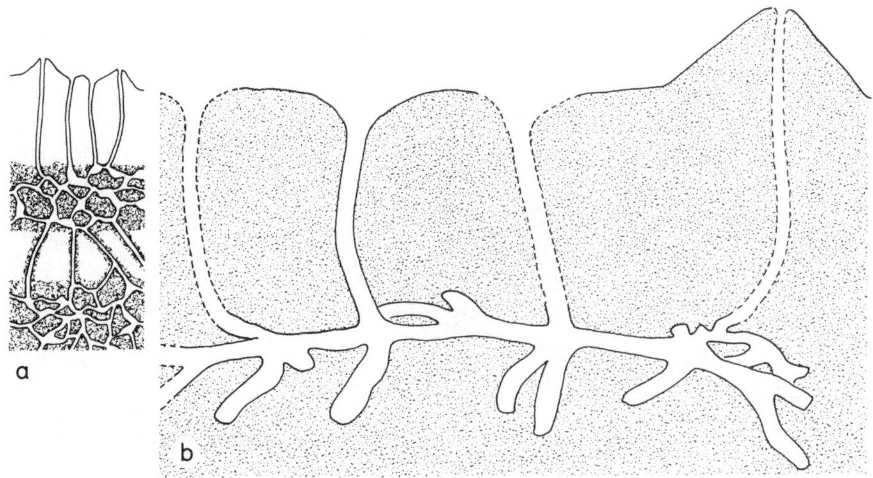

Figure 4.23 Two deposit-feeder callianassid burrow systems drawn to the same scale. (a) The tiny *Callianassa biformis*, height of section 40 cm. Modified after Hertweck (1972). (b) *Callianassa* sp. Modified after Suchaneck *et al.* (1986).

mud walls like the much larger *Callichirus major* burrows. They show lateral migration which produces spreite-like structures. In mud, deposit feeder boxwork systems are produced, showing no visible wall structure (Fig. 4.23a; Hertweck 1972). *Callianassa jamaicense*, although preferring to burrow in mud, can manage sand if there is a source of mud with which to line its sand walls (Phillips 1971).

Similarly, the European *Callianassa subterranea* applies no special walling in mud substrates, whereas sand is stabilized with mucus (Lutze 1938). *C. subterranea* produces a nodular network like that of *C. californiensis* in mud, at nearly half a metre beneath the sea floor. Atkinson and Nash (1990) could find only a single shaft to the surface in most burrows they cast, and also assumed that this species has a strong ability to withstand periods of anoxia. There is a turning chamber near the top of the shaft; perhaps this is used for respiration. A few burrows possess a long additional gallery that proceeds straight or horizontal, or obliquely downwards, from the network.

Howard and Frey (1975) and Dworschak and Ott (1993) described the work of a snapping prawn (Caridea), *Alpheus heterochaelis*, in the tidal creeks of Georgia (Fig. 4.24). The structure consists of several tiers of branched galleries, beneath which a zig-zag shaft extends deeper into the substrate. The purpose of this shaft must be different from that of the boxwork above, in which several prawns live communally. Indeed, this shaft invites comparison with similar strucures in *Upogebia* burrows (section 4.3.4).

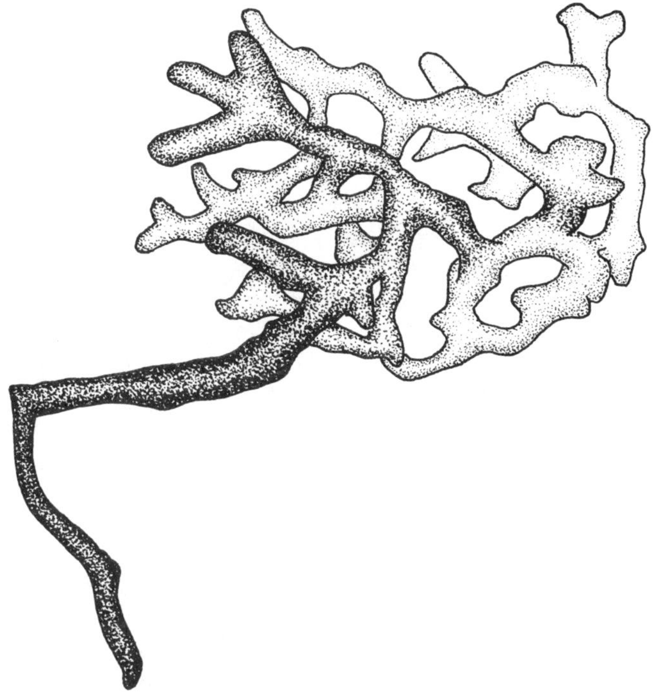

Figure 4.24 Boxwork system of *Alpheus heterochaelis* seen obliquely from below. Mazes at four levels are indicated by intensity of stippling, the palest shallowest. Modified after Bromley and Frey (1974).

4.3.3 Spiral and dendritic architecture

Burrows of some anomurans are constructed on a spiral principle, although the presence of lateral branches and flattened feeding galleries at several levels masks the helical form. For example, Dworschak (1987a) found *Callianassa tyrrhena* to construct a spiral up to 62 cm deep; a very similar burrow was illustrated by Braithwaite and Talbot (1972, pl. 3) from the Seychelles; a spiral arrangement also characterizes the burrows illustrated by Shinn (1968) and Enos (1983, fig. 4) from the Bahamas; and by Farrow (1971, figs 12B and 14A) from Aldabra Atoll (Fig. 4.25).

Spiral burrows are constructed also by the thalassinidean shrimp *Jaxea nocturna* in the Adriatic Sea (Fig. 4.26; Pervesler and Dworschak 1985). In these and the spiral *Callianassa* burrows, sediment mounds at the exit shafts indicate deposit feeding.

Colin *et al.* (1986) studied the burrowing activity of several callianassids at Eniwetak Atoll and noted the importance of resuspension of fines as excavated material was wafted out of the burrow on a water current.

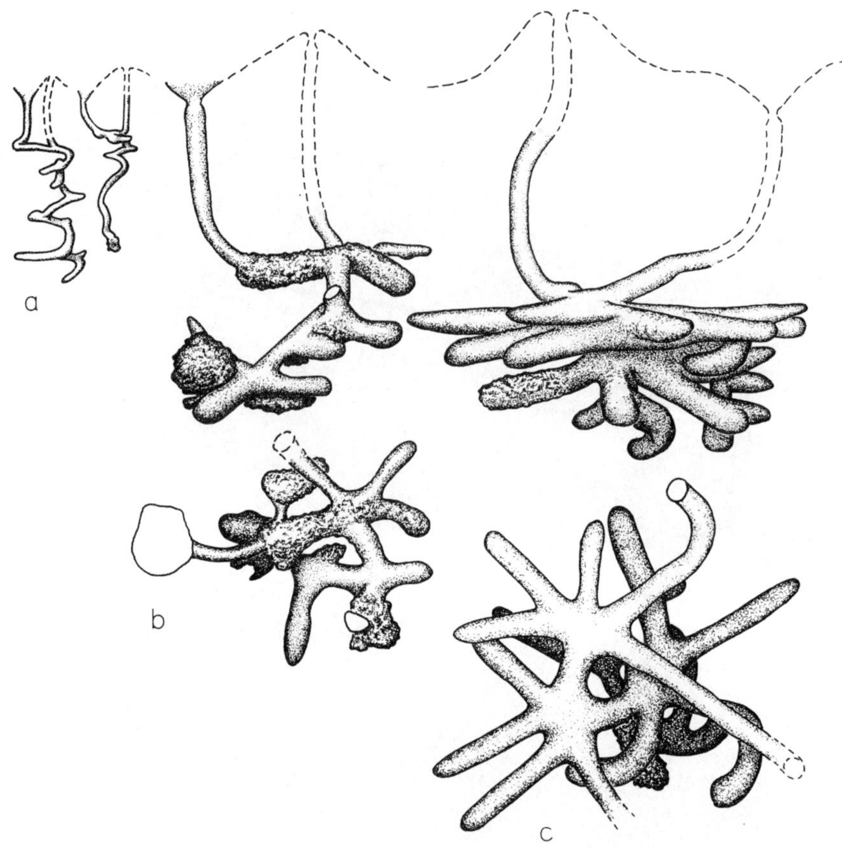

Figure 4.25 Spiral callianassid burrows, to the same scale, deeper tiers stippled dark; irregular chambers filled with plant debris are indicated. (a) Burrows 20 cm deep dug by *Callianassa* sp. cf. *C. bouvieri* (cf. Dworschak and Pervesler 1988). The slender vertical shaft leads to a mound, the spiral one to a depression. (b) Larger structure produced by *Callianassa* sp., viewed laterally and from above. (c) Similar structure from callianassid, cf. *Glypturus acanthochirus*, viewed laterally and in plan. Modified after Shinn (1968) and Braithwaite and Talbot (1972).

Other burrows of deposit-feeding anomurans have been described that have a dendritic development. *C. rathbunae* (Fig. 4.27b) processes surface sediments that are pulled into subsurface chambers. Fines are pumped out again to form mounds, where they become recharged with algal coatings. Coarse particles are selected out and dumped in deep chambers up to 3 m below the surface. These gravel-filled passages cause considerable heterogenization of the sediment.

Quite a different type of dendritic structure with mounds is produced by *Callichirus islagrande* (Fig. 4.27a).

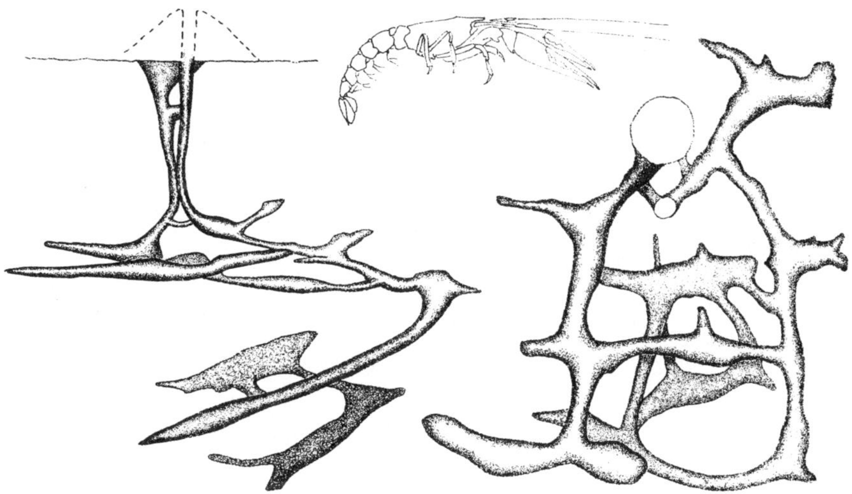

Figure 4.26 Spiral burrow of *Jaxea nocturna* seen in plan view and from the side. Modified after Pervesler and Dworschak (1985).

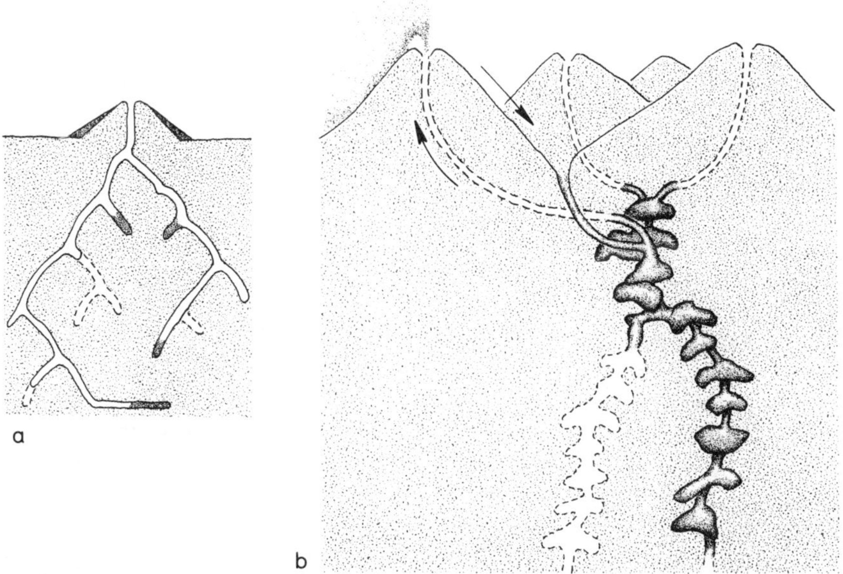

Figure 4.27 Dendritic burrows of callianassids. (a) That of *Callichirus islagrande*, showing 24 hours of sediment processing. Broken lines indicate new excavations, dark tint represents deposited material. (b) Sediment processing by *Callianassa rathbunae*. Gravel is advected down into the chambers, fines are lost in suspension. Modified after (a) Hill and Hunter (1976) and (b) Suchanek (1983).

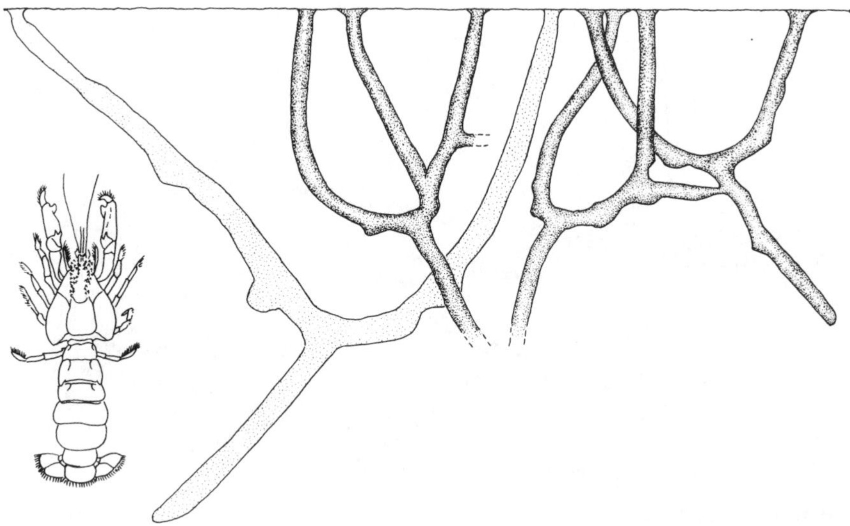

Figure 4.28 Burrows of *Upogebia pugettensis*. The larger example is 60 cm deep, modified from Thompson (1972). The more complex, smaller ones are modified after Swinbanks and Murray (1981).

4.3.4 Y-burrows for suspension-feeding gardeners

The several species of the genus *Upogebia* show remarkably uniform burrow design. Of these, perhaps *U. pugettensis* of the northwest Pacific is best known (Fig. 4.28).

Stevens (1929) cast burrows of this species in plaster of Paris, and demonstrated the characteristic Y-shaped structure. Her lively account of the activity of the mudshrimp was expanded by MacGinitie (1930), who studied the animals in glass tube 'burrows' and narrow aquaria in the laboratory. The shrimps use a hair basket assembled by the front limbs for catching seston. Significantly, he noted that although the permanent burrow was securely mucus lined, the shrimp frequently applied extra mud to the walls.

Indeed, in contrast to the excavator species of *Callianassa* and *Callichirus* in sands, the mud-burrowing *Upogebia* spp. rarely pass sediment to the outside when they construct their burrows (Swinbanks and Luternauer 1987), and mounds are not produced. They exploit the compactable properties of the substrate and create compression structures by pressing spoils into the walls (Ott *et al.* 1976).

Thompson (1972) observed that the result of all this attention was a multilayered lining topped by a smooth, mucoid surface. The external surface of the wall against the substrate is knobby, like that of *Callichirus major*. She then examined the relative resistance of this species and *Cal-*

lianassa californiensis to anoxia (R. Thompson and Pritchard 1969) and discovered the *Callianassa* to be the more tolerant of reducing conditions. The metabolic rate of *U. pugettensis* is higher and the animal is secure within a permanent burrow, isolated from the surrounding pore waters, in contrast to the deposit feeder.

Swinbanks and Murray (1981) made resin casts that displayed the lower part of the burrow. From the U-structure, a shaft descends to as much as 60 cm below the surface.

The burrow of *U. affinis* of the Atlantic coast of USA is similar to that of *U. pugettensis* (Fig. 4.29). Frey and Howard (1975) succeeded in casting the ends of several of the deep shafts. Many terminated, about 1 m below the sea floor, in unwalled chambers containing plant-leaf material. Minute branching burrows extended from the surface of these chambers (Bromley and Frey 1974; Frey and Howard 1975) and were interpreted as

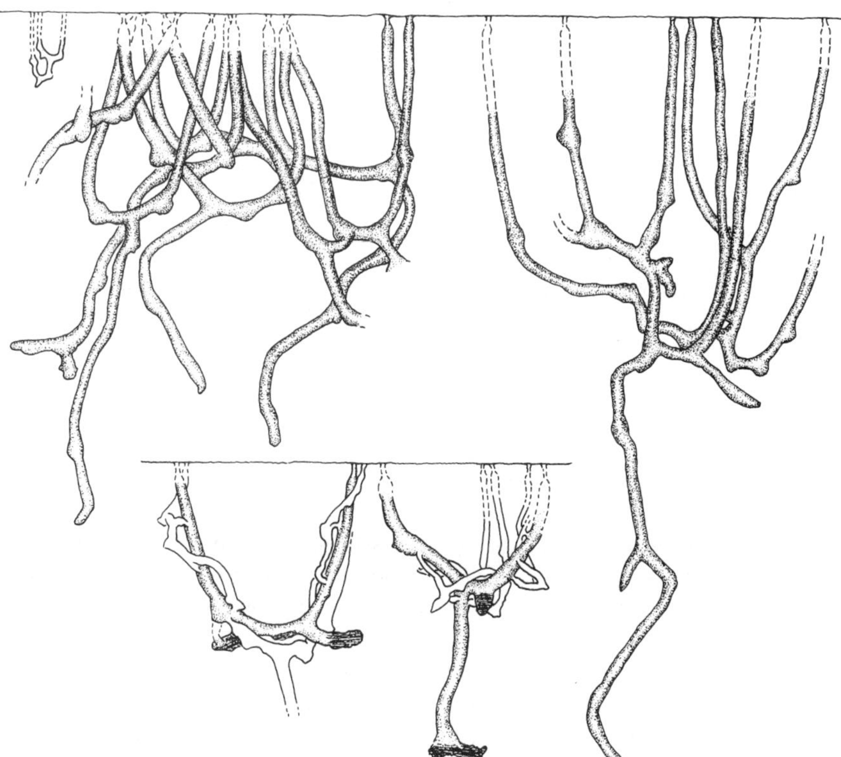

Figure 4.29 Above: burrows of *Upogebia pusilla* showing complex interference by several individuals and (left) a juvenile showing mode of extension of the U-burrow. The longest shaft reaches nearly 1 m deep. Modified after Ott *et al.* (1976) and Dworschak (1983). Below: at the same scale, burrows of *Upogebia affinis* showing interference by several individuals. Roughly-walled chambers containing plant debris are indicated. Modified after Bromley and Frey (1974).

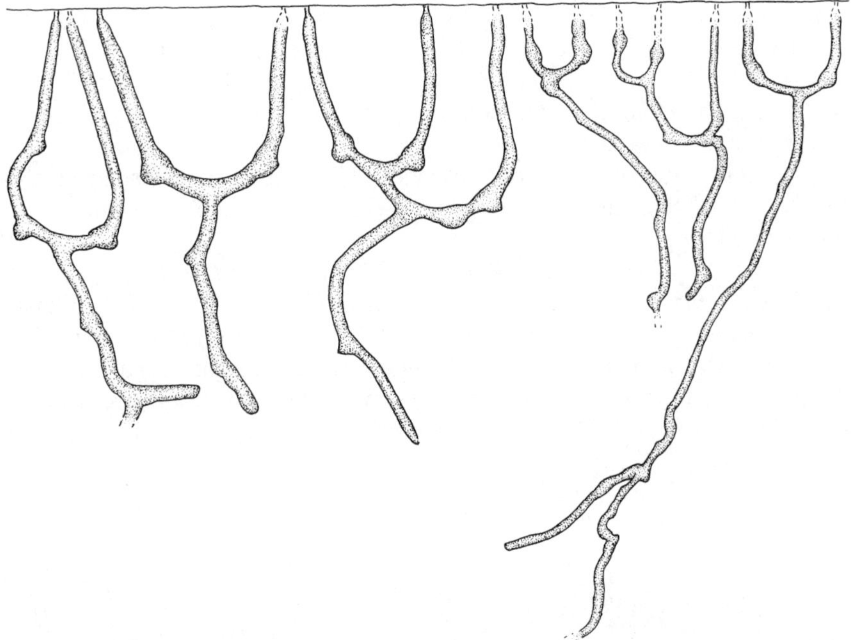

Figure 4.30 Burrows of *Upogebia pusilla*, the longest nearly 1 m deep. Modified after Dworschak (1983, the three to the left) and Ott *et al.* (1976).

the work of juvenile shrimps. It was speculated that the plant material might be used to culture bacteria for ingestion by the shrimps, adult and juvenile alike.

A putrid chamber 1 m below the sea floor is a hazardous place for a juvenile shrimp to begin burrowing. Alternatively, the small burrows may be of commensal crustaceans of another species better adapted for life in such a location, perhaps amphipods, exploiting the local food source.

Ott *et al.* (1976) showed the shafts of the Mediterranean species, *U. pusilla* to penetrate to 1.5 m. Some shafts bifurcated and many contained decomposing sea-grass leaves in the deepest parts (Figs 4.29 and 4.30). The contained water here was strongly reducing owing to the bacterial activity. These authors did not see suspension feeding take place. Instead, the shrimps were constantly occupied in shifting material from one place to another on the wall of the U-burrow. These authors therefore speculated that the shrimps were culturing bacteria for food, selecting decaying leaf material in the shaft and planting it in the oxygenated walls of the irrigation burrow.

Comparison with the Y-burrows of the bivalve *Solemya* is unavoidable (section 4.1.4b); is the shaft a sulphide well fuelled by the addition of sea-grass leaves? Where might the chemosymbiosis occur in the absence of chemosymbionts within the shrimps' bodies – on the burrow walls?

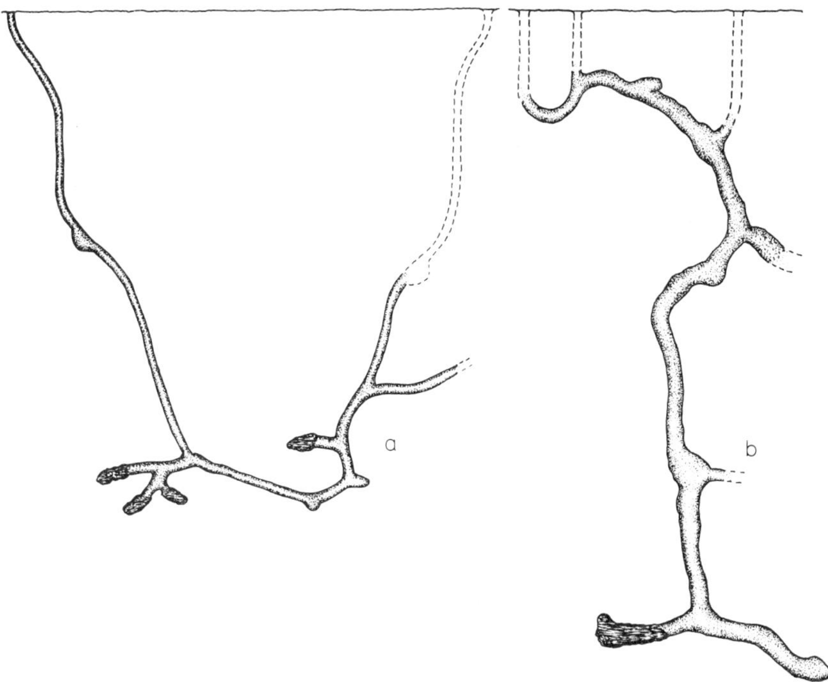

Figure 4.31 Left, the burrow of *Corallianassa longiventris* 1.5 m deep. Modified after Suchanek (1985). Right, at the same scale, the deep burrow of *Axius serratus*. Modified after Pemberton *et al.* (1976).

It is interesting, therefore, that Dworschak (1983, 1987a), when studying the same species at another locality, found no plant material in the vertical shafts. Indeed, these individuals were occupied with suspension feeding. Apparently, mud shrimps are versatile and can choose to exploit the most advantageous food source available in a given environment.

Corallianassa longiventris, among some others, apparently represents a group of species that shows similarities to the *Upogebia* life-pattern. Although the burrow is not a Y, it is basically U-shaped and contains chambers from 1 to 2 m below the surface that are used to store plant matter (Fig. 4.31). *Corallianassa longiventris* catches drifting sea-grass leaves and algae at the sea floor and carries these down to the chambers. The fragments are chopped up and some are incorporated into the burrow wall (Suchanek 1985; Dworschak and Ott 1993).

Similarities with *Upogebia* spp. are obvious and gardening is a plausible explanation of these activities. The deep, straggling burrow of *Axius serratus* (Fig. 4.31) may be part of a similar system; Pemberton *et al.* (1976) noted sea-grass leaves in the walls of this structure also.

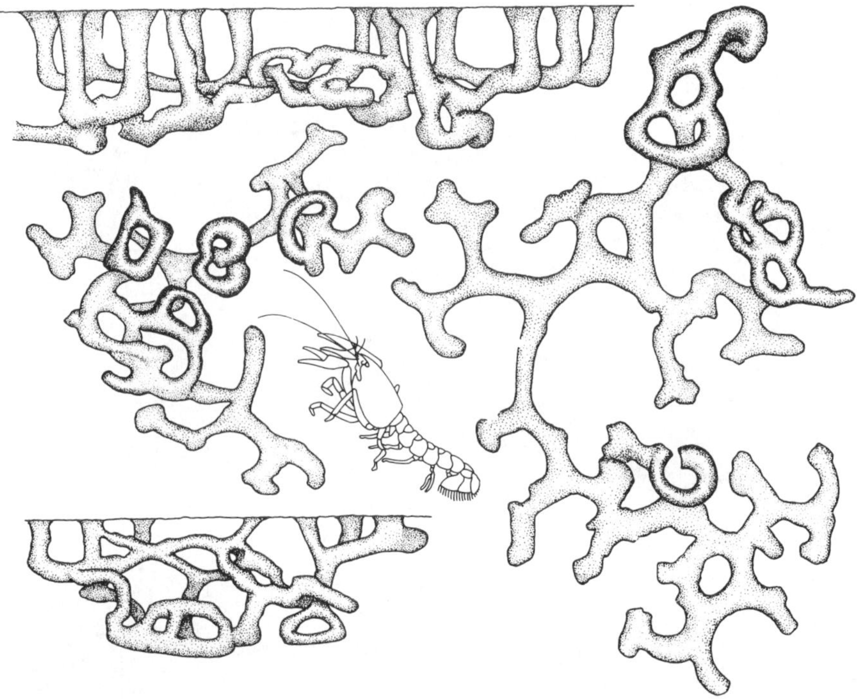

Figure 4.32 Burrows of *Calocaris macandreae* viewed laterally and from below. Maximum depth 20 cm. Data from Nash *et al.* (1984).

4.3.5 Gardening deposit feeders

Another example of gardening is provided by the thalassinidean *Calocaris macandreae* in offshore muds of the North Sea. A communal system of burrows is constructed in two tiers, at 10–19 and 14–22 cm depths, respectively. The upper level is connected to the surface by numerous shafts and is well aerated. The deeper galleries, not invariably present, are wider and comprise a series of Us interconnected by arcuate or circular galleries (Fig. 4.32; Nash *et al.* 1984). The animals eagerly seize pieces of animal material, but never eat them. Instead, these are buried in the burrow walls. Stomachs are invariably full of mud (Buchanan 1963). It was suggested that bacteria are specially cultured in the burrow walls for food: deposit feeding spiced by gardening.

How widespread such gardening is among apparently deposit-feeding shrimps is not known. Several callianassid species store sea-grass leaves in special chambers of their burrows (Figs 4.25 and 4.29). Shinn (1968), Enos (1983) and Dworschak and Ott (1993) noted this in the callianassid *Glypturus acanthochirus* in the Bahamas, Suchanek (1983) in St Croix, Braithwaite and Talbot (1972) in a callianassid in Seychelles, and Farrow

(1971) in a callianassid in Aldabra Atoll. Farrow also saw sea-grass leaves stored within burrows of the thalassinidean *Neaxius* sp.

There is the additional possibility that some of these crustaceans are using the collected organic material to generate methane or sulphides for chemosymbiosis with chemolithoautotrophic bacteria (section 1.1.8).

4.3.6 *Classifying thalassinidean burrow systems*

In this section I have tried to group those few examples we have information on, in trophic–morphological groups, following the lead of Suchanek (1985) and Griffis and Suchanek (1991). These 'ichnological groups' are poorly defined and the situation is aggrevated by the variety of trophic behaviour shown by individuals and species temporally and according to environment. Dworschak and Ott (1993) meanwhile emphasized consistency of architecture among species of architect and felt this more important than behavioural grouping. There is something for everyone in the morphology of these incipient trace fossils: trophic indications for the sedimentologist seeking evidence of palaeoenvironment; taxonomc fingerprints to help the palaeontologist identify the tracemaker.

4.4 Stomatopods

Mantis shrimps are voracious carnivores that first appeared in the Jurassic. Many species burrow and wait in ambush for passing prey (Caldwell and Dingle 1976). The basic burrow varies according to species, but is commonly a wide U having two similar openings, or one may be closed (Fig. 4.36d).

In the Mediterranean, *Squilla mantis* constructs flat U-burrows having two openings (Manfrin and Piccinetti 1970; Pervesler and Dworschak 1985). On the Atlantic coast of the USA, *Squilla empusa* produces branched burrows having extra apertures (Fig. 4.33; Frey and Howard 1969; Howard and Frey 1975). Hertweck (1972) described a thick mud lining on burrows of *Squilla* sp. in sand. *S. empusa* produces two types of burrows at the northern end of its range in Rhode Island, USA. Myers (1979) found it occupying a flat U-burrow in summer (Fig. 4.33). In autumn, however, the animal digs a shaft 50 cm deep, possibly having a side chamber, in which to hibernate in stagnant water.

4.5 More crustaceans and some fish

We cannot leave the crustaceans without mention of some other tracemakers among them, though most of the structures they produce are basically very simple.

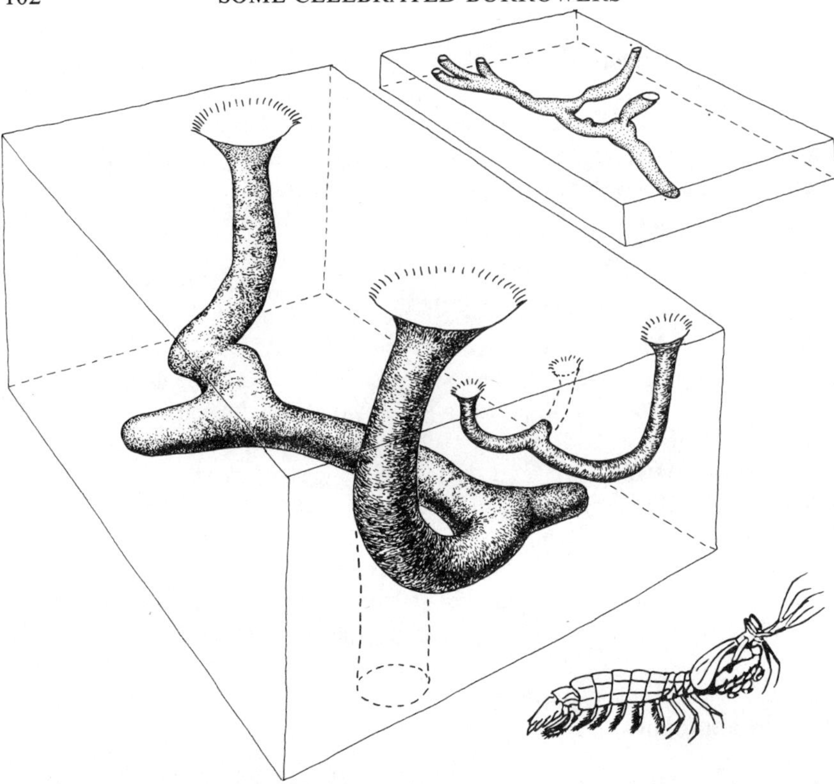

Figure 4.33 Burrows of *Squilla empusa*. The large block is 50 cm high and shows summer burrows. The top of the deep shaft constructed in winter is indicated by a broken line. The juvenile shows the mode of enlargement with growth (cf. Fig. 3.5). Data from Myers (1979). The smaller structure is modified after Frey and Howard (1969) and shown at the same scale.

Several species of lobsters construct U-burrows or J-structures, commonly having a subtriangular cross-section and a flat floor. *Homarus vulgaris* and *H. americanus* make such basic structures, but may extend these to create side branches and three or four apertures (Dybern 1973; Myers 1979).

The Norway lobster, *Nephrops norvegicus*, carries this further to produce quite complicated networks (Fig. 4.34; Dybern and Hoisæter 1965; C. Chapman and Rice 1971). Moreover, on settlement from their larval phase, juvenile *N. norvegicus* preferentially form burrows within the walls of existing adult burrows. As each juvenile grows, it extends its own burrow away from the adult burrow, so that a complex burrow system containing two or more individuals of different sizes develops. Gradually, the juvenile burrows will separate off as isolated units (Tuck *et al.* 1994).

The brachyuran crab *Goneplax rhomboides* also builds U-burrows that may be extended as mazes (Fig. 4.34). These have a flattened cross-sec-

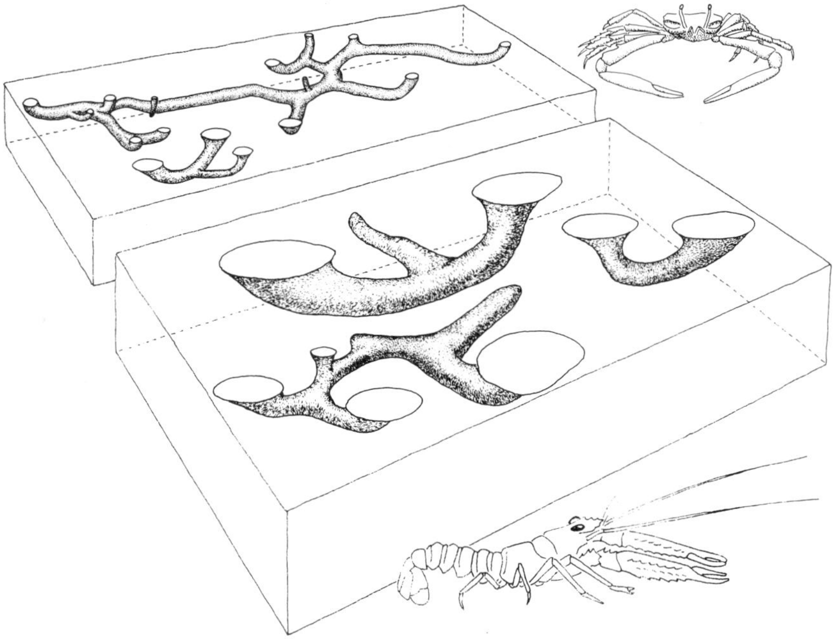

Figure 4.34 Burrows of the crab *Goneplax rhomboides* and the lobster *Nephrops norvegicus*. Blocks about 25 cm high. Data from Atkinson (1974a, b) and Rice and Chapman (1971).

tion and a smaller diameter than the burrows of *N. norvegicus*. Very comparable burrows were constructed by the closely-related Miocene crab *Ommatocarcinus corioensis*, which was found entombed within its *Thalassinoides suevicus* burrow (Jenkins 1975).

Similar structures are excavated in the same environment by a fish, *Lesueurigobius friesii* (Fig. 4.35). The basic U is inflated as a poorly-defined turning chamber, but may be extended as a small network occupied by several individuals (Rice and Johnstone 1972).

A wide variety of fish produce a broad spectrum of burrow morphologies. We must pass these by, but the interested reader should consult the excellent reviews by Atkinson (1986) and Atkinson and Taylor (1991). In particular, tilefish (e.g. Fricke and Kacher 1982), garden eels (e.g. Tyler and Smith 1992) and lungfish (e.g. McAllister 1988) produce impressive incipient trace fossils.

At a smaller scale, the amphipod *Maera loveni* also produces a U-burrow in the same sediment as the Norway lobster. Multiplication of small U-burrows and interconnections can produce an extensive maze that is not unlike a miniature edition of some callianassid systems (Fig. 5.4).

In the upper foreshore and backshore, fiddler crabs and ghost crabs are active burrowers throughout the subtropics and tropics. They live in

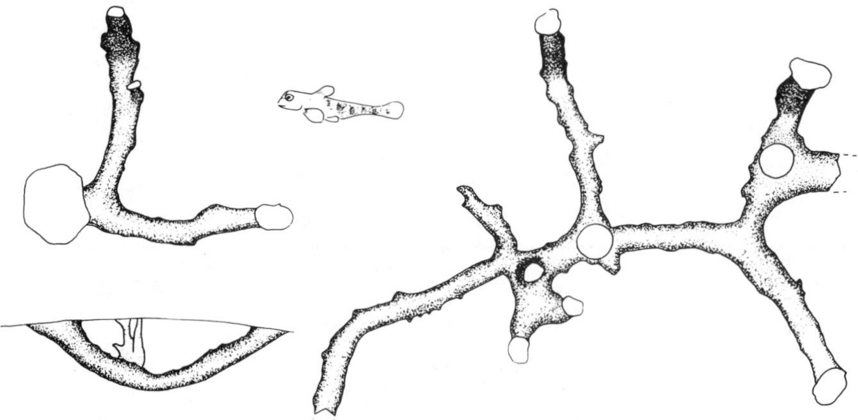

Figure 4.35 Burrow of the goby *Lesueurigobius friesii*, viewed laterally (10 cm deep) and in plan. Modified after Rice and Johnstone (1972).

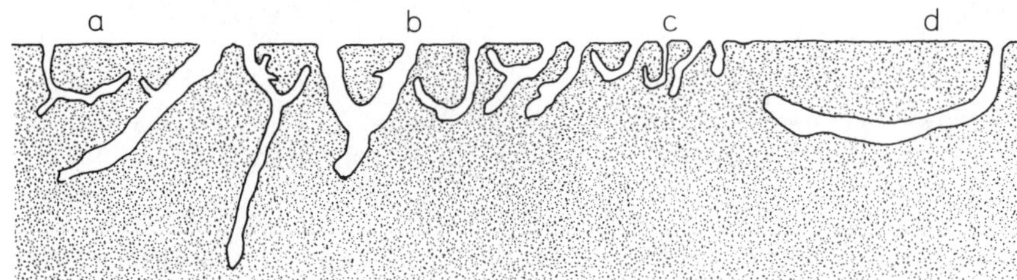

Figure 4.36 Burrows of ghost crabs *Ocypode quadrata* from (a) foredunes, (b) backshore and (c) foreshore, the longest 50 cm deep. Modified after Hill and Hunter (1976). (d) Burrow of *Pseudosquilla ciliata* at the same scale. Modified after Braithwaite and Talbot (1972).

simple J-shaped domiciles that serve to connect the crab with the tidally-fluctuating water-table (Fig. 4.36a–c; Hayasaka 1935; Frey and Mayou 1971).

Land crabs likewise are omnipresent burrowers of the terrestrial realm and backshore in warmer regions. We know very little about their burrow morphologies, despite their abundance. Hogue and Bright (1971) described the spiral shafts constructed by *Cardisoma carnifex* in Kenya. Crab burrows described by Farrow (1971) on Aldabra Atoll also had a spiral shape. In order to reach the water-table, some land crabs descend over 3 m (Fig. 4.37; Bright and Hogue 1972).

Many freshwater crayfish are known to burrow, at least producing minor domiciles. In Canada, Williams *et al.* (1974) observed *Cambarus fodiens* to be in part a deposit feeder. The form of the burrow varies at different times of year; it is branched and reaches down to the water-table 1.5 m or so below the surface. Alterations to the structure in time,

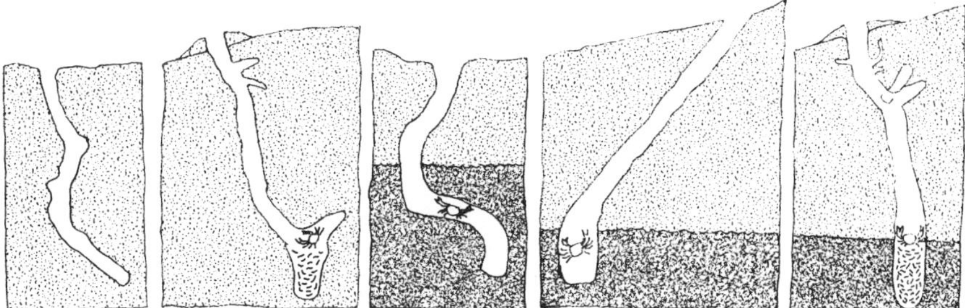

Figure 4.37 Burrows of land crabs, *Cardisoma carnifex*. The water table is indicated where recorded. The longest burrow is 1.3 m deep. Modified after Hogue and Bright (1971).

in response to water-level changes, create a complex of passages (Hasiotis and Bown 1992).

4.6 Spiral traps

I conclude this brief survey of sediment processors with a worm that in fact displaces very little sediment. It builds highly characteristic structures, however, that resemble spiral and meandering trace fossils, and adds much slime to the sediment (Schäfer 1962; Röder 1971).

Paraonis fulgens inhabits medium- to high-energy tidal flats and sand beaches. In the North Sea it occupies a zone a little lower down the foreshore than the reverse-conveyor worm *Scolecolepis squamata* (Fig. 3.24).

The burrow has two parts, each serving a different function (Fig. 4.38). The upper part, within oxygenated sand, comprises a number of spirals, one above the other, at least three per burrow; these are connected by steeply inclined or vertical shafts. The spirals are horizontal, either dextral or sinistral, and may end centrifugally with incomplete whorls producing meanders. Rarely, but significantly, a bridge may connect adjacent whorls. Burrow diameter is 0.2–0.4 mm and spiral diameter reaches 8 cm, seldom more. About 3 mm separates successive whorls. Burrow walls are soaked with copious mucus, saturating the sediment between the whorls.

The lower part of the system consists of a downward extension of the vertical connecting passages well into the reducing zone. These burrows branch and may connect as straggling U-circuits, but walls are not oxidized, indicating that water is not circulated in these anoxic regions.

At low tide, when the water-table sinks to below the level of the spirals, the worm seeks refuge within the lower passages and ceases respiration. As the water-table begins to rise with the tide, the worms gradually ascend. The passing of the breaking waves reworks the topmost sediment and destroys the upper part of the burrow system. At high tide the spir-

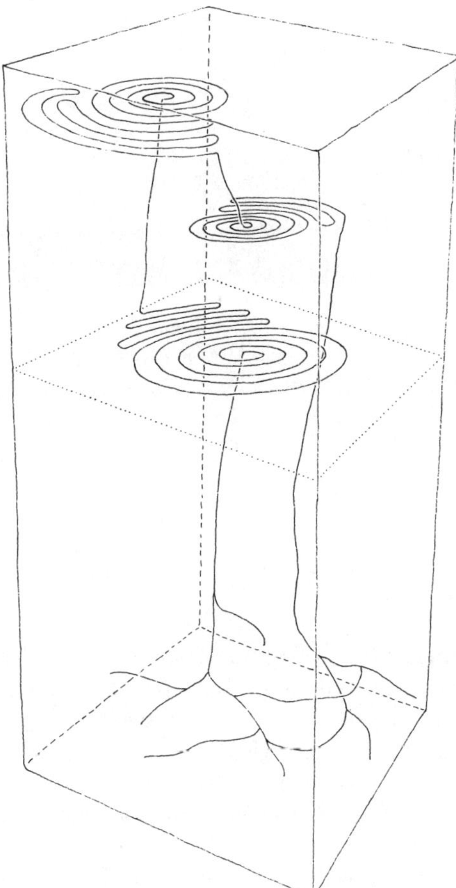

Figure 4.38 Burrow system of *Paraonis fulgens*. The stippled plane indicates the RPD approximately at the base of the zone of daily physical reworking of the sediment. Modified after Gripp (1927) and Röder (1971).

als are quickly reconstructed. As the water recedes again, however, the worm retreats and the spirals once more are obliterated. The complete burrow at high tide has not been observed; only the remaining parts that have survived the ebb are seen at low tide (Fig. 4.39). Röder (1971) found that, of the burrows he studied, 76 per cent of the spirals lay within sediment that had been reworked at the previous flood tide.

P. fulgens is not a deposit feeder; it is too small. Röder found its gut to contain benthic diatoms when collected at low tide. It is significant that these diatoms, and other interstitial organisms in the gut, are all mobile species. The mode of feeding, although not directly observed, was modelled by Röder along the following lines.

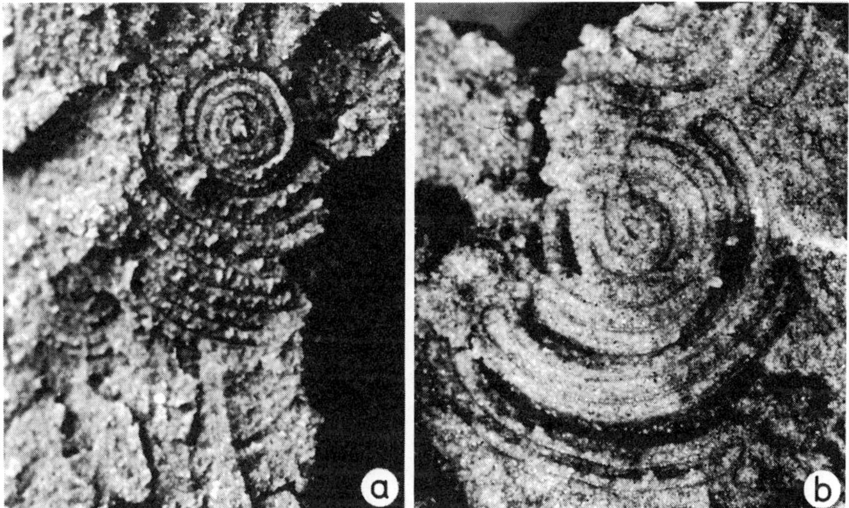

Figure 4.39 Spiral traps of *Paraonis fulgens* in a tidal sand flat, Manø, Denmark.

Benthic diatoms reproduce and grow in sunlight near the sea floor. During each ebb and flow, wave action redistributes the algae throughout the oxic, reworked zone of the beach sand. The living diatoms migrate vertically upwards towards the surface during the intervals between reworking phases. Their path is interrupted by the emplacement of horizontal, spiral slime nets, which trap the diatoms as they wander upwards. Thus, repeated visits to the spirals by the worm during their brief existence produce a continual harvest of diatoms. Vertical movement of the water-table may also help to accumulate food organisms in the slime nets.

Reworking may produce an uneven distribution of diatoms within the sediment, particularly if surface layers are buried undisturbed. The worm will naturally show preference for such horizons and the population will tend to crowd spirals at the same level under such circumstances. Risk and Tunnicliffe (1978) illustrated a crowded example from the Canadian Atlantic coast. These authors also demonstrated spirals that had proved to be rigid enough to survive gentle reworking and transport.

Clearly, the feeding system of *P. fulgens* is dependent on a narrowly defined physical environment. It will not function subtidally. Thus its chances of survival as a fossil structure seem slender (sections 6.2 and 10.7.2).

Similar but larger spiral and meander structures are well known as trace fossils, particularly from deep-water settings, but most of these structures were backfilled, not open burrows. These trace fossils (and the faecal strings photographed on the deep sea-floor) are best explained as the work of deposit feeders seeking immobile food. However, Röder's *P. fulgens* model has stimulated interpretation of some other trace fossils, the graphoglyptids, as having a trapping function (Seilacher 1977; section 9.2.6).

5 The synecology of bioturbation

The previous chapters examined the autecology of selected species. However, the biogenic reworking of sediment is rarely accomplished by one species in solitude. More often, a community subdivides the endobenthic habitat into many ecological niches, each occupied by species specialized for a particular way of life. The quality of the bioturbation process thereby becomes enormously variable. It depends on which sediment-processing species are present; the relative rates at which these work; the relative abundance of these species; seasonal variation in rate; influence of one species on another, etc. Let us examine some of these aspects of the complicated process of bioturbation.

5.1 Commensalism

The several species that comprise an endobenthic community naturally influence and interfere with each other as they share the same substrate. These species show varying degrees of dependence on the presence of others, ranging from symbiosis, involving mutual gain, to the exploitation of a resource provided by another species.

For example, coprophagy is an important element in benthic food webs (Frankenberg and Smith 1968). Faecal matter is rapidly colonized by microbes, which enhances its nutritional value (Longbottom 1970; Johnson 1977). Biodeposition by suspension feeders thus provides a food source within the sediment for deposit feeders that would not otherwise be available.

5.1.1 Combination structures

Among the closely interdependent species are those that share the same burrow. The special environment that is provided by an established burrow within the sea floor creates a habitat for other animals, offering them protection, irrigation and food remains discarded by the burrow's architect. The status of these guests is normally neutral commensalism verging on parasitism, since their presence confers little benefit on the burrower.

Examples are many. MacGinitie (1934) found seven commensals living with *Callianassa californiensis*. A copepod, a shrimp and a polychaete

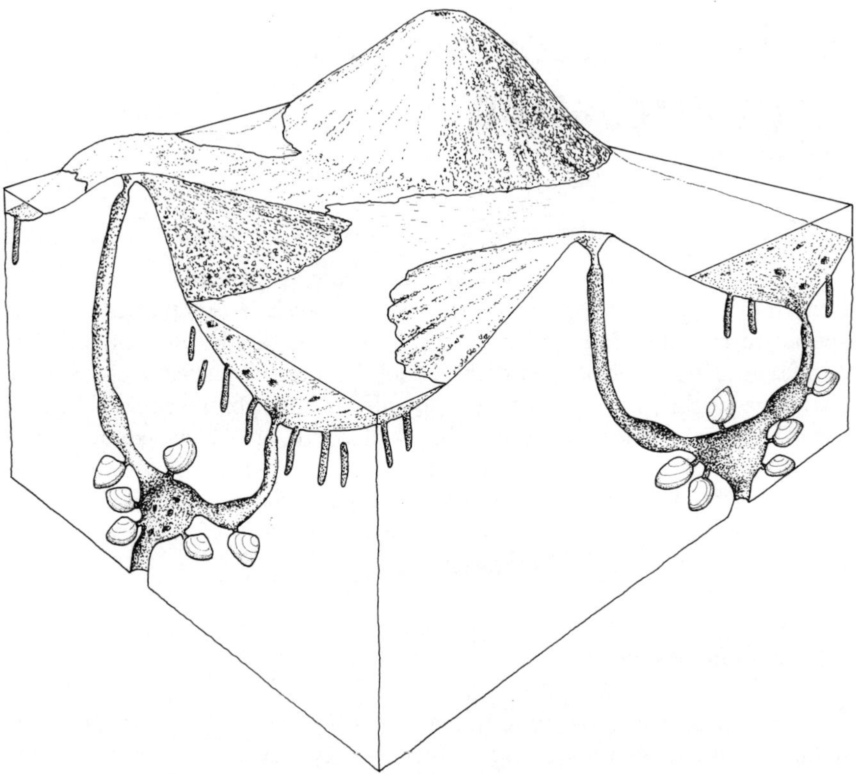

Figure 5.1 The mounds and depressions produced by the deposit-feeding activity of *Callianassa californiensis* trap standing water at low tide. The polychaete *Spio* sp. is dependent upon this topography, producing its shafts only beneath the water table. Burrowing into the walls of the upper tier of the crustacean system are many bivalves, *Cryptomya* sp. Modified after Swinbanks (1981b) and Swinbanks and Murray (1981).

worm occur only in the burrows of *C. californiensis*, whereas two pinnotherid crabs and the bivalve *Cryptomya californica* (Fig. 5.1) occur also in the burrows of *Upogebia pugettensis* and *Urechis caupo*. A gobiid fish uses burrows as a refuge at low tide.

The hospitality of *U. caupo* is famous. Fischer and MacGinitie (1928) counted three guest species: the same goby and one of the crabs that occurs in *Callianassa californiensis* burrows, plus a scale worm found nowhere else (Fig. 3.7). Phillips (1971) also found pinnixid crabs to be common guests in the burrows of *Callianassa jamaicense* and *Callichirus islagrande*. Myers (1977b) found burrows of the polychaete *Scoloplos robustus* to harbour at least one non-burrowing polychaete species and possibly several.

Examples of endobenthic symbiosis are provided by the tropical alpheid shrimps of the Pacific and Indian Oceans that cohabit with non-bur-

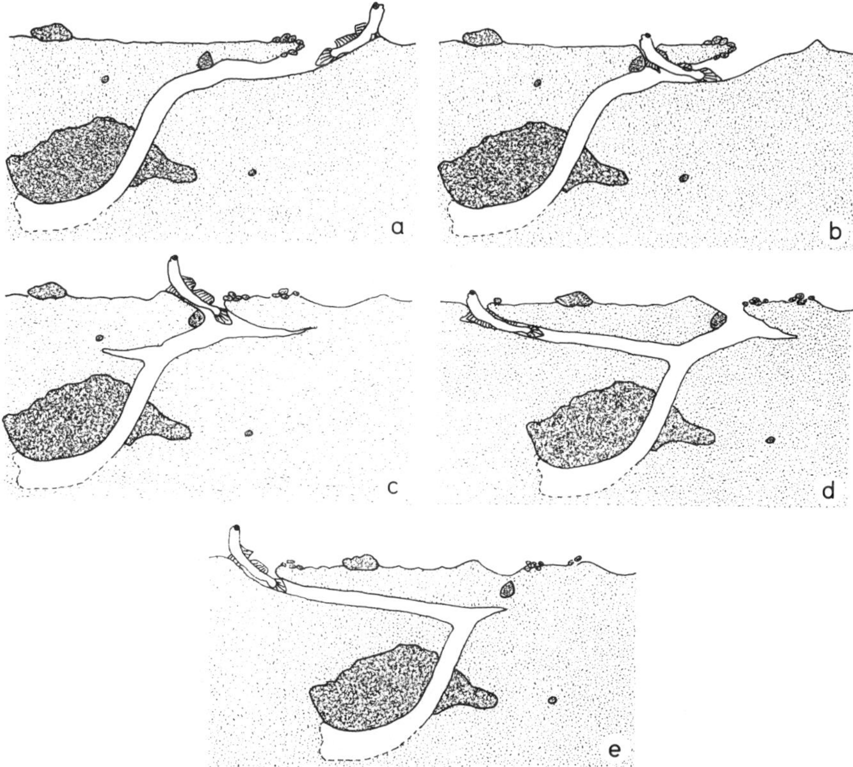

Figure 5.2 Dynamics of the upper part of the burrow of the goby *Cryptocentrus sungami* and the shrimp *Alpheus djiboutensis* (not shown) at 24 hour intervals. Modified after Karplus *et al.* (1974).

rowing gobiid fish (Magnus 1967; Karplus *et al.* 1972, 1974). The pistol shrimps have poor sight and benefit from the presence of the keen-sighted fish, which take up position at the burrow entrance during the day. The goby hides in the burrow at night and also when danger threatens in the daytime. The shrimp deposit feeds in the gravelly sediment and constantly alters the burrow (Fig. 5.2). The presence of the fish also allows the alpheid to leave the burrow safely to feed on detritus around the entrance.

In general, such guests are not burrowers themselves. In some cases, however, two or more burrowing species may share a burrow, the morphology of which comes to reflect the activity of both or all occupants. The tropical alpheid, *Alpheus crassimanus*, cohabits as described above with a gobiid fish that is itself a burrower (Farrow 1971; Karplus *et al.* 1974). The fish constructs the entrances and upper parts of the system, which have a larger diameter than the deeper regions constructed by the shrimp.

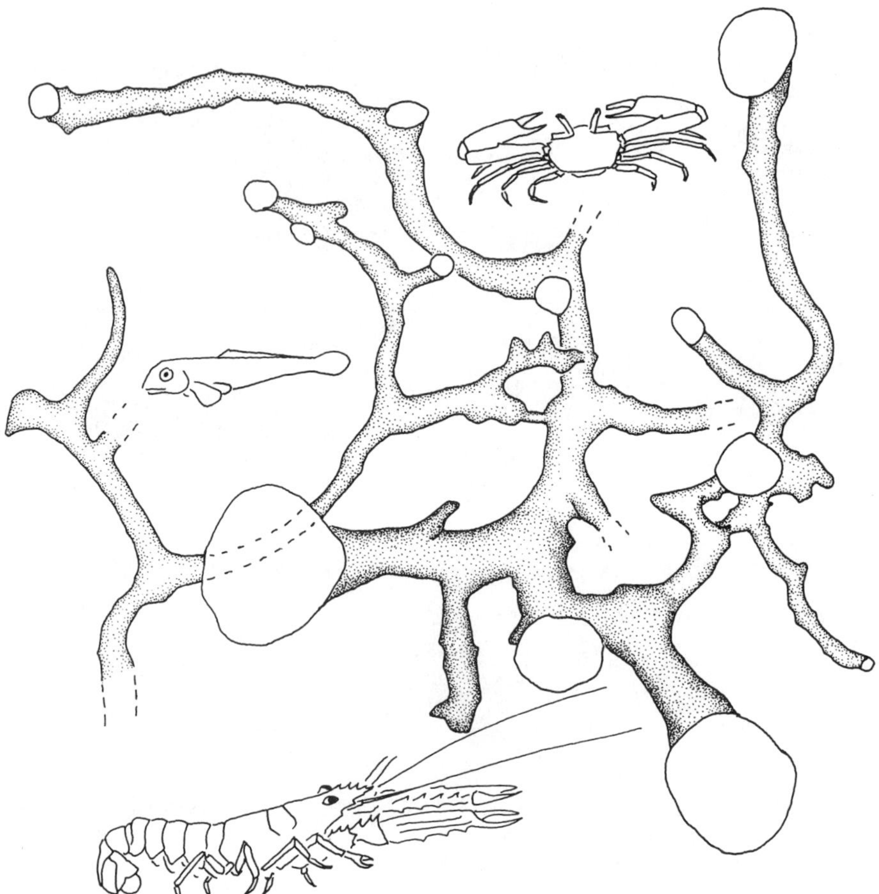

Figure 5.3 A complex single-tier system produced by interconnecting burrows of a fish, a crab and a lobster (see Figs 4.34 and 4.35). Modified after Atkinson (1974a).

A complicated case was described by Atkinson (1974a). In Scottish offshore mud bottoms the lobster *Nephrops norvegicus*, the crab *Goneplax rhomboides* and the gobiid fish *Lesueurigobius friesii* normally construct their burrows independently (Figs 4.34 and 4.35). However, combination structures also occur; one such system had 13 apertures to the surface. Various parts of this network (Fig. 5.3) had morphologies that indicated construction by each of the three species. The lobster could not enter the narrower departments of the crab and fish. Tuck *et al.* (1994) added the thalassinidean shrimp *Jaxea nocturna* and the echiuran worm *Maxmuelleria lankesteri* to the species that join burrows to such combination structures.

A greater contrast in styles is seen in the combined burrow system produced by *N. norvegicus* and the amphipod *Maera loveni* (Fig. 5.4). In this

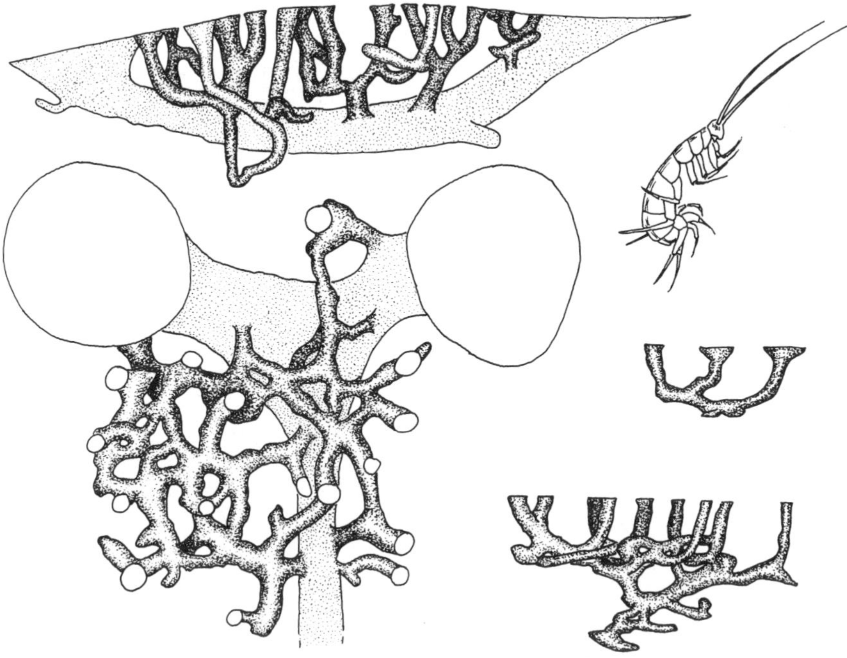

Figure 5.4 Burrows of the amphipod *Maera loveni* in connection with larger *Nephrops norvegicus* structures. Modified after Atkinson *et al.* (1982).

case, the two burrows are merely connected together, sharing water currents, and it is doubtful if there is much intercourse between the two species. The same is the case with the capitellid annelid *Notomastus latericeus* and the scalibregmiatid annelid *Scalibregma inflatum*, which connect their burrows with those of *Echiurus echiurus*, where these are sufficiently abundant, instead of directly to the surface (Fig. 5.5; Reineck *et al.* 1967).

Similarly, the short-siphoned bivalve *Cryptomya californica* obtains lodgement and irrigation far deeper and more securely within the substrate than it otherwise could, by entering the burrow walls of *Callianassa californiensis*, *Upogebia pugettensis* and *Urechis caupo* (Fig. 5.1; MacGinitie 1934). Frey and Pemberton (1987) illustrated an example from washover fans and beaches, where mole crickets intersect their burrows with those of fiddler crabs.

Combination burrows also arise through the subsequent occupation by a squatter of a structure that has been abandoned by its original constructer. Lobsters, for example, may take over the burrows of other animals and modify them to suit their own needs. Thus, *Homarus americanus* reuses burrows of *Squilla empusa* (Myers 1979).

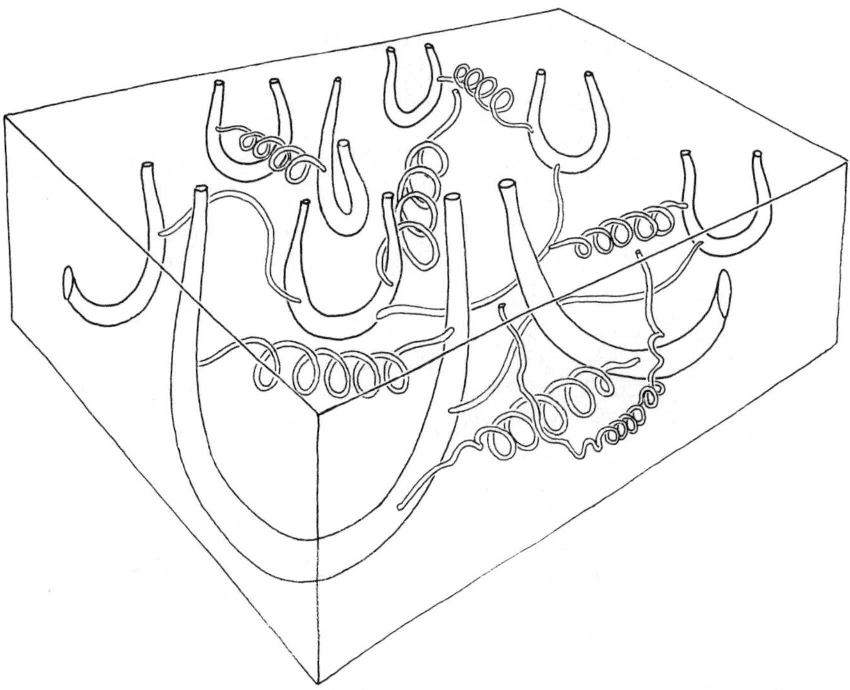

Figure 5.5 Spiral burrows of *Notomastus latericeus* and the straighter ones of *Scalibregma inflatum* using the limbs of U-burrows of *Echiurus echiurus* and thereby avoiding connection with the sea floor. Inspired by Reineck *et al.* (1967).

5.1.2 Dependence at a distance

An occupied burrow develops physical and chemical gradients around it that alter the immediate environment. This commonly creates conditions attractive to other infaunal organisms. Deep within the generally reducing zone, where a microbial food source is abundant, oxygenated haloes around irrigated burrows have a dynamic effect (Andersen and Kristensen 1991).

Reise (1981) found concentrations of meiofauna around the burrows of many tidal-flat macrobenthos. On sand flats, an abundant meiofauna is related to burrows of the amphipod *Corophium arenarium* and the polychaete *Pygospio elegans*. The inverted conveyor activity of the worm causes an increase in faecal organic material at 4–7 cm depth, at the lower end of its shaft, and a corresponding meiobenthos peak was recorded there.

The well oxygenated sediment-processing site at the head end of the pectinariid *Lagis koreni* yielded spectacular numbers of minute polychaetes (Reise 1981). Around burrows of polychaetes of the genus *Nereis*, nematodes showed increases of up to 94 per cent above ambient values.

In contrast, no meiofauna was found around the reducing walls of shafts of *Heteromastus filiformis*, which are not irrigated (section 3.5.2).

The mud walls of the U-burrow of the terebellid *Amphitrite ornata* (section 3.6) contain much organic material and are the site of intense decomposition (Aller and Yingst 1978). Microbial activity is greatest on the inside of the tube, in the presence of oxygen; outside this, respiration is anaerobic.

Thus, in the generally reducing subsurface environment, oxygenated burrows are **elite structures**, attracting more than their spatial share of abundance. After a burrow has been abandoned, physical currents may maintain an oxidizing circulation through it, creating a flow comparable to that of biogenic irrigation (Ray and Aller 1985); and even after the burrow has been filled, it remains a region of raised organic concentration, and may attract 'anaerobic' deposit feeders. This is especially so where faecal material has been buried within the fill, emplacing combustible organic matter within bacteria-rich reducing sediment.

Conveyor activity modifies the sea-floor topography and the consistency of the substrate surface. This in turn causes changes in the community structure. For example, the conical piles of sand that accumulate above the shafts of the holothurian *Molpadia oolitica* provide a substrate suitable for colonization by suspension feeding tube-worms (section 3.5.1). These sand piles are stabilized by the worm tubes and survive long after the holothurian has moved on. Between these small hills, easily reworked, pelleted fine-grained material gathers as a soupground, and is unsuitable for suspension feeders.

Another example of environmental modification by burrowers is that of forests of chimneys on the sea floor, extending from the lined tubes of worms. Woodin (1978) showed these to create a refuge for smaller burrowing species by stabilizing the floor and increasing surface roughness.

5.2 Substrate modification by bioturbation

It has long been known that the activity of endobenthic organisms profoundly changes the properties of the sediments they inhabit (Dapples 1942). The basic processes of bioturbation are concerned with the two phases of the sediments: particles and fluid. Each of these undergoes either advection (bulk transport) or diffusion (Fig. 1.1).

Physical effects mainly concern particle advection and diffusion:

1. Sediment mixing causes disruption of earlier sediment structures and leads either to homogenization of the sediment or to production of new structures.
2. The water content or compaction of the sediment may be increased or decreased as a result of different biological activities.

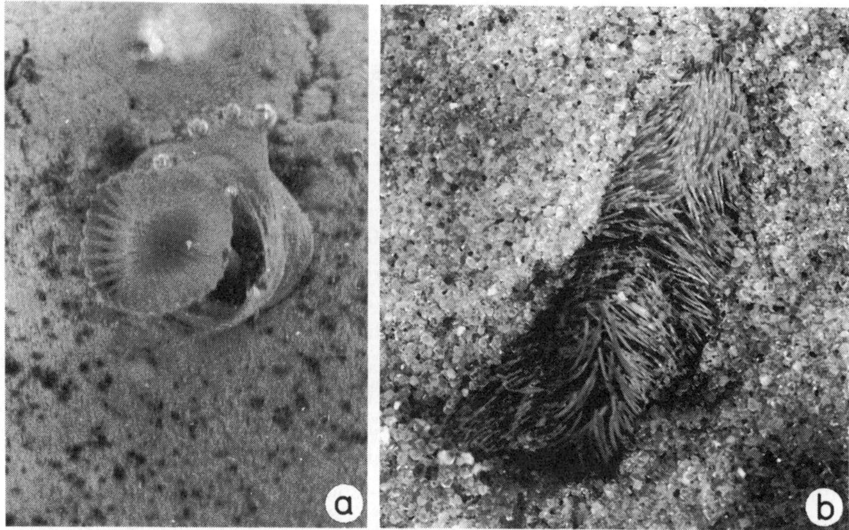

Figure 5.6 Animals and their substrates (in aquaria). (a) The polychaete *Myxicola infundibulum* succeeds in establishing itself in a watery soupground by secreting a tough mucus tube as a burrow wall (x2). (b) The last glimpse of the spatangoid *Echinocardium mediterraneum* as it disappears into loose sand. There is perfect control; not a single unwanted grain is allowed to enter the burrow. Grains in contact with the spines are cemented with invisible, elastic mucus (x5).

3. Grainsize may be increased and sorting improved by binding grains as pellets.
4. Suspension feeders may emplace much fine-grained material in the substrate.
5. Introduction of mucus by endobenthos as burrow lining, while manipulating particles and in faecal material, substantially alters the physical properties of the sediment (Fig. 5.6).

Chemical effects are highly diverse:

1. Material is translocated continually between chemical reaction zones during feeding, burrowing and the construction of tubes.
2. Burrow and faecal pellet formation alters reaction and solute diffusion geometries, creating a mosaic of biogeochemical microenvironments instead of a vertically stratified distribution (Fig. 5.7).
3. New reactive organic matter in the form of metabolic products, such as mucous secretions and dead organisms, are introduced into the deposit.
4. Feeding and mechanical disturbance influences microbial populations that mediate chemical reactions.

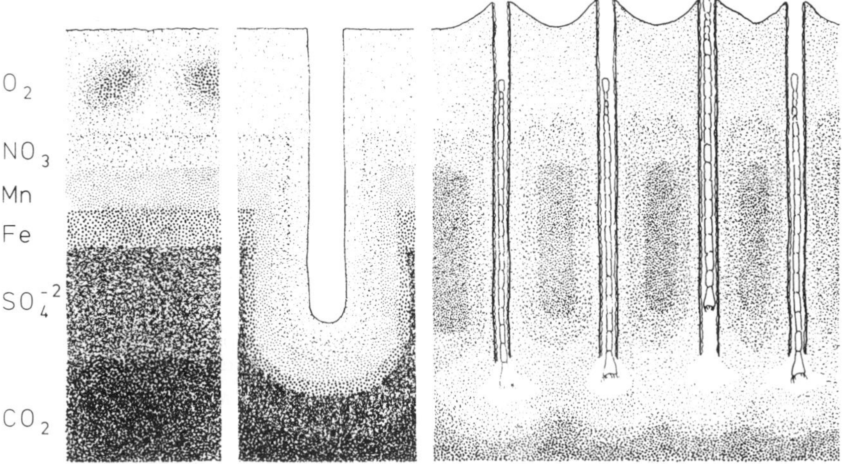

Figure 5.7 Biological disturbance of the classically assumed vertical zonation of electron acceptors in the uppermost sediment profile. (a) Just a pair of pellets in the oxidation zone. (b) Reaction zonation is deployed around an irrigated burrow. (c) Inverted oxidation–reduction stratigraphy caused by irrigation by a population of conveyor worms, *Clymenella torquata*. Modified after Aller (1982).

Biological effects chiefly centre around the ecological dependence of communities on the changing properties of the substrate:

1. Alteration of the physical and chemical properties of the substrate correspondingly alters the quality of the habitat. As the substrate environment is modified, the bioturbating community will be modified.
2. Activity of some endobenthic organisms creates conditions that exclude other activities, and species that are dependent on those activities will disappear from the community (section 5.3.2).

Let us examine these aspects of bioturbation, or 'sediment conditioning' as Levinton (1977) called them, emphasizing those that are of importance to the geologist.

5.2.1 Physical effects of bioturbation

Geotechnologists would have excellent understanding of the physical properties of lake- and sea-floor sediments were it not for the confusing effects of bioturbation (Keller *et al.* 1976; Richards and Parks 1976; Meadows and Meadows 1994). Organisms alter the bulk density and the cohesiveness of sediments in a complex way, at varying subfloor depths and to degrees that vary locally and temporally, so that their effect is extremely difficult to analyse. Studies at individual localities or of single species have produced widely differing results. Here are some examples.

Myers (1977a) and Powell (1977) agreed that the holothurian, *Leptosynapta tenuis* (Fig. 3.14), caused a tightening of the sediment in the upper 3 cm, corresponding to the funnel fillings, whereas the sediment below this became less dense owing to the production of burrows there.

Rhoads (1974) and Aller (1978) found that in Buzzards Bay, sedentary suspension feeders and deposit feeders tend to tighten the fabric, whereas mobile deposit feeders increase permeability. The production of pellets by suspension and deposit feeders creates a superficial layer 2–4 cm thick having a high water content (up to 90 per cent) and low shear strength (Fig. 5.8). Levinton and Bambach (1975) found a soupground/firmground interface 5–10 cm deep. This soupy layer is easily reworked by water currents (Yingst and Aller 1982), which can increase turbidity of the bottom water. These processes may vary seasonally, as the rates of activity of the endobenthos fluctuate. On the New England coast, McMaster (1962) found substrates to be more dilatant in summer than in winter, when bioturbational activity is much reduced. Levinton (1977) even detected a diurnal fluctuation in sediment consistency.

Tube-cementing animals obviously increase stability of the substrate, and their activity in dense populations causes organic cementation of a large proportion of the grains at the depositional interface (Fig. 4.6c; Rhoads and Young 1971; Featherstone and Risk 1977). Upward continuation of a tube as a chimney increases roughness of the sea floor,

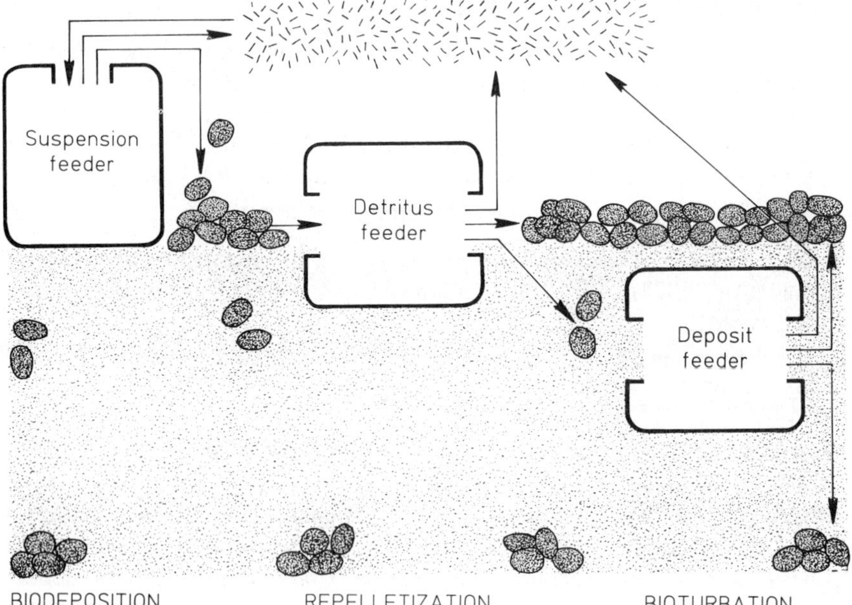

Figure 5.8 The seston–pellet cycle. Inspired by Haven and Morales-Alamo (1968) and Young (1971).

which reduces entrainment of the bottom sediments by currents (Eckman and Nowell 1984). Some tube-building animals are not as static as they appear to be. Myers (1977a) noted that the spionid polychaete *Prionospio* sp. swims out of its oblique tube and builds a new one twice a day. Predators can accelerate this process: Myers (1977a) also observed a flatworm enter tubes of amphipods, causing the crustaceans to depart hurriedly and construct new tubes elsewhere.

These autecological case histories should not be viewed in isolation. Normally the substrate is reworked by a community and the activities of its members interact in many ways. For example, Myers (1977a) found that sediment processing by the dominant bioturbators of a particular community, *Leptosynapta tenuis*, and the polychaete *Scoloplos robustus* together produce a net compaction.

In a community dominated by the conveyor maldanid *Clymenella torquata* and the terebellid *Amphitrite ornata* that lines its tube with mud (Figs 3.20 and 3.25), much of the fine-grained sediment conveyed to the surface by the maldanid worm is again transported down as wall material by the terebellid (Aller and Yingst 1978; Aller 1982).

In addition to such complications, which affect all bioturbating communities, there are relative rates of sediment disturbance by different species. Clearly, mere weight of numbers of a species does not necessarily imply that the results of its activity will dominate the final substrate fabric. Rhoads (1967) provided an example of this. In an area where relatively static tube-worms dominated the community, at two stations the rates of reworking were noticeably higher than elsewhere. At these stations only, a vagile naticid snail was present.

Thus, different species have different bioturbation values according to the amount of sediment disturbance they create. The presence of such **key bioturbators** is more important in the evaluation of the bioturbation potential of an endobenthic community than is the relative biomass of its member species.

The introduction of fine sediment into the substrate by suspension feeders alters its consistency significantly. Material that is normally in suspension mode under the given hydrodynamic conditions is trapped and collected in sand-sized pellets having a higher settling velocity than their constituent particles. Epibenthic animals are as active as endobenthic ones in this process of **biodeposition** (Figs 5.8 and 5.9b; section 4.3). Oysters, for example, may cause mud biodeposition at a rate that is seven times faster than by gravity in stationary water (Lund 1957; Haven and Morales-Alamo 1966). Tunicates also are particularly active biodepositers, an animal group having virtually no potential for preservation as body fossils.

The pellets must become buried if they are to be preserved as microcoprolites; otherwise their breakdown by bacteria or meiofauna will release the fine particles for resuspension (Haven and Morales-Alamo

Figure 5.9 Pellets are also incipient trace fossils. (a) Deposit-feeder pellets composed of aragonitic mud, Great Bahama Bank, pelleted mud facies. Length of pellets about 1 mm. SEM picture. (b) Surface of the biodeposited mud in a mussel patch on the tidal flat at Fanø, Denmark. The sediment is entirely composed of faeces and pseudofaeces of *Mytilus edulis*. A part of the shell of the bivalve is seen at left. The small round burrow apertures belong to the snail *Hydrobia ulvae* and are 0.2 mm wide.

1968, 1972; Rhoads *et al.* 1977); or, more likely, they will be reingested by detritus and deposit feeders (Fig. 5.8).

In contrast to biodeposition, **biowinnowing** also is a process of significance. Roberts *et al.* (1981) demonstrated large quantities of fines pumped out of *Callianassa* burrows as a result of the deposit-feeding process. Bottom currents transported this sediment out of the area (Fig. 4.27). Deposit feeding processes also commonly produce **biostratification** by shifting coarse-grained material to the bottom of the level of operations of the burrower, as in many crustaceans (Suchanek 1983; Dworschak and Ott 1993) and worms (Figs 1.5, 3.13, 3.20 and 3.22).

The mere production of mounds by deposit feeders is a significant sedimentary and ecological factor. Mounds comprise deeper sediment brought to the surface, burying epibenthic fauna, changing the sea-floor topography and creating new habitats. Smith *et al.* (1986) recorded depositional rates of 1–2 cm per month on large mounds in a deep-sea echiuran worm.

Finally, introduction of mucus by burrowers modifies the consistency of the substrate (section 1.3.4; Frankel and Mead 1973; Meadows and Meadows 1994; Meadows and Tufail 1994). In a fine-grained sediment containing much clay, cohesion may already be sufficient for tunnel construction, aided by compression of the walls. Even here, however, many animals secrete quantities of mucus during sediment processing, as seen in the heart urchin *Echinocardium cordatum* (section 4.2.1). In uncohesive sands, the establishment of an open burrow is virtually impossible without soaking the surrounding sediment with slime (Fig. 5.6; section 4.1.2). Deposit-feeding starfish introduce copious mucus into sediment as they digest it in bulk (Shick *et al.* 1981). This impregnation markedly alters the hydrodynamic entrainment characteristics of the sand (Fig. 4.6; Nowell *et al.* 1981).

5.2.2 Homogenization versus heterogenization

Many geologists have assumed that bioturbation leads unavoidably to homogenization and that a totally bioturbated sediment is necessarily structureless (Bayer *et al.* 1985; Droser and Bottjer 1986). Certainly, some forms of bioturbation may lead to homogenization, but it would be a sad case for ichnology if this were always so.

Animals that tend to homogenize their substrates are of several types. The meiofauna (section 2.1) obliterates the structure of the sediment yet does not impart a new one. Also, the inhabitants of soupgrounds and watery softgrounds that cause diffusive turbulance by their passage through the substrate tend to produce a structureless fabric. Gravitational sorting may accompany this activity, however, and heavy mineral grains will accumulate at the bottom of the zone of mixing (Fig. 4.15).

Mobile, unselective deposit feeders may produce a homogeneous substrate, as for instance the vertical conveyor worms. However, on close inspection, few of these animals do in fact feed unselectively, and commonly a graded bed is produced at their zone of operations (Rhoads and Stanley 1965; Jumars *et al.* 1982; Wheatcroft and Jumars 1987). Indeed, this relative movement of fine grains upward and coarse grains downward may eventually lead to an abrupt boundary between fine and coarse sediments (Figs 1.5 and 3.13; Meldahl 1987).

The opposite effect can also occur. McCave (1988) showed that diffusive bioturbation of fine-grained sediment may displace gravel-sized clasts in an upward direction. This movement may be due to reduced shear strength upwards; there is no confining pressure on large grains near the surface. This is supposedly the process that maintains the position of manganese nodules at the sea floor. Whether coarse grains are moved up or down depends on the type of bioturbation process that dominates in a given case. Thus, by using **compression**, it is easier to lift a large clast up than to thrust it down into the tighter sediment below. On the other hand, **excavating** sediment from beneath the clast will allow it to drop (Fig. 1.5); and **conveyance** will also move large grains downwards into pre-existing cavities (Fig. 4.27b).

Within a biogenically graded bed, structures may be subtle, but some sort of ichnofabric would be expected, even if hard to see. A homogeneous sediment is not the same as one in which structure is not visible (Figs 6.5 and 6.6).

In most endobenthic communities, however, there are species that fill burrows with material that differs from the surrounding sediment. Furthermore, supportive walling or constructional compression zones may demarcate the burrow. No matter how densely such structures are superimposed upon each other, the process will not ultimately lead to physical uniformity of the substrate.

5.2.3 *Chemical effects of bioturbation*

While physical effects of bioturbation are chiefly associated with the solid phase of the sediment, chemical effects are, in addition, influenced by fluid advection and diffusion.

An immediate effect of bioturbation on the chemical environment is the lowering of the RPD by the pumping of quantities of bottom water in and out of the substrate. Rhoads (1974), Rhoads *et al.* (1977) and Aller (1978) demonstrated that endobenthic pumping lowered the RPD in a sediment to 3–6 cm below the surface, whereas in the same sediment lacking endobionts, the RPD was 1 cm subsurface. A similar lowering of the RPD was observed beneath enteropneust burrow systems in deep water (Jensen 1992).

Aller (1982) found the advective transport of particles and fluids by burrowers greatly disturbed chemical gradients (Fig. 5.7). Maldanid worms, for example, can create a locally oxygenated layer beneath the general RPD.

Graf (1989) documented a striking example of this advection in the deep sea. Following the seasonal deposition of phytodetritus in early summer, key bioturbators (the sipunculan *Nephasoma* sp. and the enteropneust *Stereobalanus canadensis*) conveyed this organic material down to their activity levels to produce peaks at 10 and 6–8 cm, respectively. This advection was rapid, involving days rather than weeks.

Bioturbation also dramatically alters the flux of different elements within the benthic boundary layer. Silicon flux, for example, may be increased by a factor of ten by bioturbation (Schink and Guinasso 1977; Aller 1980), largely because of the enormous increase in area of the sediment–water interface represented by the burrow walls.

As a result of the intense decomposition processes that occur within burrow linings, several elements may be mobilized. Aller and Yingst (1978) discovered concentrations of iron, manganese and zinc in the mud walls of U-burrows of *Amphitrite ornata*. The metals are concentrated along the inner wall surface, and irrigation water passing over this surface promotes scavenging of additional metals from the sea water. Aller (1983) claimed that, by acting as molecular sieves and retaining anions, thin organic linings of burrows strongly influence the chemistry of sediments.

Paul (1977) and Jumars *et al.* (1981) emphasized another effect on sediment chemistry caused by particle advection: that of repeated exposure of grains to the bottom water. In the presence of established communities of deposit feeders, these particles pass through guts many times (Rhoads 1974). While in the gut, the chemistry and mineralogy of clay minerals may be altered. The more frequently they are ingested the further such alterations proceed (Pryor 1975).

In the deep sea, dissolution of carbonates is promoted by this process of delayed burial (Paul 1977). By the same token, in waters that are supersaturated with respect to calcium carbonate, the opposite may occur. Large, robust pellets, bound by mucoid cement and enriched in reactive metabolites, may be preferentially carbonate cemented (Fig. 5.9a). At sites of slow sedimentation, such pellets commonly act as nuclei for mineralization by glauconite and phosphate, while they are recycled within the bioturbational mixing zone.

It is clear, therefore, that bioturbation promotes the early diagenesis of sediments and individual grains by greatly increasing chemical reactivity at the sea floor.

5.3 Biological effects: amensalism and community succession

Physical and chemical changes in the substrate, brought about by bioturbation, alter the ecological effect of the bioturbating community. A fundamental biological effect of bioturbation was suggested by Reise and Ax (1979) concerning meiofauna, and Aller (1978), Yingst and Rhoads (1980) and Andersen and Kristensen (1991) concerning microorganisms. An enormous increase in the areal extent of the RPD is caused by the activity of a climax community. Microbial growth is most active at the RPD. Constant pumping of seawater into and metabolites out of the substrate, plus the continual introduction of new organic matter, creates optimum growth conditions for the microbial community. Bacteria are an essential food source for deposit feeders and they thrive under the conditions produced by the sediment-processing endobenthos (Rhoads *et al.* 1978; Branch and Pringle 1987); a form of community horticulture.

Soft-bottom community dynamics are complex, and species interact with various degrees of subtlety and intimacy. The activity of a single species can alter the structure of the habitat (substrate consistency and bottom environment) and bring about changes in the composition of the community. Commensal relationships between some organisms are treated in section 5.1. In the following are examples of the opposite effect: **amensalism**.

5.3.1 Amensal relationships

Amensalism can occur at various hierarchical levels. Individual species may be excluded by the activities of others. Owing to the close interaction of species, the loss of one may alter the structure of the community as a whole. Amensalism may also exclude **guilds** (e.g. mobile, mid-tier deposit feeders; section 10.6.3) or whole **trophic groups** (e.g. all deposit feeders).

At the simplest level, one species may exclude another by direct predation. For example, Wilson (1980) showed how the tentacles spread from its U-burrow by the terebellid polychaete *Eupolymnia heterobranchia* captured the larvae of the burrowing nereid polychaete *Nereis vexillosa*, thereby significantly reducing the nereid population.

Levinton (1977) suggested that species displace each other ecologically by two further means. By **exploitation mechanisms** a species may out-compete another by superior exploitation of a given limited resource. By **interference mechanisms** one species may directly prohibit another from exploitation of a resource. The following case histories are examples of amensalism through interference. [Further examples were reviewed by Ronan (1977), Woodin and Jackson (1979), Brenchley (1981, 1982) and Thayer (1983).]

Wilson (1981) found bioturbation by *Abarenicola pacifica* to have deliterious effects on some other endobenthic species and not to affect others. Dense intertidal populations of the lugworm were associated with decline in abundance of the spionid *Pygospio elegans* and the cumacean crustacean *Cumella vulgaris*. However, the large spionid *Pseudopolydora kempi* was not affected, nor was the amphipod *Corophium spinicornis* or total oligochaete occurrence. The smaller species cannot cope with the sedimentary disturbance created by the lugworm. On the other hand, both spionid worms significantly reduce the survivorship of juvenile lugworms by feeding on them.

Quite a different situation was described by Woodin (1977), also from an intertidal setting, in which algal gardening by tube-building worms changed the structure of the habitat and impeded the activities of deposit-feeding worms. Two nereid polychaete species catch drifting fragments of ulvacean algae and attach these to their tubes (cf. Fager 1964). The planted algae thrive and serve as food for the gardening worms. However, the flourishing algal growth, normally absent from sand areas, alters bottom conditions. It inhibits access of some deposit feeders to oxygenated water, particularly of the vertical head-down maldanid polychaetes.

Of course, it is also possible for the activity of a species to change the environment adversely so as to exclude itself. This was the case for a population of lancelets, *Branchiostoma nigeriense*, which disappeared from a Nigerian lagoon (Webb and Hill 1958). The rapid passage of lancelets through the substrate demands particularly loose packing of the sediment (section 2.4). Webb (1969) found the passage of the lancelet increased the packing of the sediment so that in time the substrate was no longer suitable for colonization. *B. caribaeum* also prefers clean sands that are subject to frequent redeposition by wave or current action (Cory and Pierce 1967).

5.3.2 Trophic group amensalism

This concept was established by Rhoads and Young (1970). It was found that suspension feeders in subtidal sediments of the New England coast are largely confined to sandy or firm mud bottoms whereas deposit feeders attain high densities in soft muddy substrates (Sanders 1958). Deposit feeding in the soft muds destabilizes the substrate and produces a pelleted surface that is easily resuspended. Suspension feeders are excluded from such substrates, not for lack of suspended food material, but because their filtering mechanism becomes clogged by the excessive turbidity, and the shifting surface tends to bury or discourage settling larvae. Thus, the activity of animals belonging to the one trophic group prevents colonization by members of the other. Comparable results were reached in other areas by Johnson (1971), Aller and Dodge (1974), Whitlatch (1977) and Tamaki (1988).

The opposite relationship may also occur. Where the cemented tubes of suspension feeders occur densely in a sediment, the mobile deposit-feeding guild (section 10.6.2) may be unable to colonize it. Reineck (1963) noted that *Echinocardium cordatum* is not found in areas where the substrate is riddled with the shell-lined tubes of the polychaete *Lanice conchilega*. In fact, tube mats have been shown significantly to inhibit the activity of several mobile burrowers (Brenchley 1982).

5.3.3 Community succession

The community modifies its environment in such a way that eventually the habitat becomes less suitable for the existing community. Its members can no longer successfully compete with invading species of the next stage. In this way, pioneer communities give way to later-stage communities (Johnson 1972).

It is difficult to distinguish between community changes that are 'spontaneously' caused by internal, ecological evolution of a maturing community as opposed to those imposed by changes in the external environment (Johnson 1972; Levinton and Bambach 1975). The first is 'community succession', the second 'community replacement'.

A classic study of community succession was that of McCall (1977). An area of shallow sea floor was defaunated so as to resemble a catastrophic ecological disturbance. Within ten days the floor had been colonized abundantly by a single species of polychaete. Two more polychaete species arrived after 30 days and were soon joined by two bivalve species. The worms produced abundant small shafts and U-burrows. These pioneers were rare or absent from the background community. It was not before months had elapsed that species from the mature community began to appear, and a tiering structure developed. It was a year or so before the climax community was restablished.

Ecological disturbance repeats this succession of communities and thereby, in the long-run (**time-averaged**) assemblage, brings about an increase in diversity (Thistle 1981).

A different sort of community succession was suggested by Kidwell and Aigner (1985). Under conditions of slow deposition, the skeletons of successive generations of infaunal bivalves accumulate in a softground and gradually convert the substrate into a firmground containing abundant hard skeletal grains (Fig. 5.10), unsuitable for the original softground community.

Finally, here is an example of possible community replacement, in which two communities having contrasting key bioturbators occur in contiguous subenvironments (Fig. 3.24). Vertical shafts of the polychaete, *Scolecolepis squamata* occur on slight topographic rises and the spiral traps of *Paraonis fulgens* are produced in nearby depressions on a foreshore profile. Small environmental fluctuations could cause these two facies to alternate in a geological sequence. A hasty interpretation of such a sequence might lead to a suggested alternation of intertidal and abyssal environments on the basis of traditional evaluation of individual trace fossils!

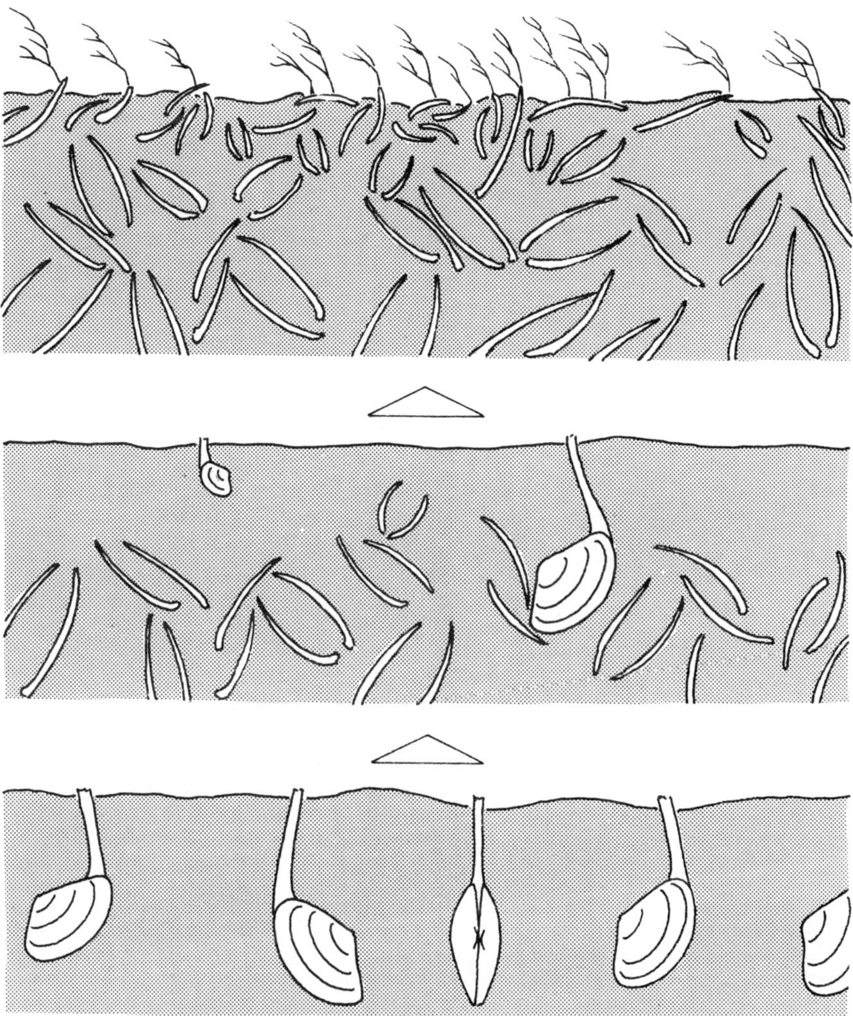

Figure 5.10 Community succession in a stable mud floor. Accumulation of generations of skeletons changes the consistency of the substrate. Based on Kidwell and Aigner (1985).

5.3.4 Community replacement

Changes of the environment imposed from outside also cause a change in the bottom community. Such environmental variations may include temperature, salinity, oxygen or sedimentary changes (Miller 1990).

Reineck *et al.* (1968) and Hertweck (1970a) provided an example of community replacement in the North Sea south of Helgoland. The cold

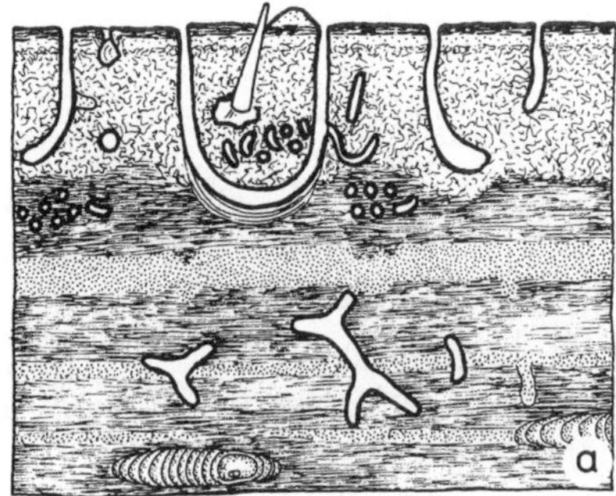

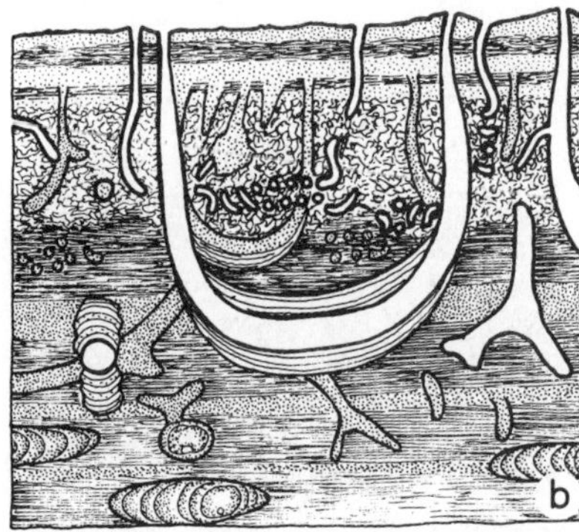

Figure 5.11 Development of an *Echiurus echiurus* community near Helgoland over 5 years. (a) 1964: a dense population of juvenile echiurans, together with pectinariid worms and some bivalve species, totally bioturbate the upper part of the substrate. Lower tiers are occupied by *Notomastus latericeus* spirals (Fig. 5.5), *Callianassa subterranea* in boxwork systems and, deepest, backfilling *Echinocardium cordatum.* (b) 1967: sparse adult echiurans, shifting less than the juveniles but producing spreiten, cause incomplete bioturbation. (c) 1968: the echiuran population is dead. The homogenized horizon produced by the juvenile echiurans, and its colonization surface, remain distinctive. In the historical layers (section 5.4.4), evidence is seen of recurring echiuran colonization events. Modified after Hertweck (1970a).

winter of 1962/3 killed off the benthos in the mud bottom at –20—30 m. The pioneer community that developed in 1963–64 was dominated by a dense population of juvenile *Echiurus echiurus*, which completely bioturbated the upper 10 cm (Fig. 5.11a). By 1965 the population of *E. echiurus* had reduced markedly.

By 1967, further decline had affected the echiuran population. There having been no recruitment, all individuals now were adult, their burrows associated with spreite structures. The bed of total bioturbation caused in 1964 by the juvenile echiurans was now being reworked by the polychaete

Notomastus latericeus (Fig. 5.11b). All echiurans were dead by 1968, but the horizon of their activity, now buried beneath laminated sediments, was distinctive (Fig. 5.11c) and deep box cores showed the existence of similar horizons at 20–30 cm intervals. So what did the cold winter do? Kill off the predators and open a **colonization window** for the echiuran larvae?

A more classic example of community replacement is that which occurs during periods of non-deposition. The sediment is gradually dewatered by physical compaction and bioturbation and the substrate changes in consistency. Examples are numerous where trace fossil assemblages are compound (section 10.1.3), the work of a softground community being cross-cut by that of a firmground one. If the sea floor becomes cemented as a hardground, the final colonization is by a community of boring organisms (e.g. Goldring and Kazmierczak 1974; Bromley 1975; Fürsich 1978; Mángano and Buatois 1991; Wilson and Palmer 1992).

Another effect can lead to the same relationship of firmground structures cross-cutting softgound burrows. This is produced by relative depth of burrowing – tiering (section 5.4) and the increasing firmness of the sediment with depth – concealed firmground (section 1.3.5). Thus, similar ichnofabrics may indicate the ordering of structures reflecting community replacement (time), or tiering of a single community in an aggrading sea floor (space). These alternatives may not be easy to distinguish (Walker and Diehl 1986).

5.4 Tiering

Subaqueous sediments are vertically zoned with respect to physical, chemical and biological parameters. This creates a vertical partitioning of the endobenthic environment.

5.4.1 Environmental gradients

Gravitational compaction within the uppermost metres causes dewatering and changes in consistency. This creates a downward increase in shear strength through regions that are within reach of endobenthic activity. Bioturbation by the endobenthos modifies and generally accelerates this firming-up process (e.g. Myers 1977a).

There is normally a steep gradient in the distribution of organic matter within the uppermost metre, showing an extreme peak at the top (Carney 1989). Whitlatch (1980) found the detritus to contain 300 per cent of that of the subsurface sediment. Biogenic mixing causes the organic content of the interval from 2 to 20 cm down to be rather uniform (Johnson 1977).

Decomposition of this organic material produces gradients in oxygen pressure and pH. The distribution of organic material and oxygen within the sediment in turn lead to a vertical zonation of microbial communities (Rhoads *et al.* 1976); and as bacteria are the main food source for deposit feeders, this gradient further accentuates the vertical polarity of resources.

5.4.2 Vertical habitat partitioning

Biologists have long studied the phenomenon of differential vertical structure in communities; in forests or aquatic benthic communities it is referred to as 'ecological stratification' (Seilacher 1978). This term is confusing in geological contexts and geologists have adopted the equivalent

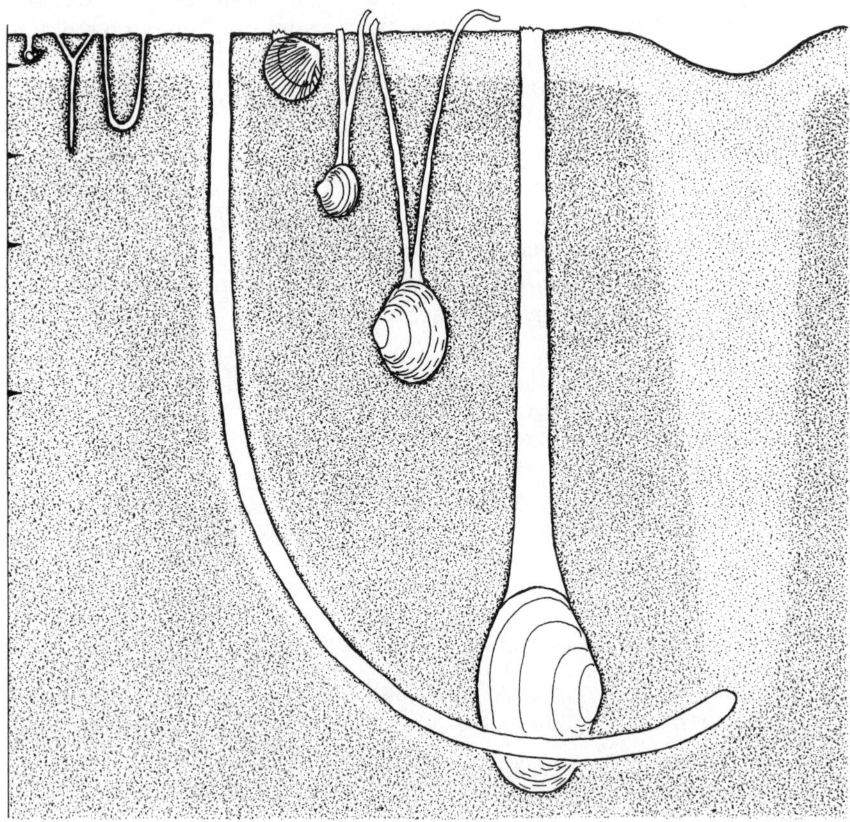

Figure 5.12 An early tiering diagram for tidal communities of the Danish North Sea. Thamdrup (1935), after whom the figure is modified, recognized five tiers, 0–1, 1–4, 4–7, 7–12 and 12–25 mm deep, respectively. From left: the snail *Hydrobia ulvae*, the Y-burrow of the polychaete *Pygospio elegans*, the U-burrow of the amphipod *Corophium volutator*, the J-burrow of the lugworm *Arenicola marina*, and the bivalves *Cerastoderma edule, Macoma balthica, Scrobicularia plana* and *Mya arenarea*.

term **tiering** as proposed by Bottjer and Ausich (1982). In one of the breakthrough papers on tiering, Werner and Wetzel (1982) referred to 'storeys'.

Endobenthic tiering has been documented in greatest detail in the intertidal and shallow subtidal zones. An early tiering diagram was published by Thamdrup (1935) for Danish tidal flats (Fig. 5.12). This work has been extended to the subtidal communities of Danish coastal waters (Fig. 5.13). On the American side of the north Atlantic, Whitlatch (1980) studied the vertical distribution of endobenthic polychaetes in an intertidal environment; in California, Ronan (1977) did likewise for polychaetes and Levinton (1979) for bivalves.

5.4.3 Some reasons for vertical restrictions

Purely physical factors restrict the distribution of many endobenthic animals. The cheapest method of locomotion is to swim or walk epibenthically from feeding site to feeding site. Thus some animals burrow only to feed or hide and these generally disturb the sediment surficially (Fig. 9.3). Likewise, the work required to burrow may limit the depth distribution of some animals to the topmost sediment where shear strength is minimal. Myers (1977b) considered this reason as largely responsible for the fact that 85 per cent of the species of a community were restricted to the topmost 2 cm.

Numerous studies have indicated the concentration of activity at and just beneath the sea floor. Whitlatch (1980, 1981) showed the importance of organic matter in this respect. Diversity may be highest where there is an abundance of organic food material. He found polychaete species to feed at specific levels, many species near the surface, progressively fewer downward. Aller and Cochran (1976) found in a shallow sea floor a 4 cm zone of total sediment mixing overlying a 12 cm zone of irregular activity. Figures vary from place to place and among communities, but the pattern generally applies where a lack of physical perturbation allows climax communities to develop (section 10.5).

Despite the attractions of the uppermost levels, this is a danger zone with regard to physical stress, predation, disturbance and competition for space (Fig. 6.3). Many suspension feeders, therefore, being less mobile than deposit feeders, seek security deeper in the substrate.

A few specialist deposit feeders and bacterial gardeners occupy deeper tiers. The RPD constitutes an attractive level within the substrate for these deeper burrowers. Construction of irrigated passages within the anoxic sediment enormously enlarges the redox interface and, with this, the potential for microbial activity (sections 5.1.2 and 5.3).

In dysaerobic settings, where the RPD lies at the sea floor, this critical balance of oxygenated waters in contact with reducing pore water may support anomalous shelly epibenthos that have symbiosis with sulphur-

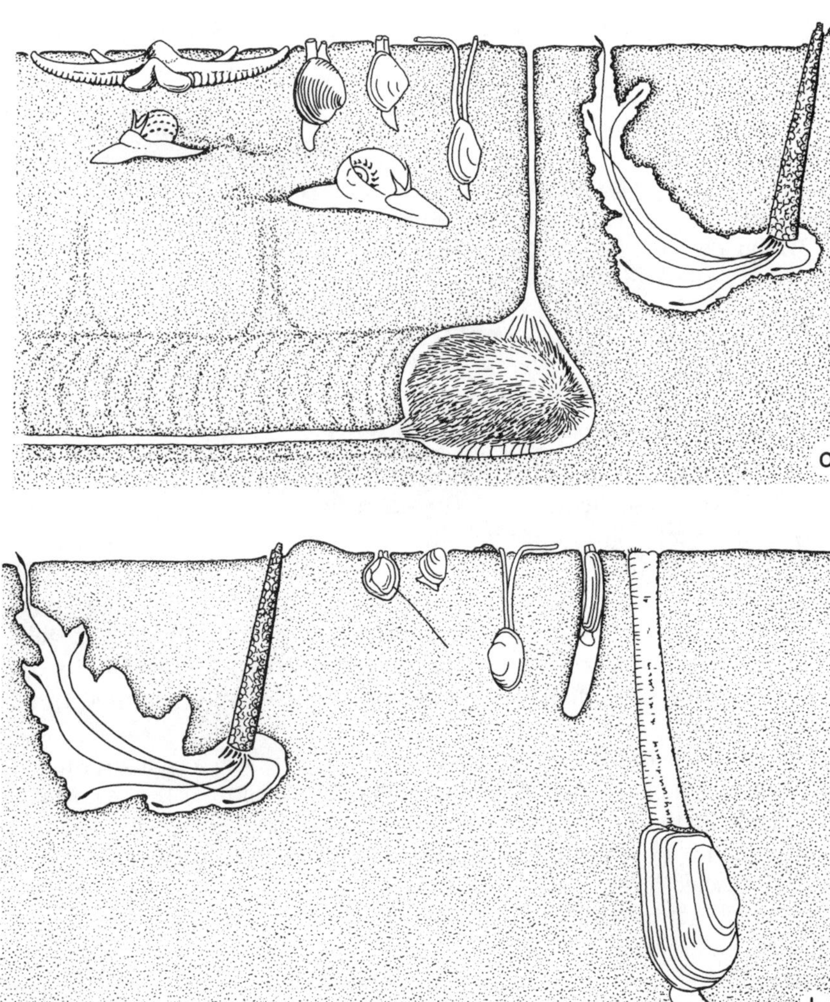

Figure 5.13 Tiered communities from the Danish sublittoral, based on Thorson (1968). (a) The *Venus* community, in sand at –15 to –20 m. From left: the starfish *Astropecten* sp., the bivalves *Venus gallina, Spisula subtruncata* and *Tellina fabula*, with two carnivorous snails, *Natica nitida* and *N. catena*, the heart urchin *Echinocardium cordatum* and the pectinariid worm *Lagis koreni*. (b) The *Syndosmya* community, also in sand at –15 to –20 m. From left: *L. koreni*, and the bivalves *Corbula gibba* (anchored by a stout byssus thread), *Nucula tenuis, Syndosmya alba, Cultellus pellucidus* and *Mya truncata*. (c) The *Amphiura* community, in mud at –20 to –30 m. From left: the heart urchin *Brissopsis lyrifera*, the carnivorous polychaete *Nephthys ciliata*, the brittle-star *Amphiura filiformis* suspension feeding, the snail *Turritella*, also suspension feeding, the detritivorous snail *Aporrhais pespelicani* and the sea pen *Pennatula phosphorea*. (d) The *Haploops* community, deeper than –30 m in mud. From left: the amphipod *Haploops tubicola* feeding through a slit near the top of a tough parchment-like tube, the enterpneust *Harrimania kupferi* and the polychaete *Polyphysia crassa* in U-burrows, the bivalve *Nuculana pernula* and the hagfish *Myxine glutinosa*.

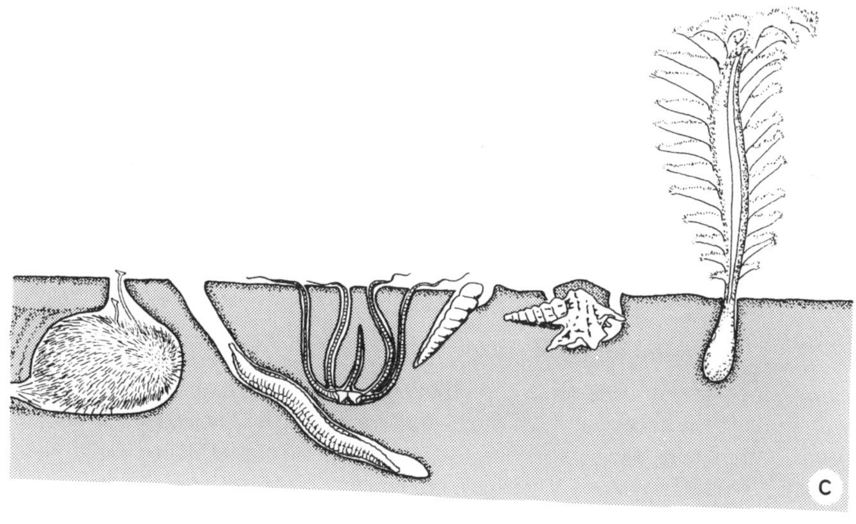

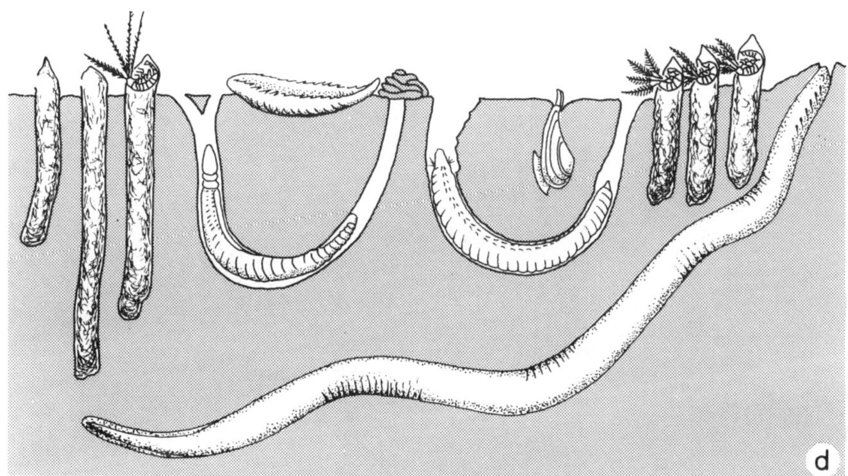

oxidizing bacteria. Savrda and Bottjer (1987a) introduced the term **exaerobic biofacies** to cover this situation.

In fact, this exploitation of the redox interface is also present in many communities, including chemosymbiotic species in oxic sea floors, where the RPD is subsurface. However, these species must maintain contact with the sea floor, and thus are in danger of suffering interference with the busy upper tiers. Levinton (1977) showed how the deeply burrowing bivalve *Solemya velum* avoided contact with shallow-burrowing bivalves by shifting the upper part of its burrow (Fig. 4.11) (section 4.1.4b).

As we have seen, many species exhibit more than one trophic habit, simultaneously or alternately. For example, the burrows of *Upogebia* spp. (section 4.3.4) have an upper U-burrow for suspension feeding and respiration and a lower shaft for culturing microbes. Compare also *Heteromastus filiformis*, feeding in a deep tier, but respiring in a shaft at the surface (section 3.5.2).

Levinton (1977) reminded us that animals may shift their level of operations during ontogeny (Figs 1.6 and 5.11). Adult, vertical conveyor species feed at deeper levels than juveniles. *Cistenides gouldii* (Whitlatch 1974) and *Scoloplos armiger* (Reise 1979) likewise feed at deeper levels with growth and, as reproduction follows the season, the tier activity is also seasonal.

5.4.4 Deep-sea endobenthic tiering

Ever since gravity and box cores began to provide intact samples of the sea floor from deep water, authors have remarked the vertical segregation of the faunas or their biogenic structures. In turbidites from the abyssal ocean floor, Griggs *et al.* (1969) recognized six types of burrows, each having its characteristic depth of occurrence, from 10 to 50 cm below the sea floor. Jumars (1978) found a marked vertical zonation of species of polychaetes in cores from the bathyal Pacific. An ichnological study by

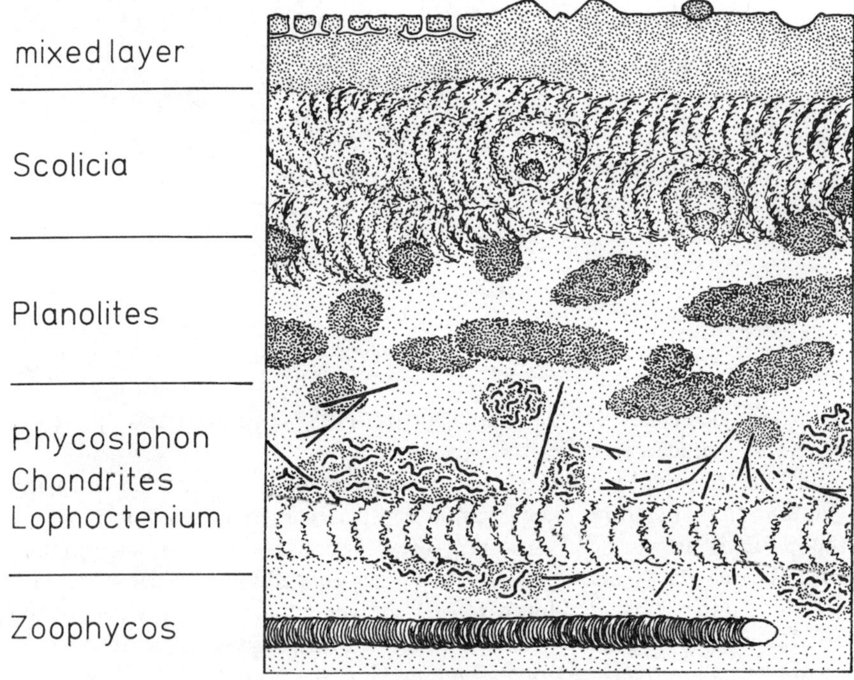

Figure 5.14 Tiering diagram of traces seen in box cores taken in –2—3.5 km water depths off northwest Africa. Surface trails and *Paleodictyon* isp. are indicated in the mixed layer. Modified after Wetzel (1984).

Wetzel (1981, 1983b) revealed a five-tier structure of endobenthic activity in the bathyal sea floor off West Africa (Fig. 5.14).

Berger *et al.* (1979) established a generalized model for burrow stratigraphy on the basis of box cores from bathyal and abyssal depths: further detail was added by Ekdale *et al.* (1984b). Beneath a **mixed layer** a few centimetres thick, where total bioturbation homogenized the sediment (Thistle *et al.* 1985), a **transition layer** showed a highly heterogeneous texture, owing to the descent of a few deeper burrows to this level. Colour contrast of the transition layer is high, but fades in the level beneath, the **historical layer**, which generally lies beneath the zone of active bioturbation (Fig. 5.15). Swinbanks and Shirayama (1984) pointed out that the uniformity of the mixed layer was due to oxygenation and the colour contrast peak of the transitional layer was due to heterogeneous manganese reduction, which became uniform in the reducing historical layer below. The boundaries between the layers are, strictly speaking, diagenetic, but the diagenesis is controlled by bioturbation.

Animals also burrow deeply in the deep sea. Thomson and Wilson (1980) and Weaver and Schultheiss (1983) found open burrows extending 2 m vertically into deep-sea sediments.

Figure 5.15 Deep-sea box core of foram–nannofossil ooze from –4 km water depth on the East Pacific Rise, just above calcite compensation depth. Mixed and transitional layers are indicated. Photo courtesy of A. A. Ekdale.

5.5 Modelling bioturbation

Many authors have attempted to construct models that will explain and predict the effects of bioturbation. These models fall into two categories: descriptive and mathematical.

5.5.1 Descriptive models

An early descriptive model of bioturbation by Moore and Scruton (1957) was based on a transect of shallow-water deposits in a series of cores from the Gulf of Mexico. Nearshore, the sediment was coarse and homogeneous, and at the seaward end of the transect the sediment was fine and homogeneous. In between, there was a passage from layered structures to increasingly mottled and finally homogeneous sediment. Bioturbation was considered the cause of this structural gradient, causing not only homogenization of layered fabrics, but also heterogenization of structureless sediments (Fig. 5.16). Hill and Hunter (1976) showed that some of the homogeneous layers that were considered to be primary by Moore and Scruton actually had been homogenized by bioturbation.

The descriptive model of Berger *et al.* (1979) operates essentially with two tiers (section 5.4.4). The upper, mixed layer, is totally bioturbated [although Berger *et al.* (1977a) subdivided this as homogeneous above and lumpy below]. This overlies a lower tier of incomplete reworking, the transitional layer. A similar two-tiered system was mentioned for shallow

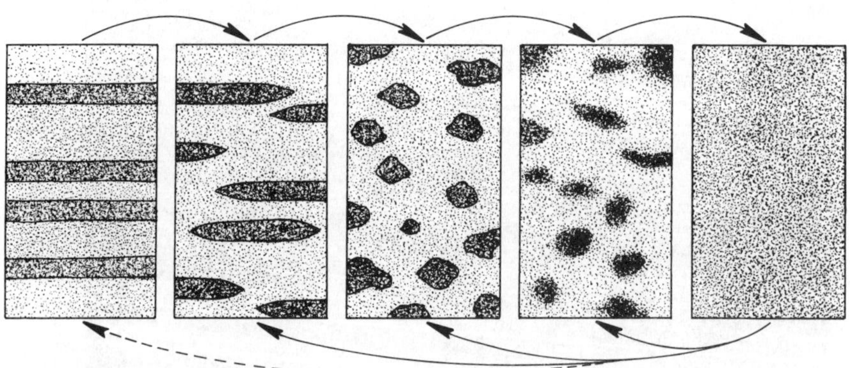

Figure 5.16 An early bioturbation model, modified after Moore and Scruton (1957). Laminated, unbioturbated sediment (left) may be progressively altered by the work of burrowing animals to interrupted bedding, mottles, indistinct mottles and homogeneous fabric (upper arrows). However, bioturbation may also unmix a homogeneous sediment (lower arrows) and even lead to biostratification (broken arrow). Cf. Figs 4.15 and 6.8b.

water by Aller and Cochran (1976), but our knowledge of the vertical distribution of animals in sediment would suggest that the tiering system of bioturbation is much more complicated than this.

5.5.2 Mathematical models

Impetus for mathematical modelling of bioturbation is provided by the blurring effect that the process has on stratigraphical markers. In cores of sea-floor sediments, sudden events such as extinction levels of stratigraphically important organisms, microtektite and ash falls, or radioactive fall-out horizons, were found to be gradual instead of sharp. This record-smearing was attributed to bioturbation (Bramlette and Bradley 1942; Ericson *et al.* 1963; Glass 1969) and it was naturally considered necessary to model the process of bioturbation in order to be able to compensate for it. Eventually it should be possible to 'unmix' the record by running the model backwards (Berger *et al.* 1977a).

The disadvantage of these models, however, is their simplicity in comparison with the extreme variability of the actual process. Usual requirements for such models (Berger and Heath 1968) are: (1) constant sedimentation rate; (2) constant mixed-layer thickness; (3) complete homogenization within the mixed layer; and (4) no burrow activity beneath the mixed layer. As Ruddiman and Glover (1972) pointed out, the last two requirements are rarely met. Guinasso and Schink (1975) modified the Berger and Heath model to allow for different rates of mixing and deposition, and various thicknesses of the mixed layer.

On the whole, there is not enough biological realism in these models to make them useful for the palaeoecologist or ichnologist. Most are based on the unrealistic assumption that bioturbation involves diffusive processes only (Boudreau 1986a). But some models do emphasize advective processes (Fisher *et al.* 1980; Robbins 1986). Having discovered the paper by Rhoads (1974) describing 6 cm upward conveyor activity of maldanid worms, Boudreau (1986b) presented a general theory based on advection. It is a pity that he had not yet encountered the 5 m upward advection activity of *Callichirus major* (section 4.3.1)! Robbins (1986) generously allowed animals to make occasional excursions to 25 cm, although Kershaw *et al.* (1983) had expressed misgivings about dumping plutonium on sea floors where echiurans were advecting surface material 40 cm down.

Carney (1981) and Robbins (1986) pointed out that the nature of the redistribution of a marker event through the bioturbation zone depends entirely on the type of sediment processing in progress. Conveyor processing works selected particles upwards, or accumulates them as a residual layer at the base level of activity. Passive fill may displace particles metres downward. Active filling and inverted conveyance displaces selected grain-types decimetres to a metre downwards. Horizontal backfill commonly causes total bioturbation yet only slightly displaces grains in a vertical sense. Coarsest grains may even be worked up to the sea floor (section 5.2.2).

I believe that a bioturbation model based on tiering should be of more use to the ichnologist.

5.5.3 A tiering model

We have seen in the preceding chapters that not only is bioturbation a highly complex process, the quality of which depends on the composition of the endobenthic community, but it is also subject to a distinct vertical repartition. Thus, although there is commonly a double zonation of bioturbation of the sea floor, an upper level totally reworked and a lower one incompletely so, the layering of biological activity is usually more complicated.

If we assume that a given sea floor has, at a given time, a five-tiered biological system, we may label these tiers A–E (Fig. 5.17). In the uppermost tier (A), superficial disturbance from mainly epibenthic animals, and homogenization by mobile burrowers, intruders and meiofaunal activity will totally mix the sediment and obliterate primary depositional structures. The large majority of the endobenthic species will be active within this layer, which may be 2–3 cm thick.

In tier B, more specialized, deposit-feeding, vertical conveyor worms

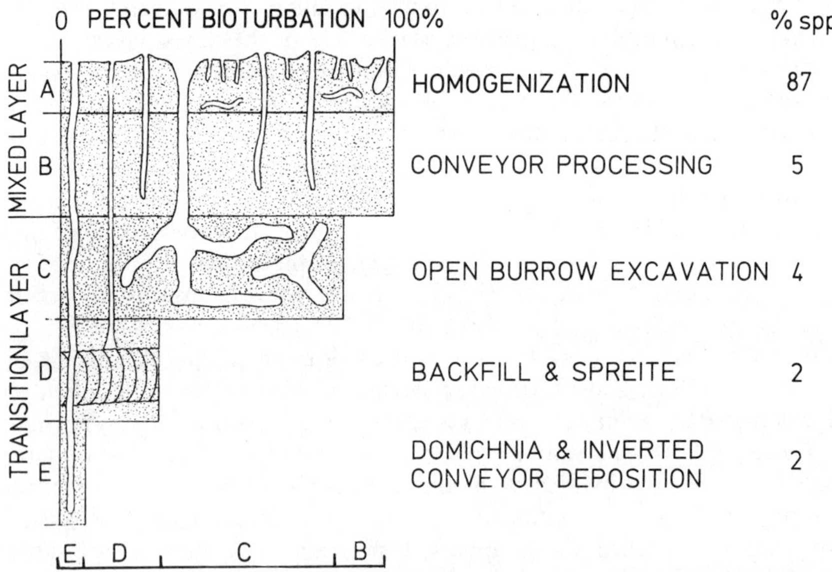

Figure 5.17 A generalized tiering bioturbation model, indicating five levels of activity. For each tier, the usual quantity and type of sediment turnover are suggested, together with the proportion of the whole community that may be active in each tier. The scale at bottom indicates the expected representation of the tiers in the final ichnofabric (historical layer).

may dominate the subcommunity. If bioturbation is complete within this tier, the work of these relatively few animals will obliterate any structure created by tier A activity.

In tier C we may suggest an open burrow network of a species of deposit-feeding crustacean, working perhaps 20 cm down. Yet again, the activity of these animals will largely rework the sediment and leave little of the fabric of the shallower tiers for posterity.

Beneath this, a tier (D) may exist where deep deposit-feeding worms win nutrient from below the RPD, possibly in symbiosis with chemolithoautotrophic bacteria. Finally, security-seeking, immobile suspension feeders may construct shafts to form, at their distal ends, a bottom tier (E), and here too we find the sulphide-well structures that fuel chemosymbiosis.

It will be seen that the occupants of the deeper tiers (C–E) will have difficulty in maintaining contact with the sea floor, and will suffer much interference from the crowded upper tiers. Indeed, for this reason the deeper tiers may be only locally and temporarily accessible. Yet it is the fabric produced in these lower tiers that will be preserved in the fossil record, and the appearance and disappearance of these ecological niches will significantly alter the ultimate fabric of the rock (Fig. 5.17; section 10.3).

Direct communication between benthic ecologists and palaeoichnologists is rendered impossible by this tiering system. One might say that the biologist looks at it from above and the geologist from below. The box-coring biologist is confronted by a totally bioturbated sea floor, and for him/her the deeper tiers are a matter of remote sensing. In contrast, the geologist sees rock dominated by trace fossils created in the deeper tiers whereas the topmost tier normally is not even preserved.

But we are now touching on matters of taphonomy and ichnofabric, and these belong to the geological half of this book. First we must cross and examine the **fossilization barrier**.

PART TWO

Palaeoichnology

6 The fossilization barrier

In general, geologists make too little use of information derived from modern environments in their analyses of biogenic structures. This is partly because the biological information cannot be applied directly to trace fossils and the geological record of bioturbation. Even between such patently similar structures as the burrow of *Callichirus major* and the trace fossil *Ophiomorpha nodosa*, a one-to-one comparison cannot be made.

The two realms are quite distinct and are separated by what Seilacher (1967a) called the 'fossilization barrier' and Curran (1994) the 'modern to fossil transition'. This is more than the taphonomic filter that separates living organisms from equivalent body fossils. The ichnological fossilization barrier is complicated by several features that are not found in the conversion of a skeleton into a body fossil. The work of an animal produces cumulative structures by the continuous processing of sediments in the course of time (section 6.3). Bioturbation modifies the substrate in ways that directly influence diagenesis. Furthermore, the process of bio-

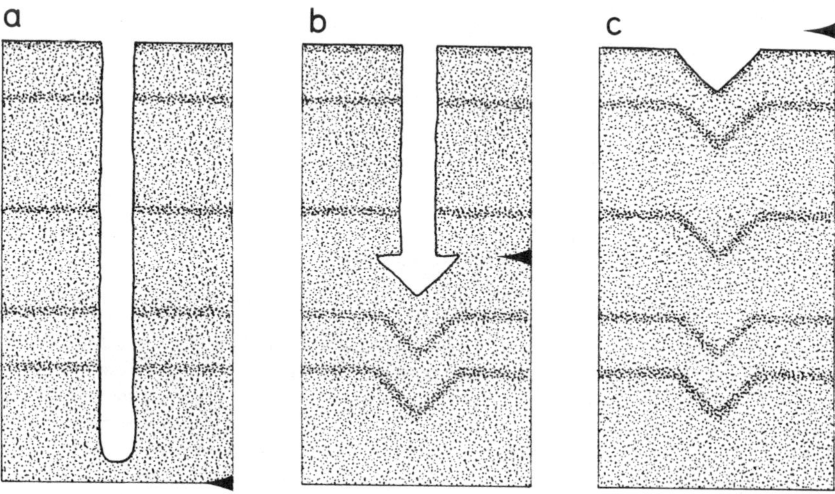

Figure 6.1 Pattern of sediment collapse seen in vertical section in a tidal aquarium. The dark arrowhead indicates the water table. (a) A simple, unlined shaft is excavated in sand at low tide. (b) With the rising tide, water percolates upward. In the saturated zone, the sand collapses as a conical depression. (c) Continued upward progress of this system produces, at high tide, a column of conical structures in place of the burrow. Data courtesy of Claus Heinberg.

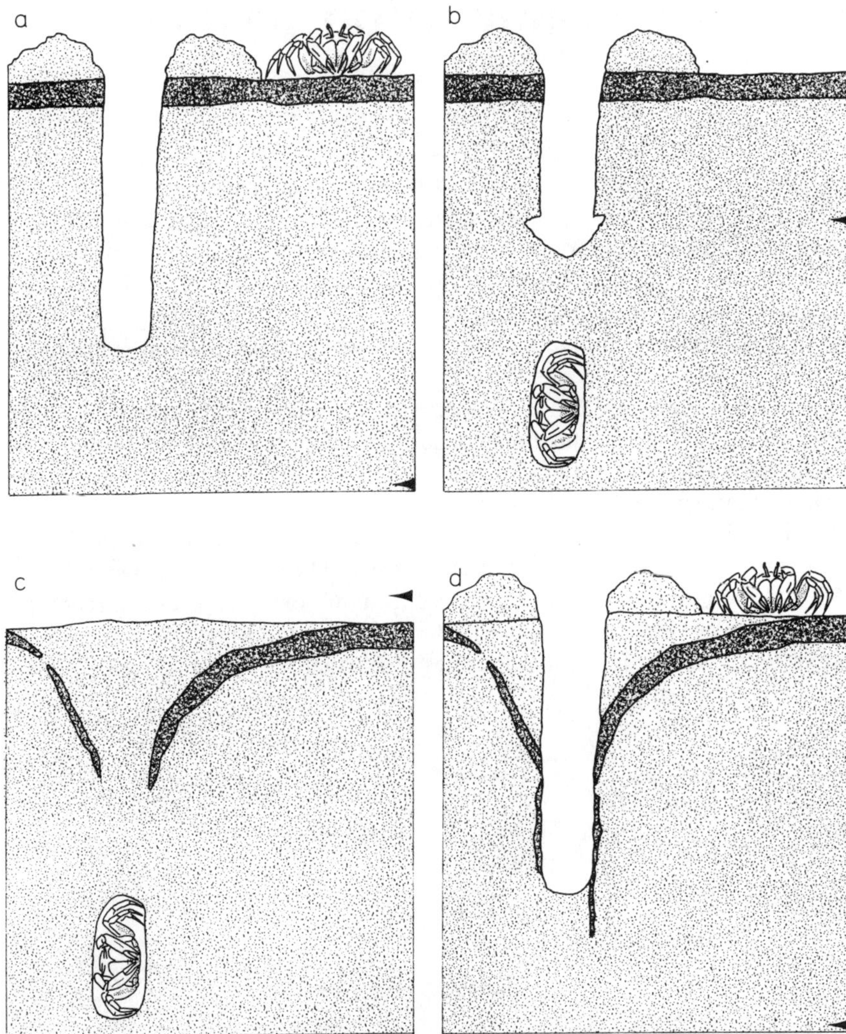

Figure 6.2 Structures associated with the burrowing activity of the soldier crab *Dotilla myctiroides* in the tidal zone (Thailand). (a) The small crab feeds on detritus beside its open burrow at low tide. (b) At the approach of high tide, the crab digs into the base of its shaft, where it maintains a small air-filled chamber. The burrow above is obliterated gradually by the rising water table as seen in Fig. 6.1. By high tide the detritus layer has collapsed as a conical depression. (c) After several high tides, the detritus cone works deeper. Sediment from beneath the detritus, advected up during the repeated re-excavation of the shaft, fills the depression. (d) Low tide after many tidal cycles. The detritus layer has been worked down, both by physical collapse and by the activity of the crab, as a 'lining' around the burrow. Tidal aquarium observations, data courtesy of Claus Heinberg.

turbation is destructive as well as constructive; the construction of burrows obliterates earlier structures.

In many environments, the actual biogenic structure is not preserved as such; nevertheless, its transitory presence produces a recognizable fabric in the rock. Biodeformation structures (section 1.2.1) fall into this category, as do collapse structures in intertidal sands (Figs 6.1 and 6.2).

In general it may be said that trace fossils are relatively easy to observe but are difficult to interpret; modern burrows, on the other hand, are difficult to observe but are comparatively easy to interpret. Therefore, the better we understand this discontinuity in the nature of the data, the more accurate our interpretations of fossil biogenic sediment structures are likely to be.

6.1 Taphonomy of trace fossils

Taphonomy is the study of processes that lead to the loss of information incurred as sediments pass from the active benthic boundary layer into the geological record. In ichnology, however, there may be as much gain from this passage as loss. For example, the geologist can study the whole life's work of an animal, preserved in detail and easily accessible in the form of a trace fossil, whereas the biologist may manage to make a cast of 'today's burrow'. Trace fossils that originated in deep water have supplied information on abyssal sub-sea-floor animal behaviour that is not available to the biologist.

This is the reverse of the situation in body fossil palaeontology, and led Frey (1975) to speak of reverse uniformitarianism in the trace fossil world, or of 'the past as a key to the present' (Frey and Seilacher 1980).

But, of course, it is not universally the fate of burrows to receive such favoured treatment. As with body fossils, an infinitesimal proportion of the biogenic structures that are produced actually become preserved in the geological record. Goldring (1965) wrote of the 'preservation potential' of such structures, which is highly variable for different types of trace, and depends on numerous contingencies. As a result, the preserved record of endobenthic activity is a gross distortion of the work of the original community. Certainly, it gives us little evidence of the relative biomass of the different species. It does, nevertheless, provide us with a wealth of information on the sediment processors whose work is preserved, and the conditions under which they were working.

6.2 Preservation potential

Many authors have stressed the vastly different potential the several types of biogenic sediment structures have for survival. Hertweck (1972), for example, after describing the distribution of the traces produced in sediments of a coastal region of Georgia, USA, confronted the problem of their possible preservation in the geological record.

In turbulent coastal environments, Hertweck found no uninhabited burrows. Physical reworking of the upper layers of the sediment destroys the structures as fast as they are produced. Thus, in high-energy settings, only deep-tier structures have good preservation potential.

Locally, coastal sand is densely populated by *Donax variabilis* (Fig. 6.3), which is active in the topmost 3 cm.

Although these bivalves may dominate the endobenthic community in biomass, any structure that their activity imparts to the sediment is normally destroyed by the next passing of the surf zone and has a negligible chance of survival. In contrast, the deeper parts of the burrows of *Callichirus major* (section 4.3.1) in the same environment have a high preservation potential (Weimer and Hoyt 1964), being strongly constructed and generally lying well below the depth of biological or physical sediment reworking.

It would not be true to say, however, that the work of *Donax variabilis* never could be preserved. The fact that the narrow uppermost portions of (presumed) *Callichirus major* shafts have been found in nearby Pleistocene sands (Fig. 4.20) shows that, under rare circumstances of sudden and permanent burial, the *Donax* tier is spared destruction.

No biogenic structure can be said to have zero preservation potential; however, structures such as those of *Donax variabilis*, lancelets (section 5.3.1) and *Paraonis fulgens* (section 4.6), which function only when emplaced in sand that is constantly being reworked physically, certainly have small chance of survival. Risk and Tunnicliffe (1978) need not have expressed surprise that, despite our extensive experience of Quaternary beach sands, the eye-catching spirals of *Paraonis fulgens* do not appear to have been observed in the fossil state (section 10.7.2).

It would seem, then, that structures in deep tiers stand a better chance of survival than do those of shallow tiers. Indeed, in cases of complete bioturbation in the uppermost, mixed-layer tiers, only those structures that extend beneath this level will pass into the geological record. Thus, we have a grand rule: 'To be preserved, a burrow must reach below the mixed layer and into the transition layer' (Berger *et al.* 1979, pp. 205, 215).

6.2.1 Semirelief preservation

This rule holds true while there is gradual accretion of the sea floor. But, like all rules, this one has exceptions. A sudden event, such as **obrusion** (sudden burial) or defaunation (e.g. anoxia) will interrupt the gradual trace-destruction process. Savrda and Ozalas (1993) demonstrated how oxygenation event beds can preserve the fabric of the mixed layer. Exam-

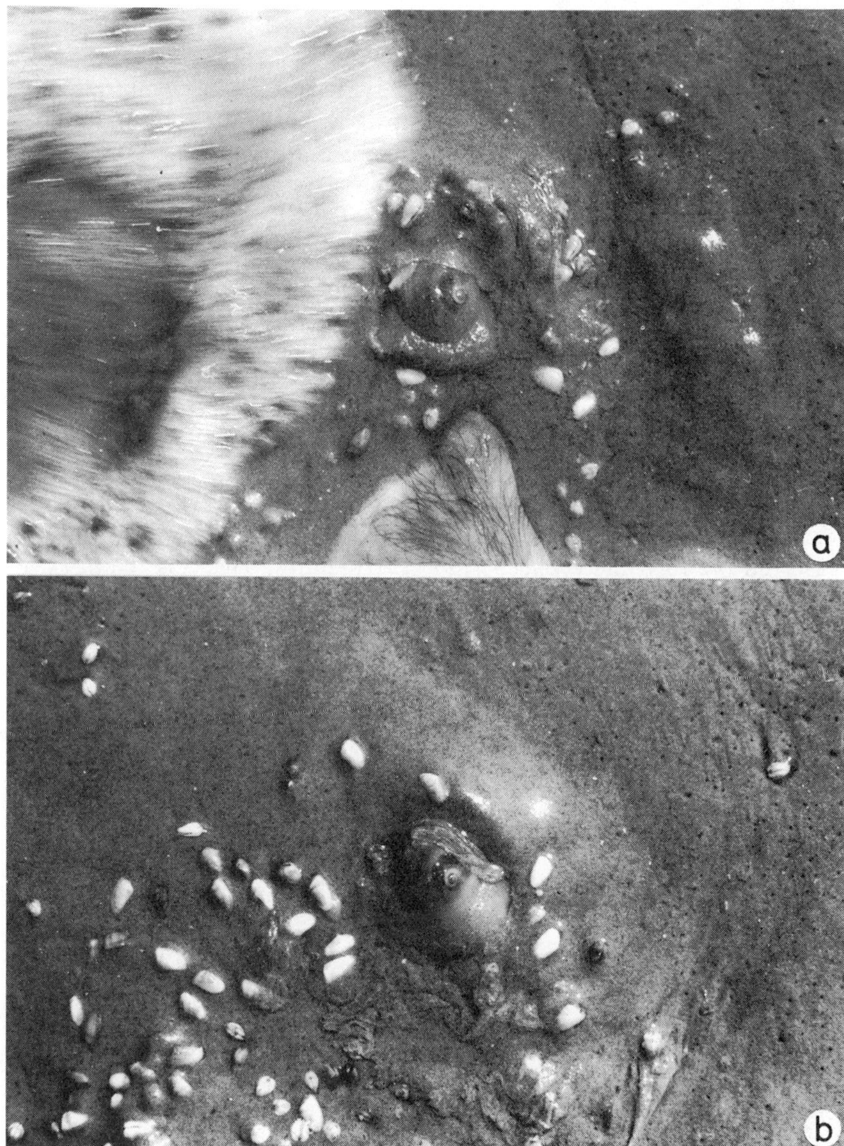

Figure 6.3 Shallow-tier bioturbation on an exposed Atlantic beach, Sapelo Island, Georgia, USA. (a) Intruding my foot into the sand in the surf zone of the rising tide exposes numerous individuals of *Donax variabilis* and a predaceous snail, *Polinices duplicatus*. (b) A few seconds later, the snail is rapidly re-entering the sand and many of the bivalves are already in vertical position, actively digging in. To the right, the undisturbed area shows numerous small apertures of bivalves living immediately under the surface.

ples are rare, but Watling (1991) and Brodie and Kemp (1995) have both illustrated mixed-layer ichnofabric.

Sudden burial of the muddy deep-sea floor by a sandy turbidite can fill and cast the open burrows within the mixed layer and elegantly preserve the uppermost tier of structures. In this way, sole surfaces of turbidite beds may have semirelief preservation of structures such as the ornate graphoglyptid trace fossils that are common in many Flysch sequences (Fig. 9.5). Ekdale (1980), Ekdale *et al.* (1984b) and Gaillard (1988) demonstrated these delicate structures in the topmost tier of the mixed layer, being visible in box cores from the ocean floor only where still inhabited or, at least, still empty. Adolf Seilacher was particularly happy with that discovery; he had predicted it on the basis of graphoglyptid taphonomy (Seilacher 1977)!

Of course, the turbidite bed must be sufficiently thick to bury the previous surface to a depth beneath the base of the mixed layer that will develop at the new sea floor (Fig. 10.6). In fact, the thickness of the turbidite will determine which, if any, tier of the next (**post-depositional**) community will reach the buried sand–mud interface and mingle with the preserved pre-turbidite (**pre-depositional**) structures. This was the basis on which Seilacher (1962b, 1964) determined the relative depths within the substrate at which the different burrows of turbiditic trace fossil assemblages were produced (Bromley and Ekdale 1986).

Various degrees of erosion immediately prior to burial by a turbidite will expose and cast different tiers of burrows. The same phenomenon

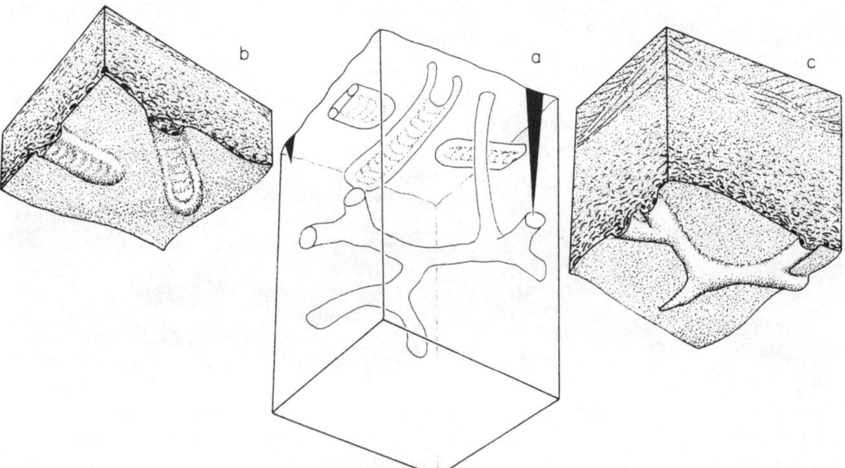

Figure 6.4 In a sea floor containing a community working at several levels, different amounts of scour and fill will cast different types of biogenic structures. (a) A two-tiered deposit-feeder community. (b) Sole surface of a tempestite or turbidite that involved only slight erosion; *Rhizocorallium* isp. is preserved in semirelief. (c) Deep erosion has reached the *Thalassinoides* tier. Inspired by Wetzel and Aigner (1986).

occurs in shallow water. Wetzel and Aigner (1986) pointed out that trace fossils in semirelief on soles of storm beds belonged to different tiers, and that according to the known depth of that tier, they could judge the amount of erosion represented by the storm bed (Fig. 6.4). This process has recently been exercised by Droser *et al.* (1994) in Cambro–Ordovician sediments.

6.2.2 *Full-relief preservation*

In contrast to storm or turbidite events, steady deposition tends to produce full-relief preservation. Such trace fossils may be conspicuous where the original burrow has a wall sediment that contrasts lithologically with the substrate, or has been filled with contrasting material (Fig. 8.12a).

Even where lithological contrasts are lacking, biogenic structures may nevertheless be well preserved because of constructional differences in compaction, grain orientation and organic content of fill and walls

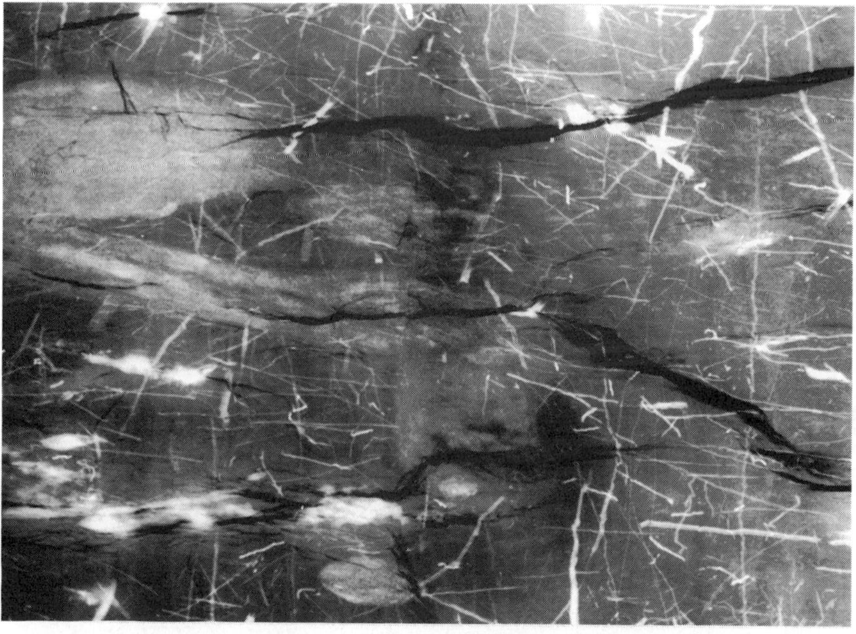

Figure 6.5 X-radiograph of a vertical slice 1 cm thick of Gram Clay (Miocene, from Gram, Denmark), slightly reduced. The rock itself is almost black and structures are very poorly visible in reflected light, whereas in the radiograph the fabric is distinct. The background is very faintly mottled and is totally bioturbated (at other horizons, unbioturbated intervals appear finely laminated). The white lines are *Trichichnus* ispp., pyrite tubes or burrow linings in several size classes (cf. Thomsen and Vorren 1984). Cutting these are pale burrow fills 1 and 3 cm in diameter. (The black structures are cracks in the specimen.) Courtesy of Peter Laugesen.

introduced by the burrower. These textural and chemical differences may initiate localized diagenesis that will enhance the visibility of the trace fossil (Figs 8.3 and 9.10).

In monotonous lithologies, visibility of biogenic sediment structures is often poor. Clays and fine mudstones are particularly difficult in this respect. In many cases, a technique must be developed to enhance the visibility of the structures (Farrow 1975).

In box cores of deep-sea muds, normal shipboard procedure, whereby the sample is cut with a wire, may produce smeared surfaces that reveal only vague burrow mottling. Far greater detail of bioturbation fabric was obtained by gently spraying the cut surface with a jet of water (Berger *et al.* 1977b). Similarly, deep burrows in abyssal red clay were not seen on core surfaces using routine methods, but were revealed unexpectedly by cleanly fracturing the sediment (Thomson and Wilson 1980).

In clay units, X-radiography can yield good results in rock that appears to be homogeneous (Fig. 6.5; e.g. Wetzel 1984; Fu and Werner 1994). Computed tomography is now being applied increasingly for

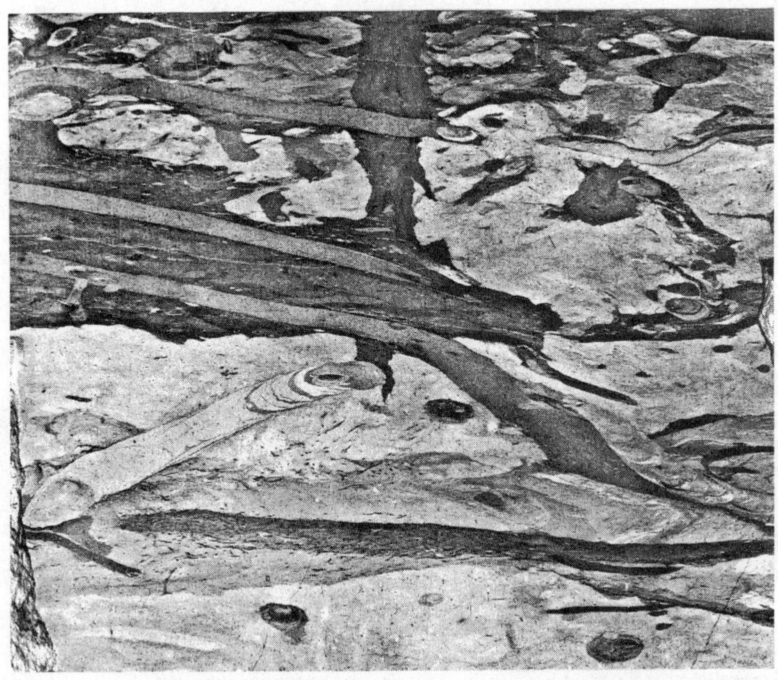

Figure 6.6 Vertical section of marly chalk, smoothed and oiled, and photographed on lith film (Bromley 1981a). A pale, *Planolites*-mottled background is overprinted by black *Thalassinoides* isp. These are, in turn, cut by *Zoophycos* spreiten. Minute white and black *Chondrites* isp. represent the last biogenic events. Upper Maastrichtian (Cretaceous), Marl M12, Dania Quarry, Denmark. Natural size.

revealing full-relief structures in fine-grained sediments and other substrates (Fu *et al.* 1994; Genise and Cladera 1995). Again, by applying oil to smooth, dry surfaces, ichnological details can be studied in chalk that appears to be structureless (Fig. 6.6). Thus, trace fossils and ichnofabric may well be preserved even where they are not immediately visible.

6.3 Cumulative structures

Another troublesome aspect of the fossilization barrier is the non-equivalence of trace fossils to extant burrows. Burrowing animals show a great range of mobility, but even 'stationary' deposit and suspension feeders periodically shift their location (Ronan 1977; Wilson 1981; section 5.2.1). Mobile deposit feeders and many carnivores move more or less continually. Thus, although a burrow at any one time may have a well-defined construction, its frequent or continual alteration or migration through the substrate produces cumulative structures having an entirely different morphology (Figs 3.6, 3.8a, 3.12e, 3.21, 4.2, 5.2 and 6.7; Bromley and Frey 1974).

The J-shaped burrow of the lugworm (section 3.4.2) may, in time, develop into quite a different cumulative structure having radial symmetry. The entire structure may be preserved as a single trace fossil like, for example, the ichnogenus *Dactyloidites* (Figs 6.7c and 6.8). Even after sudden burial of the sea floor, which might 'freeze' the structure, it may not be clear which part of it comprises the burrow that was active at the time of burial. It is not easy to compare such cumulative structures with recent counterparts.

Another example is the spreite. This cumulative structure is hard to see in sediments that have undergone little diagenesis. Spreiten have been glimpsed in connection with the work of only a handful of modern burrowers: *Echiurus echiurus* (section 3.3.3), *Corophium volutator* (section 3.4.1), *Thyone briareus* (section 4.4.4), *Heteromastus filiformis* (section 3.5.2) and *Nereis diversicolor* (Seilacher 1957). Yet the trace fossil record is rich in spreiten from Cambrian to Holocene and they must be as abundant in the modern sea floor; indeed, radiography of cores reveals them clearly (Wetzel 1984).

Part of the problem is that spreite structures such as *Zoophycos* ispp. occupy the deeper tiers and are difficult to sample in the deeper-water sea floor. We may assume, however, that animals that can produce *Zoophycos* are well known to taxonomists and reside already in phials of spirits in the collections of zoological repositories, as did the worm *Paraonis fulgens* for several decades before its sediment structures were first observed. Levinsen (1884) caught the adult worms while swarming at the surface of the sea, and had no idea that this was a burrowing worm. Gripp (1927) found the spiral burrows but not the tiny worm. It was still longer before

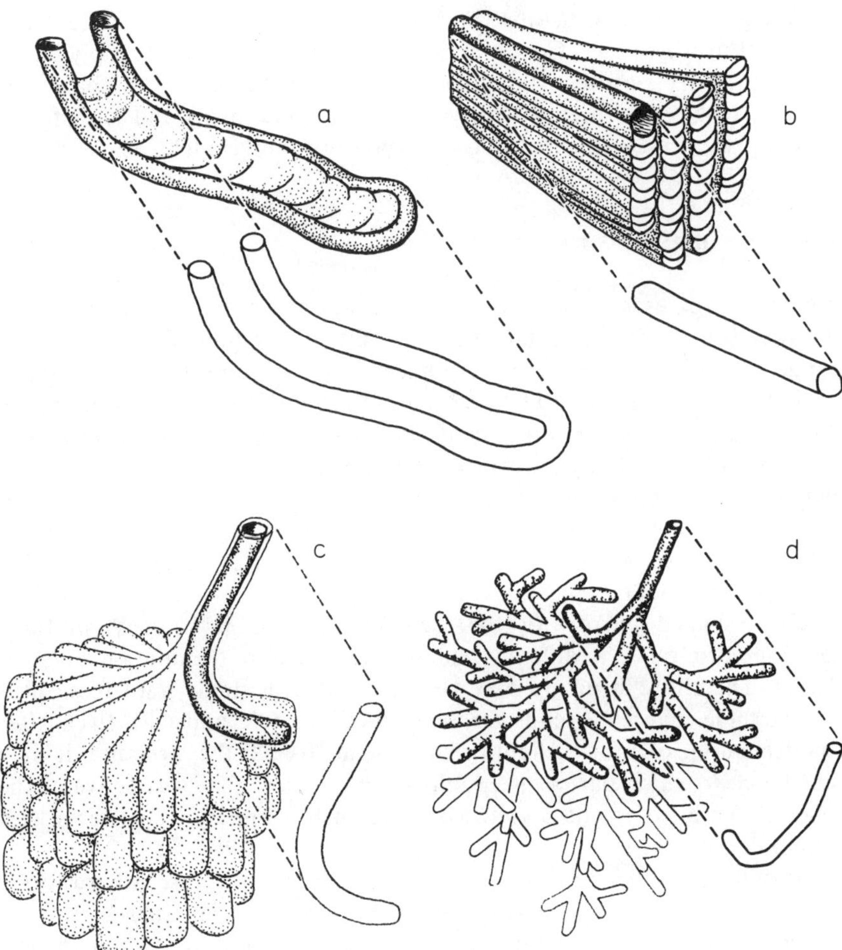

Figure 6.7 Crossing the fossilization barrier. Four cumulative structures produced by stationary deposit feeders, as preserved in the rock. The causative burrow is extracted, representing 'today's burrow', as would be found in the sea floor by the biologist. (a) *Rhizocorallium irregulare*. (b) Clustered *Teichichnus rectus*. (c) *Dactyloidites ottoi*. (d) *Chondrites* isp. Inspired by Seilacher (1957).

the worm and its traces were identified as being genetically connected (Remane 1940).

6.4 Key bioturbators and elite trace fossils

Endobenthic species that are highly mobile and bring about sediment disturbance out of proportion to their size (section 5.2.1) are of paramount

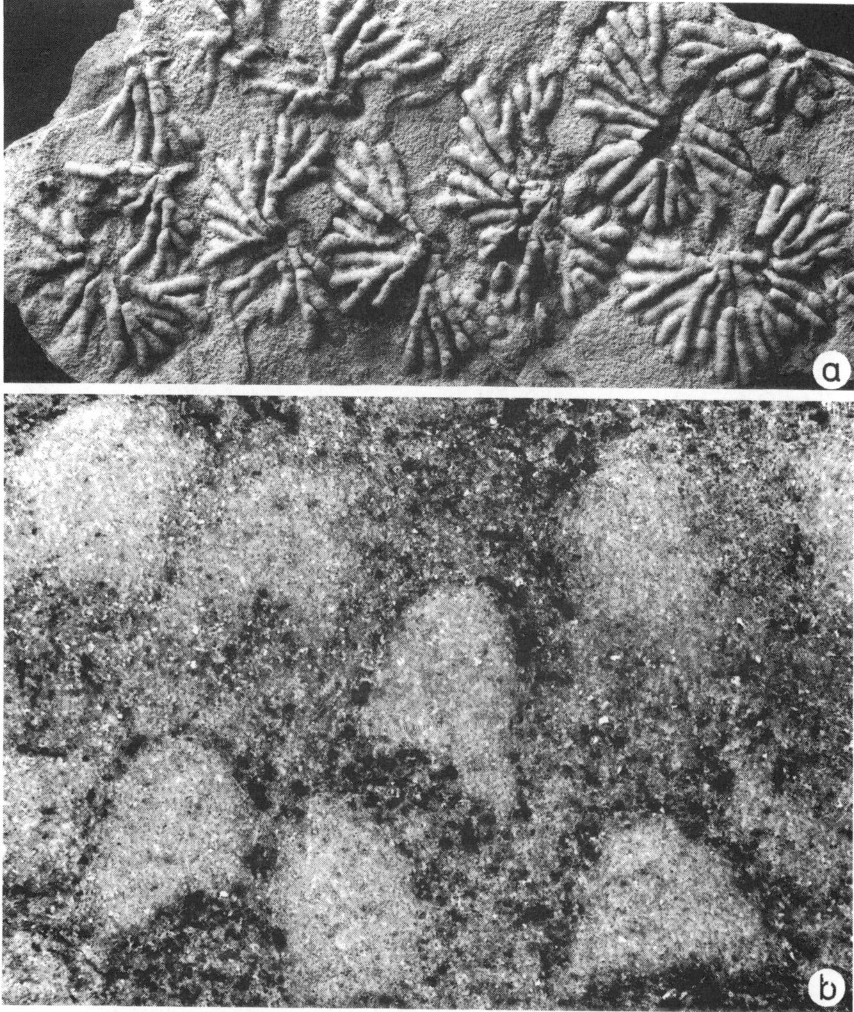

Figure 6.8 The rosetted trace fossil *Dactyloidites ottoi*, Upper Cretaceous, Nûgssuaq, West Greenland (see Fig. 6.7c). (a) Weathered upper surface of a sandstone horizon showing several cumulative rosettes, possibly all the work of a single animal. Half natural size. (b) Enlarged vertical section through the trace fossil. The animal has unmixed a poorly sorted sand, excluding plant fragments from the processed sediment of its spreite (Fürsich and Bromley 1985). Cf. Pickerill *et al.* (1993).

importance. Key bioturbators of this sort, e.g. active, prey-seeking carnivores, normally inhabit shallow tiers and contribute to the diffusive stirring of the mixed layer. However, when relentlessly mobile species work at depth, as do some heart urchins (section 4.2), the structures they produce obliterate shallow-tier structures, and come to dominate the preserved fabric of the rock.

Structures that, in the fossil state, totally dominate the fabric may be called **elite trace fossils**. Special diagenesis may also create elite trace fossils by providing them with disproportionate emphasis. In the European Cretaceous and Danian chalk, *Thalassinoides suevicus* has received this special treatment, acting as the nucleus of flint concretions. This is probably due to the continuous nature of the filled boxwork passages, which acted as permeability conduits through the sediment (Bromley and Ekdale 1984b; Clayton 1986; Zijlstra 1994). In this way, the black silica renders *T. suevica* highly conspicuous at many levels of the chalk, but detracting attention away from other trace fossils in the rock, which remain inconspicuous.

From the top of the Round Tower in Copenhagen you can see the white Danian limestone cliffs of the coast of Sweden. Using the public telescopes provided, for only two crowns you can see the flintified *Thalassinoides suevicus* at a distance of 13 km. Their visibility has been enhanced enormously by the fossilization process since the time they existed as inconspicuous open networks 30 cm beneath the sea floor!

7 Some ichnological principles

Biogenic sedimentary structures originate through the behaviour of animals and clearly do not represent the burrowing animals themselves. Seilacher (1967b) described trace fossils epigrammatically as 'fossil behaviour'. The special features of the taphonomic filter through which the structure must pass in order to be preserved in the fossil record further emphasize the differences between **trace fossil** and **body fossil**.

Nevertheless, many biologists and palaeontologists still nurse the belief that, when the fossil record is better understood than it is today, each trace fossil will be attributable to its trace-making organism and separate names for the trace fossils will become redundant. Since Seilacher (1953a) first tried to dispell this biotaxonomical attitude to trace fossils, many workers have repeated the attempt, but the idea persists. 'Which animal produced that trace fossil?' remains typical of the first question put to an ichnological sample. This is not the right way of addressing trace fossils; they tell us much more interesting things than that, while rarely divulging the nature of their architects.

When the International Code of Zoological Nomenclature (**ICZN**) was due for revision, Bromley and Fürsich (1980) tried to outline the basic concepts of ichnology as a set of principles, in the hope of influencing the revisers' attitude to ichnotaxa; the principles have been expanded since by Ekdale *et al.* (1984a). I shall restate these principles here and illustrate them with examples drawn from the first half of this book.

7.1 The same individual or species can produce different structures corresponding to different behaviour patterns

The several activities that are essential for life, i.e. respiration, feeding, retreat, breeding, may take place within specifically designed parts of the burrow. The form of the different parts may reflect their functions (section 8.5.1).

The burrows of *Heteromastus filiformis* (section 3.5.2) and *Upogebia* spp. (section 4.3.4) are examples of this. In addition to burrows, the faecal pellets of deposit feeders and most suspension feeders are also composed of sediment and have a reasonable preservation potential as trace fossils.

The chief significance of this principle for trace fossil analysis, however, is that different activities receive different emphasis, according to the environment. Thus, *Callianassa biformis* (section 4.3.2) produces an unlined burrow system in mud that is referable to the ichnogenus *Thalassinoides*; in loose sand, however, the same species supports the wall of the burrow with a mud lining, thus producing a structure referable to ichnogenus *Ophiomorpha*. The spreite U-burrow of *Corophium volutator* in mud resembles protrusive *Diplocraterion parallelum*, but in sand this amphipod constructs a single shaft, ichnogenus *Skolithos* (Fig. 3.11).

Many other animals, such as *Arenicola marina* (section 3.4.2), *Leptosynapta tenuis* (section 3.4.3) and perhaps *Echiurus echiurus* (section 3.3.4), can feed in several different manners and produce correspondingly different structures. Clark and Ratcliffe (1989) gave examples among the insects. They emphasized that structures produced by larvae and adults of the same species may be very unalike.

7.2 The same burrow may be differently preserved in different substrates

This principle involves substrate consistency and stratinomy, and has been illustrated by the classical example of the ichnogenera *Nereites*, *Scalarituba* and *Neonereites*. Seilacher and Meischner (1964) showed that these distinctive forms actually are representatives of the same activity pattern preserved under different stratinomic conditions. There is disagreement among ichnologists as to how this problem should be handled ichnotaxonomically: one ichnogenus or three? The everyday human footprint is a good example of this (Fig. 7.1).

Likewise, a burrow that is constructed in mud firmground is likely to bear ornament (**bioglyph**) on its walls that reflects the digging activity: a network of scratches, annulations or longitudinal striation. In a mud softground such an ornament may not be preserved, even though the basic activity of the burrower remains unchanged. Wall ornament is considered important in trace fossil taxonomy, but just how important it should be is again a matter of debate (section 8.5.1).

7.3 Different tracemakers may produce identical structures when behaving similarly

'Diverse organisms perform similar functions to solve similar problems' (Forbes 1989, p. 172). Behavioural convergence guided my organization of Chapters 2–4. Similar, if not identical, trace fossils may be produced by phyletically widely divergent species. Vertical unlined dwelling shafts (ichnogenus *Skolithos*) are a good example because of their simplicity and abundance. Today they can be produced by certain eel-like fish (Klausewitz 1962; Fricke 1973), some sipunculan worms (MacGinitie and MacGinitie 1949), most phoronid worms (Ronan 1978), numerous polychaete

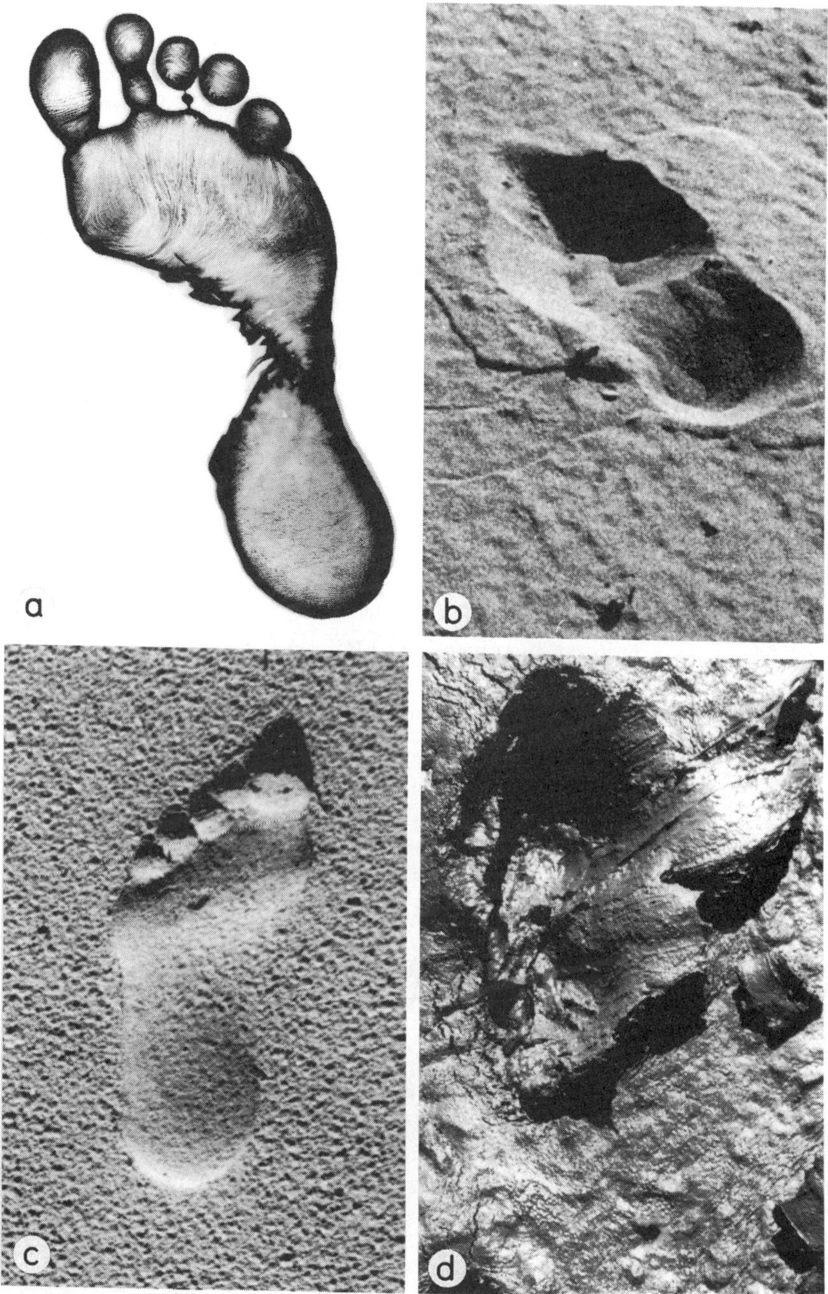

Figure 7.1 Four examples of my footprints, representing similar activity, but on substantially different substrates. The resulting traces have radically different morphologies. (a) On photographic paper. (b) In dry sand. (c) On wet sand after a shower of rain. (d) On an intertidal mud flat.

worms (Myers 1972), some actinian anemones (Mangum 1970), many insects and spiders (Ratcliffe and Fagerstrom 1980) and so on. Similarly, the extremely thin, commonly deep and branched shaft *Trichichnus* is produced by sipunculans in today's ocean floors (Thompson 1980; Romero-Wetzel 1987). However, similar structures are made by polychaetes and, presumably, pogonophores.

While this convergence of burrow form diminishes any hope of identification of the causative animals for fossil material, it adds to the ecological value of the trace fossils. *Skolithos* ispp. are useful environmental indicators from late Precambrian to Quaternary, a far longer period than the range of any of the *Skolithos*-creating species.

Morphological convergence is seen, of course, in more complicated structures also. Seilacher (1953b, 1960) provided vivid examples among resting traces.

Compare also the common U-burrow, and the cumulative structure that develops when this is enlarged in the manner of a W. This is a mode of growth shown by numerous unrelated taxa (Figs 3.5, 3.6, 3.8, 3.9c, 3.14, 4.29, 4.33 and 5.4).

7.4 Multiple architects may produce a single structure

As mentioned in section 5.1, commensalism or mere tolerance of proximity may result in a single burrow being constructed by two or more tracemakers simultaneously (Figs 5.2 and 5.3). Alternatively, one species

Figure 7.2 The glauconized lining of the spiral structure *Gyrolithes davreuxi* almost invariably contains densely-packed *Chondrites* isp. Lower Campanian smectite (Cretaceous) from near Visé, Belgium (Bromley and Frey 1974). x3.

may take over a burrow after another has abandoned it and modify its form to suit its own needs (section 5.1.1).

Subsequently, a long-abandoned and filled burrow may function as an elite structure that attracts deposit feeders when its contained organic matter has produced a harvest of microbes. Alternatively, the bacterial breakdown of organic matter in the fill or wall of the burrow may produce a suitable environment for chemosymbiosis (Fig. 7.2). Pickerill (1994) referred to such trace fossils, containing two distinct and unrelated ichnotaxa, as 'composite specimens' (cf. section 8.5).

7.5 Non-preservation of causative organisms

Many of the advantages conferred on an animal by its skeleton are duplicated by the burrow: physical protection, protection against desiccation, enclosure of a respiration or breeding chamber, etc. Consequently, burrowers tend to save expenses and reduce their skeleton. Among bivalves, most deeply-burrowing species cannot withdraw completely into their thin shells; arthropods such as callianassids have greatly reduced the calcitization of the skeleton; deeply-burrowing echinoids have thin tests.

Add to this the fact that, for the same reasons of protection, the endobenthic environment is a refuge for innumerable representatives of the naked phyla, and we have to admit that the burrowing world is populated by chiefly soft-bodied animals having low potential for preservation as body fossils. Put another way, the ichnological record offers the palaeoecologist a wealth of information on that part of the community that is not generally preserved as fossil.

There is a strange twist to this story, because the open burrow offers a microenvironment of significantly higher preservation potential than the rest of the sea floor in general. Any skeleton that falls into a burrow, being buried in a deep tier, escapes the most destructive phases of early diagenesis.

There are striking examples of this. Voigt (1959, 1974) described exquisitely preserved bryozoans that were found only within *Thalassinoides paradoxicus* fills in Danian and Cretaceous hardgrounds, where they had escaped the physical destruction or degradation processes that reigned on the sea floor above.

In Cretaceous chalk environments, bodily remains of nektic fish stood little chance of preservation; the small bones and scales were recycled by scavengers. However, those bones and scales that were used as wall support material by *Thalassinoides survicus* builders received excellent preservation (Fig. 7.3). Whether the architect collected the scales from the sea floor, or ate their owner, or shared its abode, we are unlikely ever to know. Collection is favoured by the fact that mistakes were made: rare, tiny, scale-like lingulid brachiopod shells are included in the linings.

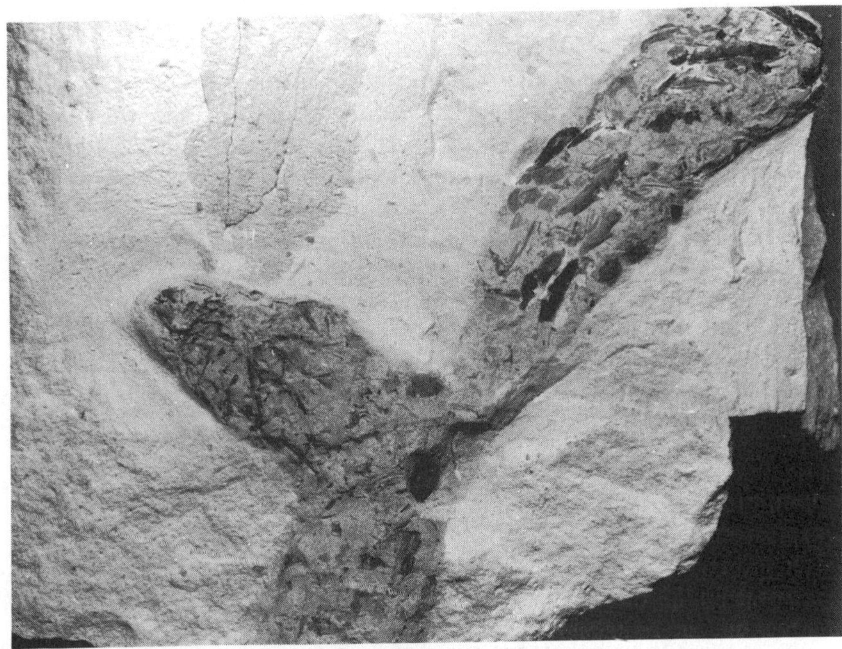

Figure 7.3 Fish bones and scales in the walls and floor of *Thalassinoides suevicus*. Maastrichtian chalk (Cretaceous), Møns Klint, Denmark. Natural size.

MacGinitie and MacGinitie (1949) observed that the echiuran worm *Urechis caupo* would have almost no chance of producing a recognizable body fossil. However, in a U-shaped Pliocene burrow attributable to the work of an echiuran worm they found the remains of the crab *Scleroplax granulata*, a common commensal that shares the worm's burrows today (Fig. 3.7).

Despite the general rule that the tracemaker is never preserved in its burrow, there are the usual rare exceptions. The giant spiral shafts, ichnogenus *Daemonelix*, were identified as the burrows of Miocene beavers, genus *Paleocastor*, on the basis of complete skeletons of these mammals preserved in the fill (Voorhies 1975). (See the exceptionally fine example on exhibition in the National Museum of Natural History, in Washington, DC!) Similarly, in the Permian playa sediments of the Karoo Basin, Smith (1987, 1993) found many helically spiral burrows to contain one or more articulated skeletons of a therapsid reptile, curled up in the terminal chamber.

After a long search of many kilometres of Pleistocene *Scolicia* isp., a few individuals of *Echinocardium cordatum* were found at the ends of their backfill structures, their spines still in an active digging position (Bromley and Asgaard 1975). Also, in rare cases, Mesozoic *Thalassinoides suevicus* contain remains of palinuran shrimps of the genus *Glyphea*, a fossorial form that clearly had constructed the burrow in the fill of which it was preserved (Sellwood 1971; Bromley and Asgaard 1972b). Likewise, Jenkins (1975) found some Miocene *T. suevicus* to contain the burrowing crab *Ommatocarcinus corioensis*.

Mikulás (1990) and West and Ward (1990) have described Palaeozoic ophiuroids that have been preserved within their resting trace, *Asteriacites*; Pickerill and Forbes (1987) found a Palaeozoic polychaete at the end of its trail; and other 'exceptions to the rule' could be cited. However, the overwhelming trend is that tracemakers escape being caught 'red-handed', and their identity becomes a fixation in the minds of many exasperated palaeontologists.

It is clear, then, that not only are trace fossils, to a large extent, biologically anonymous, they are nomenclatorially unconnected to the zoological taxonomical system, and must be treated separately according to ichnological principles.

8 Ichnotaxonomy and classification

We have inherited a set of names for trace fossils the most venerable of which originated in a series of misconceptions and misidentifications. Most early workers named trace fossils of invertebrates as remains of algae, sponges and other organisms, i.e. as either botanical or zoological body fossil taxa. More recently erected names have been based on more reasonable interpretations but, owing to the laws of priority, it is the oldest available names that are the basis of a taxonomy. Such ichnogeneric celebrities as *Cruziana*, *Zoophycos* and *Chondrites* were erected as algal taxa, *Nereites* as a worm.

We owe an immense debt to Häntzschel (1962, 1965, 1975) for working out the synonymies within a plethora of names and misspellings. His labours created a workable taxonomy out of a jungle of synonyms.

Only two ranks of ichnotaxa are in general use: **ichnogenus** and **ichnospecies**. These terms are usually abbreviated as ichnogen. and ichnosp., but the neater forms **igen.** and **isp.** are gaining in popularity. Higher-ranking groups are used informally by some workers, e.g. graphoglyptids, and the introduction of ichnofamilies is at present under debate (section 8.6).

8.1 The development of trace fossil nomenclature

The early history of ichnotaxonomy has been recorded by several workers (Osgood 1975; Teichert in Häntzschel 1975; Pemberton and Frey 1982) and need not be repeated here.

At the 15th International Zoological Congress in 1961, the Commission on Zoological Nomenclature made a strange decision. It ruled that names based on the work of an animal that were established after 1930 were to be 'accompanied by a statement that purports to give characters differentiating the taxon' [Article 13(a)(i) of the 1964 edition of the ICZN]; i.e. the causative organism had to be identified. Names established before 1931, on the other hand, were to continue to be treated on the same basis as body fossils. Since the specific affinity of a trace fossil taxon is anonymous, post-1930 names became essentially unavailable.

So began the Dark Age of ichnotaxonomy. Most ichnologists maintained order simply by applying ICZN rules even though they were not bound by them. In the Treatise on Invertebrate Paleontology, Häntzschel (1962, 1975) treated valid and invalid names alike.

Neither were the pre-1931 ichnotaxa satisfactorily provided for. Having similar status to body fossil taxa meant that names of trace fossils and names of causative organisms could compete under the law of priority. The ineptitude of this situation was illustrated by Osgood (1970), who documented a body fossil of the trilobite *Flexicalymene meeki* Foerste, 1910, preserved within the resting trace *Rusophycus pudica* Hall, 1852, which it had clearly excavated. It would cause more than a little chaos in trilobite taxonomy were the trace fossil name to be considered the senior synonym!

When the next revision of the ICZN became due, ichnologists began to agitate for improvement. Two courses of action were debated. Either the trace fossil names should be entirely removed from the ICZN, or the zoological code should be adapted to accommodate them satisfactorily.

The first procedure was advocated by Sarjeant and Kennedy (1973), who wrote a draft for an entirely separate code for trace fossil nomenclature. This was republished by Sarjeant (1979), but had no legal standing. The draft was based on the venerable rules of the ICZN, incorporating some modifications borrowed from the ICBN (the botanical code).

Many zoologists also advocated complete removal of trace fossils from the domain of the ICZN (e.g. Lemche 1973). After all, trace fossils do not perpetuate themselves in natural populations by sexual or asexual reproduction, as do the other objects covered by the code. Nor are they parataxa in the sense of names applied to parts of an animal, that ultimately can be worked out when the anatomy becomes fully known. Eventually, zoologists admitted that a single animal could produce many different sorts of work, each of which may be named (Melville 1979). Although having a biological origin, then, trace fossils really are 'different'.

The Sarjeant and Kennedy (1973) Code might have been adopted. However, the alternative procedure was undertaken by Häntzschel and Kraus (1972), who proposed amendments to the existing code. This proposal was largely implemental in the revisions of the code that followed (Teichert in Häntschel 1975; Basan 1979; Melville 1979). The Dark Age was over.

8.2 Status of trace fossil names in the ICZN

All ichnotaxa are now covered by the present ICZN (Ride *et al.* 1985), the 1930/31 limit having been removed. Ichnogenus and ichnospecies have received genus-group and species-group status, respectively (Article 10d). A type ichnospecies is not required and, where one has been designated, this should be disregarded (Articles 42b, 66). This is most unfortunate, as it is desirable to have an ichnospecies labelled as 'typical' of the ichnogenus when creating other ichnospecies; otherwise the nature of the ichnogenus may drift away from its original conception as new ichnospecies are added.

Most important, the names based on the work of an animal do not compete in priority with names given to causative organisms (Article 23g). The present position of trace fossil nomenclature was succinctly put by Kelly (1990) and Rindsberg (1990).

8.2.1 Fossil or not fossil?

The code states clearly that (after 1930) only fossil traces are covered (Articles 1a, 1b). The reason for this is the lingering belief that one trace, one animal is the normal situation; 'traces of living animals can always be related to their causative organism, and there is no need to name them separately' (Melville 1979). This applies to some individual traces, of course, but not to individual ichnotaxa. Ichnologists can live with this restriction although, as pointed out by Bromley and Fürsich (1980), it carries with it several problems, one of these being the definition of the fossilization threshold.

In the case of the tracemaker, there are fairly distinct boundaries between living organism, dead body and body fossil. Such boundaries are far from obvious, however, in the case of trace fossils. When does the work of an animal become fossilized? At what stage does the backfill of a heart urchin become a trace fossil?

In deep-sea box cores, Wetzel (1984) demonstrated that in some *Zoophycos* just over 1 m below the sea floor the marginal burrow was still empty and was probably still occupied by the unknown tracemaker. Wetzel was surely correct in applying an ichnotaxon to these recent structures; how else could they be named? Likewise, Ekdale (1980) and Gaillard (1988) applied the ichnogenera *Paleodictyon*, *Spirorhaphe* and *Cosmorhaphe* to apparently active burrows at the recent ocean floor.

This is probably merely a semantic problem. However, Bromley and Fürsich (1980) suggested that, when revising or erecting ichnotaxa, type specimens should be chosen from unequivocally fossil material. These authors further proposed that, when describing material deemed to be unfossilized, but which nevertheless can be referred to an ichnotaxon, the adjective 'incipient' should preceed the name.

8.2.2 Dual nomenclature

Whether we like it or not, we have to operate with a dual nomenclature. Sediment structures caused by biological activity are named as ichnotaxa, the causative organisms as biotaxa. The two nomenclatorial systems run parallel but are in no way interchangeable or duplicating, as will be seen in the following examples.

The burrow system constructed by *Callichirus major* may either be called 'the burrow of Callichirus major' or be referred to an ichnotaxon as 'incipient *Ophiomorpha nodosa*' (Frey *et al.* 1978). The pellets produced by the shrimp may be called 'the pellets of *Callichirus major*" or be referred to an ichnotaxon as 'incipient *Palaxius* isp.' (Fig. 4.21).

A crustaceologist would regard it as laughable to consider the pellets still lying within the hindgut of *Callichirus major* as requiring a name, let alone a name different from that of the shrimp. The palaeontologist views the situation differently, however. As soon as these pellets are voided and lose contact with the causative organism the identification of their origin begins to involve speculation.

If we find similar pellets preserved in Pleistocene deposits the amount of speculation greatly increases. Do other callianassid species make similar pellets? The details of the internal canal system in pellets of anomuran Crustacea may well be species diagnostic (Fig. 4.21b), though we shall never be able to demonstrate this in fossil material. In every case, therefore, a trace fossil name is desirable.

Should similar pellets be found in association with *Ophiomorpha nodosa* in Jurassic deposits there would be no justification for ascribing these to *Callichirus major*, the family Callianassidae having first appeared in the Late Cretaceous and *Callichirus major* much later.

In the Upper Jurassic of England, pellets of the ichnogenus *Palaxius* were found within *Thalassinoides* fills (Kennedy *et al.* 1969). A single specimen of the shrimp *Glyphea udressieri* was found in a *Thalassinoides* fill at the same locality (Sellwood 1971). This is a fair example of an intimate association of trace and body fossils in which a genetic connection obviously suggests itself, but is difficult, if not impossible, to prove. That the *Palaxius* isp. in this occurrence probably are the faecal pellets of *Glyphea udressieri* in no way establishes that such pellets may not have been produced by other species at other times. Moreover, the burrow structure of *Thalassinoides suevicus* is certainly not produced by that species of shrimp alone. Neither, perhaps, was it the only type of burrow that that species was capable of making. Thus, ichnotaxa should be used for all fossil material, regardless of the degree of certainty with which the structures can be related to specific tracemakers.

8.3 Ichnotaxobases

Surprisingly little has been written on what exactly is being named when an ichnotaxon is applied to fossil material. Many new names are applied in a completely random way, according to the appeal the specimen has on its describer. A motor-cycle enthusiast, for example, has named a fossil arthropod trackway *Hondichnus* isp. (Ausich 1979). In the final analysis, it is the morphology of the trace fossil as an expression of animal behaviour that is the basis of the name, but this unavoidably becomes coloured by the interpretation placed upon the structure by its discoverer.

Small, round sediment bodies within the fossil are described as sporangia by those who interpret the object as an alga (e.g. Sternberg 1833, for his genus *Muensteria*), but as faecal pellets by an ichnologist. Cross-partitions may be regarded as cell walls of an alga (Heer 1877, for his species *Taenidium serpentinum*) or as backfill meniscae by an ichnologist. However, as long as type material is available or the morphology is adequately described and illustrated, interpretations based on it may be modified.

Subjectively assuming, then, that the objects to be named *are* biogenic sediment structures, we can list the following characters as being the most commonly used for the bases of ichnotaxa (i.e. **ichnotaxobases**): (1) general form; (2) wall and lining; (3) branching; and (4) fill.

8.3.1 General form

Naturally, the general shape and orientation of the structure are important; whether it is a vertical shaft, a network, a spiral meander, a spreite structure, etc. Major groups of trace fossils are separated on this basis, which may be suggestive of higher-ranking ichnotaxa (section 8.7). General size is not normally considered important.

8.3.2 Details of the burrow boundary

Many ichnotaxa are based on the structure of the wall. This may vary from a hardly noticeable or preservable film to a massive structure that amounts to a substantial part of the trace fossil. Seven main categories of wall structure may be recognized.

(a) No lining

In some trace fossils the fill abuts directly against the surrounding sediment at a clean discontinuity surface (Figs 8.1 and 9.10).

(b) Dust film

Irrigation of a burrow lined with mucus will introduce suspended dust that will collect passively on the walls. The mucus itself may have little preservation potential, but the film of dirt may be recognizable in the trace fossil. In fine-grained rocks, extremely thin films commonly demonstrate the presence of structures such as *Palaeophycus tubularis* (Figs 11.4a and 11.14c).

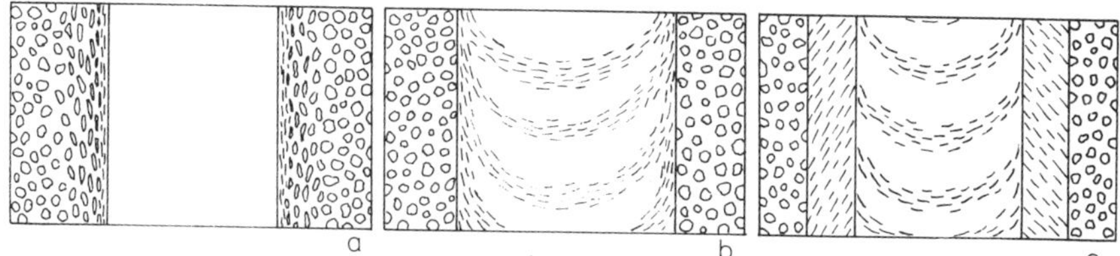

Figure 8.1 Burrow boundaries of unlined, cylindrical structures; inspired by Heinberg (1974). The trace fossils are seen in horizontal axial section; the grains are mica flakes, lying flat where undisturbed by bioturbation, steeply inclined elsewhere. (a) A sharp boundary showing compaction of the adjacent substrate. (b) Meniscus backfill; the nested menisci overlap and merge along the boundary, giving a false impression of a lining. This situation is common in *Taenidium serpentinum*. (c) Zoned backfill having a mantle and a meniscate core, as seen in *Ancorichnus ancorichnus*. Some lined structures are shown in Fig. 1.7.

(c) Constructional lining

An organic tube may be carbonized during fossilization. Sediment and special grains applied to the burrow walls as building material are more easily recognizable in trace fossils. Variations in the material and morphology of these tubes have been used to differentiate ichnotaxa. Mud lining having a knobby interface against the substrate characterizes *Ophiomorpha* ispp. (Frey *et al.* 1978); shell-lined vertical tubes may be

Figure 8.2 A thick laminate lining around a horizontal burrow. Where these bodies are spindle shaped they are usually considered deposit-feeder structures in which a probing burrow has wandered concentrically. This is the *Asterosoma* model of Chamberlain (1971). Where they are cylindrical, as in this case, they may be compared to the burrow construction seen today in *Amphitrite ornata* (Fig. 3.25). Jurassic, offshore Norway, North Sea, x2.

called *Diopatrichnus* isp. (Kern 1978); several ichnospecies of *Palaeophycus* are distinguished on the basis of lining material (Pemberton and Frey 1982).

During long occupancy of a stationary burrow, repeated additions to the wall and growth of the animal can result in concentrically multilaminate lining of disproportionate thickness (section 3.6). Such structures are common trace fossils and have puzzled many ichnologists (Fig. 8.2).

Unwanted material that enters burrows may be pressed into walls rather than laboriously carried outside. The fish scales in some *Thalassinoides suevicus* systems may be constructional support material, as these structures are deposit feeder systems (Fig. 7.3). However, scales that thickly line walls of unbranched burrows may represent remains of prey of carnivores such as stomatopods.

(d) Zoned fill

The preceding wall types all belong to open burrow structures, occupied and irrigated over long periods of time. Other types of boundary structure can arise by the sediment processing of mobile deposit feeders

Figure 8.3 Weathered bedding plane in the Middle Jurassic Vardekløft Formation of Jameson Land, East Greenland, showing *Ancorichnus ancorichnus*. The sectioned mantle is weathered as a groove on either side of the core. Natural size.

moving continuously through the substrate. In these cases, the apparent wall lining is in fact no such thing, but is the outermost layer of a concentrically zoned fill.

The trace fossil *Ancorichnus ancorichnus* consists of a two-layered cylindrical backfill (Fig. 8.3). The central zone or **core** has meniscus backfill structure, and is surrounded by a thick outer layer or **mantle**. Core,

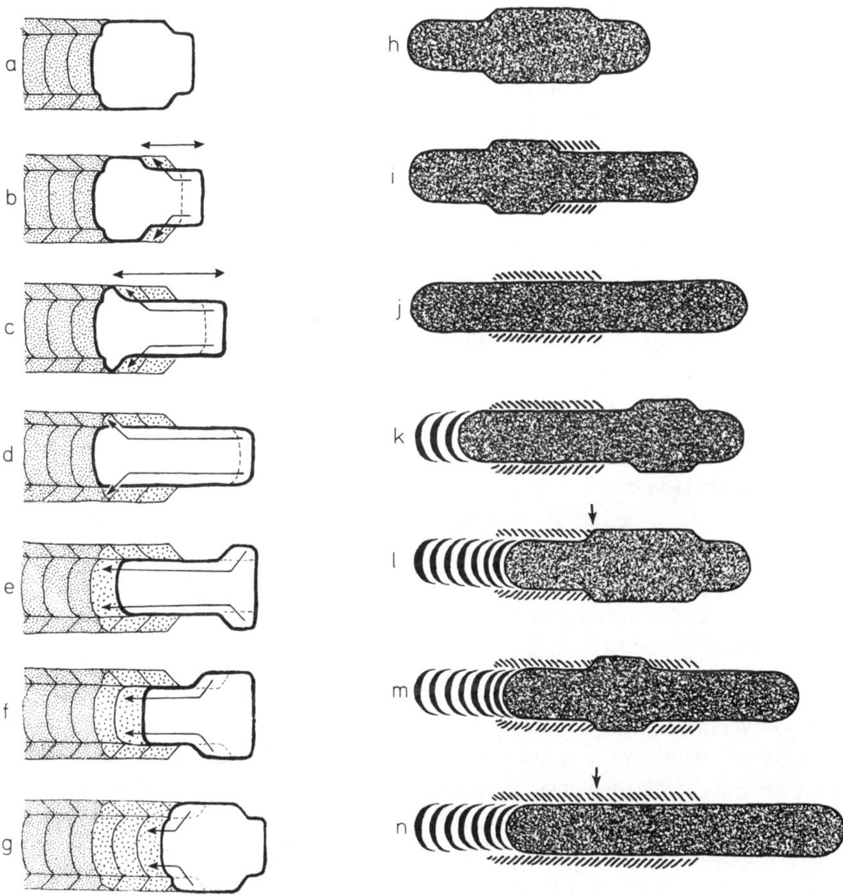

Figure 8.4 Models suggesting the mode of origin of *Ancorichnus ancorichnus*, modified after Heinberg (1974). (a–g) Dynamic model showing mode of emplacement of sediment packages in the annulated mantle and meniscate core of the zoned backfill. The double-headed arrows in (b) and (c) indicate the region where contraction of ring muscles causes elongation. The functions of penetration and terminal anchors are combined in a single anchor that moves by peristalsis along the body. (h–n) Model using a longer worm, showing the relationship between mantle and core, and successive anchor cycles. At the arrow in (l) the backward advance of the anchor has reworked mantle deposits. This reworking would be disadvantageous, and may limit the length of the worm or of the anchor. The arrow in (n) shows the apparently conformable contact between the two mantle cycles.

mantle and surrounding sediment are separated by sharp discontinuities. Heinberg (1974) provided a model to explain the burrowing technique that has given rise to this special structure (Fig. 8.4), which is clearly conceptually distinct from a true burrow lining.

A similar structure occurs in the trace fossil *Phoebichnus trochoides*, though the general morphology is quite different (Figs 8.5 and 8.6). The two-layer fill in this case was modelled by Bromley and Asgaard (1972c) as having arisen through the double passage of the tracemaker; making radiating excursions into the surrounding sediment from a central shaft, the animal returned along the same path. The ring-meniscate structure of the mantle shows that it was deposited by the animal on its outward journey from the shaft. The central core was laid down on its return journey. A similar double journey must apply to *Jamesonichnites heinbergi*, a behavioural variant of *Ancorichnus ancorichnus* (Fig. 12.9h). These two trace fossils represent a compound ichnotaxon (section 8.5).

Less distinct outer zones of fill can occur where the inner discontinuity against the core is not sharp (Fig. 8.1b). In such cases the nature of the mantle is less likely to be mistaken for the wall lining of an open burrow. In the ichnogenus *Macaronichnus segregatis*, the boundary layer is thin, sharp and dark; in *Phycosiphon incertum* the mantle is usually pale and the core dark (Fig. 11.11).

(e) Wall compaction

The sediment immediately external to the burrow wall commonly shows some disturbance resulting from burrowing activity (Fig. 8.1a). Reineck (1958) called this zone the 'Wühlhof'. This distortion may result from eddy turbulence, compaction or shear, depending on the consistency of the substrate (section 1.2.2).

(f) Diagenetic haloes

A zone of oxidation, mucus impregnation, chemical adsorption and compaction may initiate special diagenesis in the wall. Colour differences and mineralization render such structures disproportionately conspicuous as elite trace fossils (section 6.4). In DSDP cores, Chamberlain (1975) called such structures **halo burrows**.

Diagenetic bleaching or mineralization can also affect a discrete wall material differentially. In fact, after diagenetic modification it may be hard to tell if there originally was a special wall structure. The most widespread (or just the most conspicuous?) trace fossils in DSDP cores are **rind burrows** (Chamberlain 1975). Ekdale (1977) considered these bleached rings, as seen in cross-section, to represent modified wall material, but the discoloration extended inward some distance into the fill.

These diagenetic effects render some trace fossils spectacularly visible, but are unsuitable as a taxobase for formal nomenclature. Vernacular

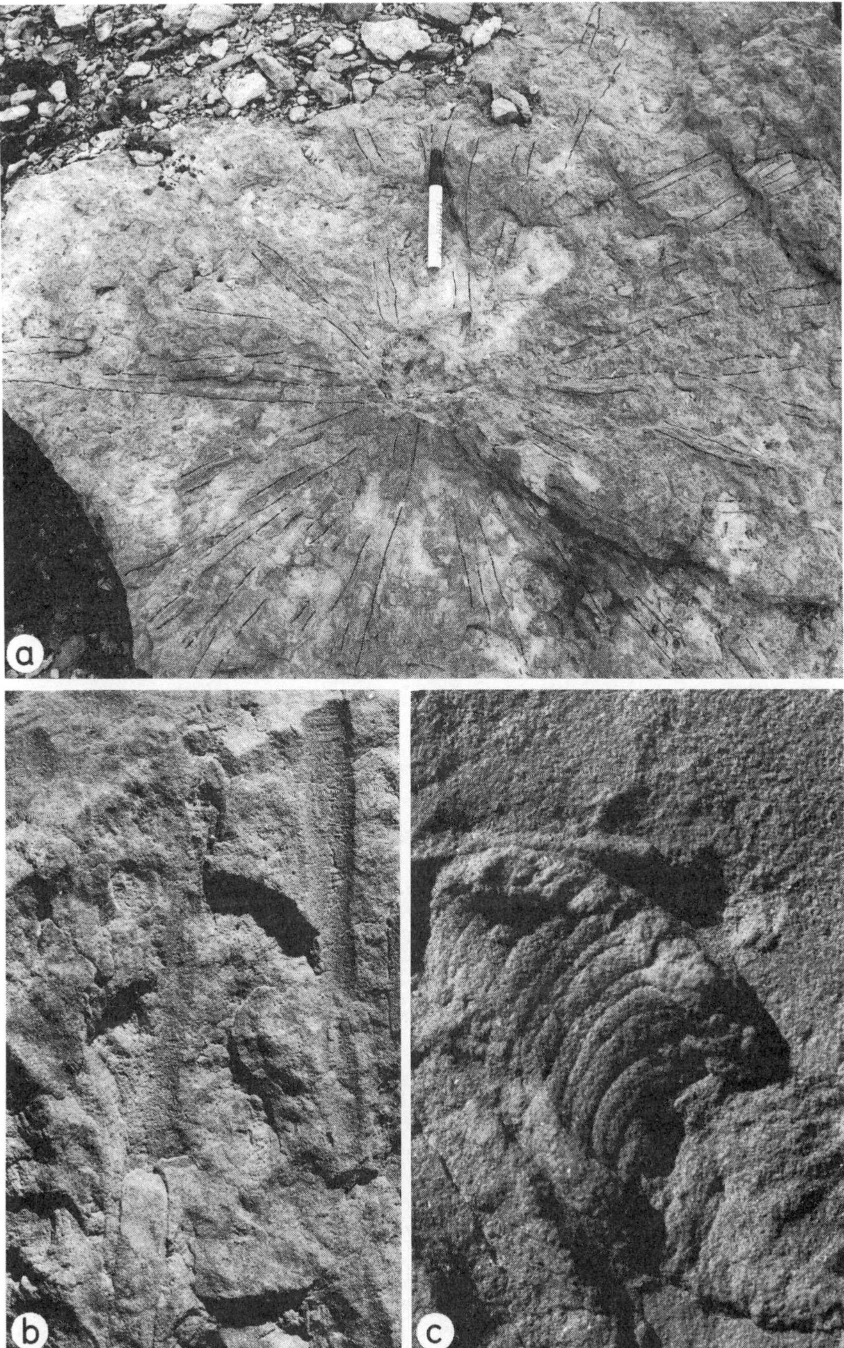

Figure 8.5 *Phoebichnus trochoides*. (a) A system of *P. trochoides* in a giant concretion within the shaly Vardekløft Formation (Middle Jurassic) of Jameson Land, East Greenland. A marker pen (scale) has been used to outline the radii. (b) Two radii showing annulation of the outer interface of the mantle. (c) Close up of the mantle on a weathered surface, x2.

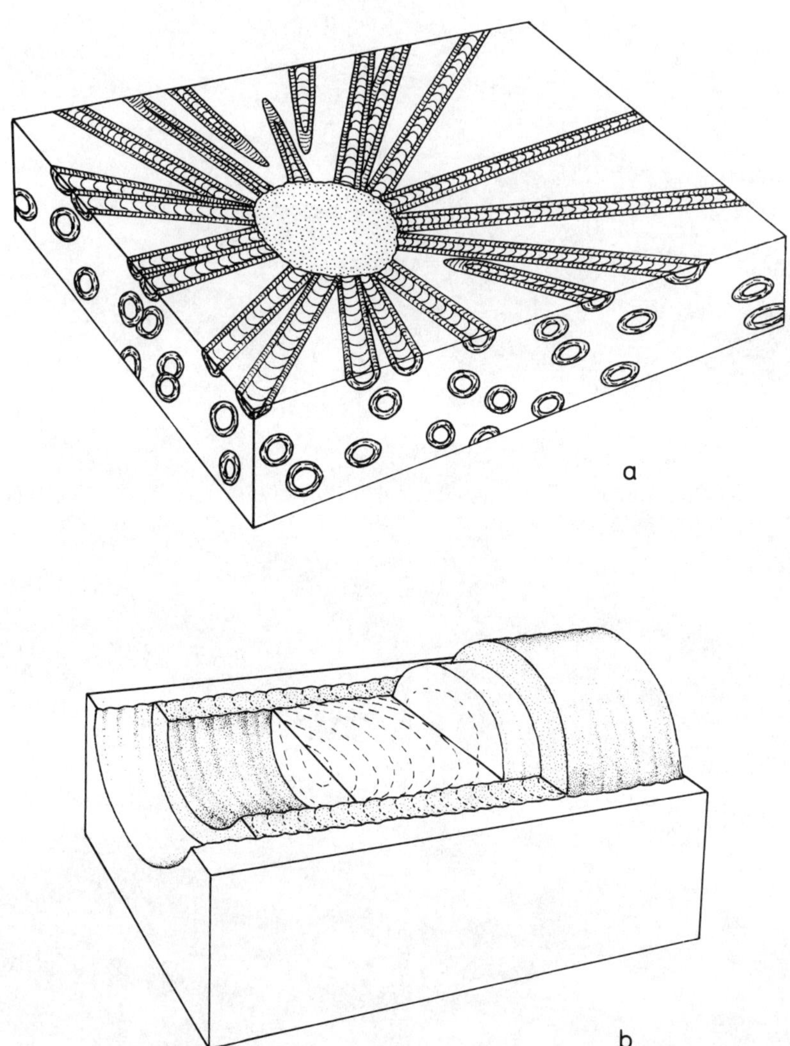

Figure 8.6 The structure of the giant radiating trace fossil *Phoebichnus trochoides*. (a) The middle part of the structure; the radii may be twice as long, reaching a length of 1 m. The central shaft is poorly defined. (b) Construction details of a radiating arm; the meniscate backfill comprises a central core and surrounding mantle. Modified after Bromley and Asgaard (1972c).

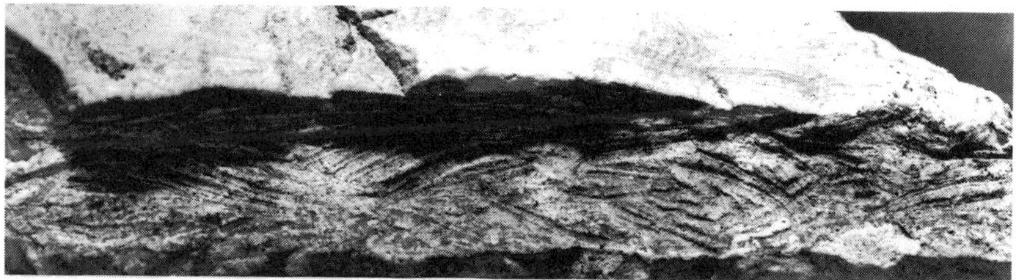

Figure 8.7 A burrow wall constructed in firm chalk at the basal Eocene unconformity of the London Basin. The trace fossil, *Glyphichnus harefieldensis*, is a wide U-burrow having a powerful bioglyph indicating the work of a crustacean. Harefield, Middlesex, England, x2.

names such as rind burrow and halo burrow are adequate for descriptive purposes.

(g) Wall ornament

Another and quite independent aspect of the burrow wall that is of taxonomic importance is the presence or absence of a bioglyph. This is dealt with elsewhere (section 7.2 and Fig. 8.7).

8.3.3 Branching

Branching of biogenic sediment structures can arise in several ways, and it is necessary to distinguish different types of branching when analysing a trace fossil. The fossilization barrier plays a role here, and it is vital to state whether it is the burrow, the cumulative trace fossil or both that are branched.

An open burrow system, repeatedly visited by the occupant, can be extended as a branched structure (e.g. Fig. 4.22). In contrast, a mobile deposit feeder, progressing steadily through the sediment (e.g. Fig. 4.16), cannot produce a branched structure unless it turns round or backs up, retracing its path. Such manoeuvres leave recognizable structures in the trace fossil, as in *Rutichnus* ispp. (Fig. 9.9).

It must be remembered, however, that an unbranched burrow can be shifted so as to produce a cumulative structure that is branched (Fig. 6.7). Therefore, although the branched form of the trace fossil may not be related to the shape of the original burrow, it is directly related to the behaviour of the causative organism, and therefore is of paramount significance in nomenclature.

Bromley and Frey (1974) attempted to classify branching in trace fossils with reference to the original burrow. Three types were recognized: (1) **true branching**, in which the original open burrow was branched; (2) **false branching**, where an animal reworked an earlier burrow fill and then deviated from that course; and (3) intersection, which is merely a crossing

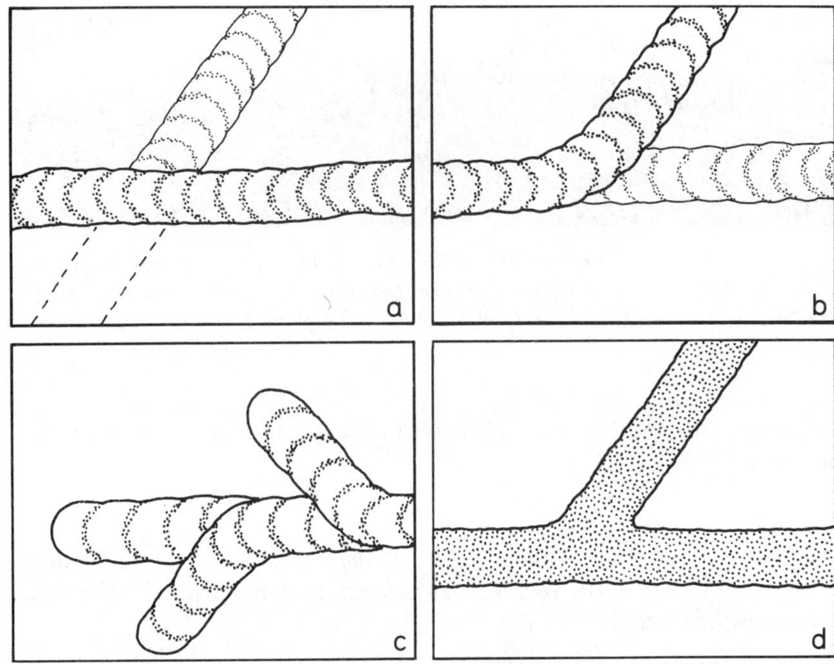

Figure 8.8 Four distinct styles of branching in trace fossils. (a) **False**: an apparent branching is commonly caused by accidental intersection and incomplete preservation, particularly in compacted mudrocks (Fig. 9.10). (b) **Secondary successive**: an unbranched burrow causes a branched structure where an animal enters and follows along an earlier fill (or later deviates from it). This trait is characteristic for some ichnotaxa. (c) **Primary successive**: a branched, cumulative structure is produced by an unbranched burrow by successive probing movements (Fig. 6.7c and d). (d) **Simultaneous**: a network of passages in which the branches were simultaneously open. Modified after D'Alessandro and Bromley (1987).

of two unbranched burrows but which, in the fossil state, can appear to be a ramification (Fig. 8.3). D'Alessandro and Bromley (1987) refined this classification so as to include probing forms such as *Chondrites* ispp. (Fig. 8.8). Recognition of these different aspects of ramification provide valuable clues to the behaviour of the causative animal and allow more precise definition of the trace fossil names (Keighley and Pickerill 1995).

8.3.4 Filling material and structure

The basic dichotomy in the nature of fill is whether it resembles or is different from the surrounding sediment. If different, this may be original or diagenetic, although the localized diagenesis must have emphasized some original contrasting factor, e.g. permeability or organic content.

The fill may be said to be passive or active. **Passive fill** enters a burrow gravitationally. It may consist of material similar to the surrounding sedi-

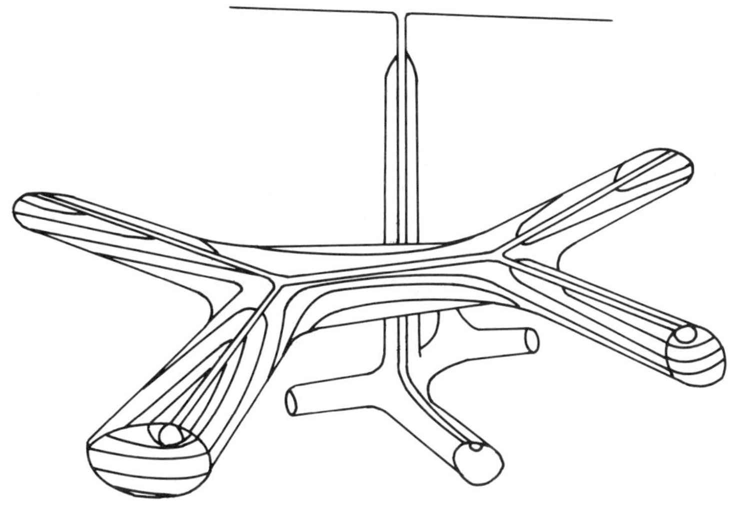

Figure 8.9 Passive fill in *Thalassinoides* or *Ophiomorpha*. Slightly inclined laminae make closures against the cylindrical wall and can be mistaken for meniscate backfill. The final filling material is deposited in the draught canal, the size of which reflects the diameter of the burrow aperture at the sea floor.

ment or, if derived from above a change of sediment type, may contrast with the ambient sediment. The last case is known as **bed junction preservation** (Simpson 1957) and, while important in the stratinomic classification of the trace fossil (Fig. 9.1), and improving preservational quality and visibility, is not of taxonomic importance.

Similarly, a draught fill canal also indicates passive filling, and is not of taxonomic importance, beyond the fact that it indicates the existence of an open burrow. Seilacher (1968) described how such a draught fill can occur in any filled cavity having a narrow entrance (Fig. 8.9). The narrow draught canal can be mistaken for a burrow itself, reworking the fill of the larger, and there have been cases where this purely physical structure has been designated a trace fossil name, e.g. *Thalassinoides ornatus (pars)* by Kennedy (1967).

Active fill is that emplaced by the burrower. This material usually contrasts with the surrounding sediment as it has been subject to biological processing; some or all of it may have passed through the animal's gut. Active fill therefore normally possesses a characteristic structure. It is generally enriched or depleted in certain constituents present in the surrounding sediment; commonly it is pelleted; and it may be packed as backfill menisci (Fig. 9.10). An active fill may be zoned (section 8.3.2d) or contain canals as in *Scolicia* ispp. Thus the characters of active fill are of the highest significance in trace fossil nomenclature.

8.3.5 Trackways

Trackways of repetitive footprints and other impressions of walking, crawling and bottom-swimming invertebrates and vertebrates have their own set of problems (Fig. 9.4). Mostly classified as repichnia, this group of trace fossils is the basis for a large number of names (e.g. Walter 1983; Walker 1985).

A particular problem is the lateral variation in morphology. This may be due to lateral variation in substrate consistency, for example, water content in marginal lacustrine settings. Where an animal walks out of a lake onto dry land it will cross a succession of different substrates (Fuglewicz *et al.* 1990; Gradziński and Uchman 1994). Another reason for lateral variation is the undertrack problem and the depth within the substrate at which the impression is preserved (Lockley 1991). There is a temptation to apply several names along the course of a single trackway under these circumstances.

Trewin (1994) suggested defining a series of taxobases on which these trace fossils could be described and named. These ichnotaxobases include: (1) trackway width; (2) morphology of individual imprint; (3) repitition modules: a ten-legged animal will leave five pairs of imprints in a characteristic grouping, a 'set' of imprints; (4) repeat distance of these sets; (5) symmetry: the animal may walk obliquely, or rake one row of legs sideways; (6) continuous traces such as tail-dragging traces. These taxobases might possibly be used to establish a Trackway Data System that was machine readable (Trewin 1994).

8.4 Ichnogenus and ichnospecies

Because trace fossil nomenclature has grown haphazardly in increments of a few ichnotaxa at a time, instead of according to a pre-existing plan, the quality and extent of different ichnogenera show great variation, and use of ichnotaxobases in different ichnotaxa is non-uniform. Fürsich (1974a, b) tackled this problem by attempting to classify the taxobases according to their degree of significance in terms of behaviour of the tracemaker. Those characteristics that relate to major behavioural traits should be used to found ichnogenera, whereas more periferal, but nevertheless distinctive, features should be used for subdivision as ichnospecies.

As an example, vertical U-burrows having associated spreite structures have been described under several names of generic rank. This morphology reflects a fundamental behaviour pattern, and Fürsich (1974b) placed these names in synonymy, the oldest being *Diplocraterion* Torell, 1868. Variation in the mode of production of the spreite, or the general morphology of the U-tube, use of wall materials, etc., which derive from dis-

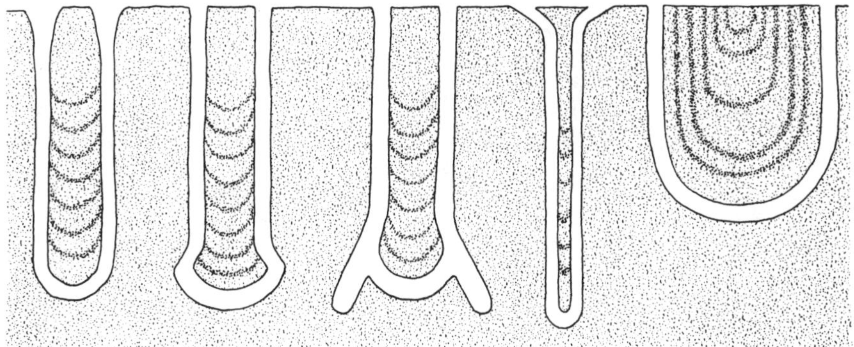

Figure 8.10 The ichnospecies of *Diplocraterion*. From left: *D. parallelum, D. helmerseni, D. biclavatum, D. habichi* and *D. polyupsilon*. Only *D. parallelum* and *D. habichi* are common. The two funnels indicated in *D. parallelum*, although the basis of the ichnogeneric name, are not considered diagnostic, even at ichnospecies level. Likewise, *D. habichi* virtually never has divergent apertures. Modified after Fürsich (1974b).

tinctive behaviour patterns, serve to support several ichnospecies (Fig. 8.10).

There are, however, aspects of the morphology of *Diplocraterion* ispp. that are unsuitable as taxobases, despite the fact that they are behavioural in origin. The two basic types of spreite, for instance, fall into this category: **protrusive** and **retrusive**. These conspicuous traits of the trace fossil are related to equilibration of the tracemaker at an unstable sea floor. The inherited ability of the tracemaker to cope with such an environment is reflected in the presence of the spreite, but its degree of protrusion or retrusion is related to vagaries of the local physical sedimentary regime. Although these features are vitally important in reconstruction of palaeoenvironments (Goldring 1964), they cannot be used in taxonomy.

Likewise, growth is reflected by a downward expansion of the structure (Fig. 8.11) when the environment remains stable for a sufficient period of time, but this feature may be modified by equilibration movements and cannot be used in the nomenclature of the trace fossil (Bromley and Hanken 1991).

Finally, Fürsich (1974b) considered that the presence or absence of funnels at the apertures of the U could not be used as an ichnotaxobase either, because in most cases this feature may merely reflect presence or absence of pre-burial destruction by erosion.

8.5 Compound trace fossils

Owing to the great variation in behaviour exhibited by tracemakers, it is common to find trace fossils that reflect such variation, and a single trace

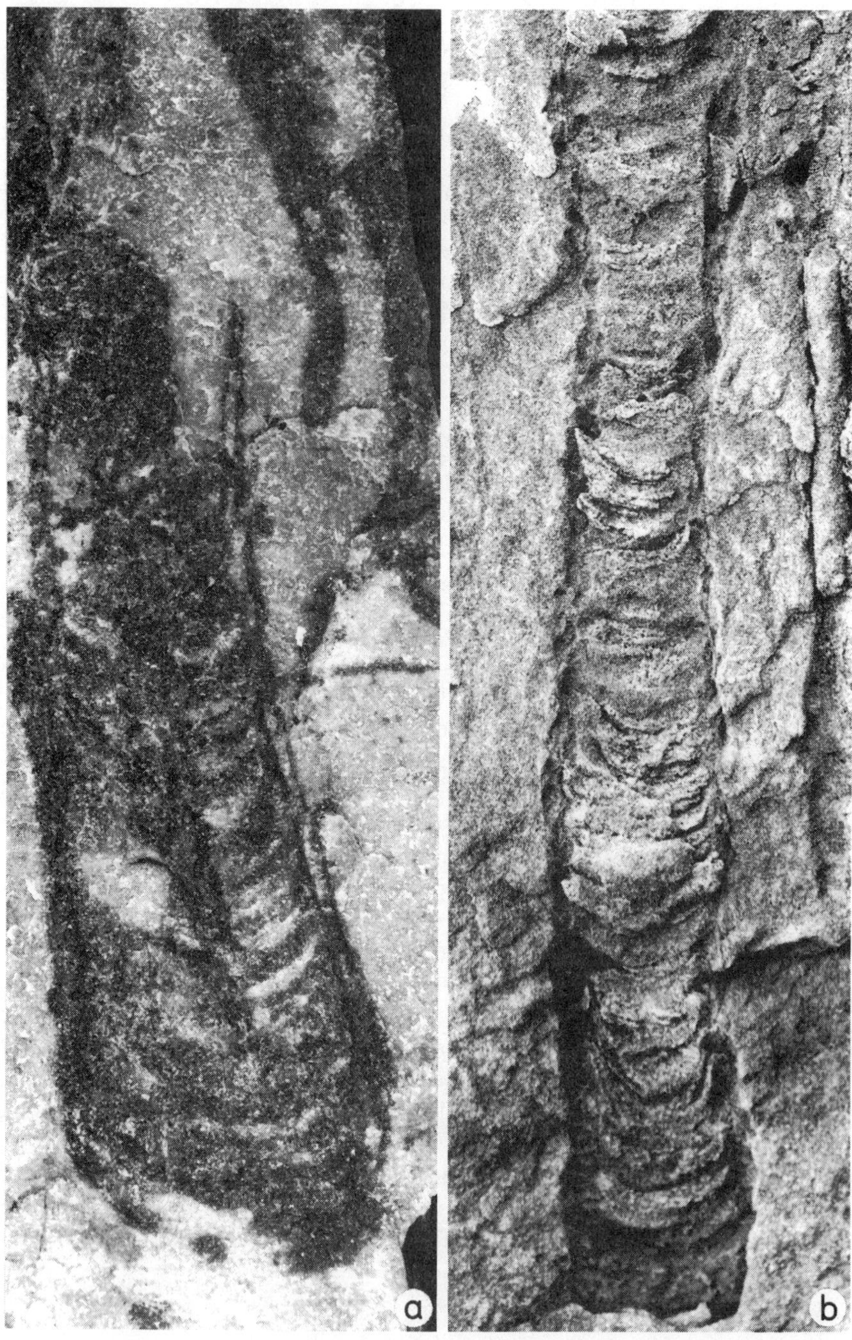

Figure 8.11 Two contrasting examples of the equilibrichnion *Diplocraterion parallelum*. (a) A Lower Cambrian form, protrusive and showing downward widening representing growth. Dividal Group, Finnmark, Norway, x1.5 (Bromley and Hanken 1991). (b) A large example from the Jurassic of Boulonnais, France. The long spreite is protrusive, representing gradual downward adjustment to slow erosion. The erosion surface is just above the top of the picture, x0.5. See Ager and Wallace (1970).

fossil may represent several ichnotaxa in lateral continuity. Pickerill (1994) and Pickerill and Narbonne (1995) called such trace fossils 'compound specimens' (cf. section 7.4).

Examples are numerous. *Ophiomorpha nodosa* commonly looses its wall lining locally and becomes *Thalassinoides suevicus*, in response to change in substrate consistency (Bromley and Frey 1974). *Thalassinoides paradoxa* locally develops scratch ornament for the same reason and becomes *Spongeliomorpha* isp. (Bromley 1967). The echinoid pascichnion *Scolicia prisca* contains local cubichnial developments where the burrower 'paused' and produced *Cardioichnus planus* (Smith and Crimes 1983). The backfilled pascichnion *Ancorichnus ancorichnus* may abruptly show primary successive branching as a fodinichnion called *Jamesonichnites heinbergi* (Fig. 12.9h).

8.6 Some problematic ichnogenera

Fürsich's (1974b) revision of the ichnogenus *Diplocraterion* has received general acceptance, as has the basic theory behind it (e.g. Pemberton and Frey 1982). However, the concept allows ample range for subjective evaluation of the relative significance of behavioural traits. Furthermore, each group of trace fossils has its own problems. In the following section, two of these groups and their systematic problems are exemplified.

8.6.1 Ophiomorpha–Thalassinoides–Spongeliomorpha

One suggestion that Fürsich (1973) made for increasing the rigour of naming trace fossils has not received general acceptance. This was to synonymize the forms that may occur connected together within a single system (sections 7.1 and 8.5). He used the network ichnogenera *Ophiomorpha, Thalassinoides* and *Spongeliomorpha* as an example.

Dichotomously or T-branched boxworks, mazes and shafts, unlined and unornamented, are referred to *Thalassinoides* Ehrenberg, 1944. Burrows showing a comparable morphology but having a thick, externally knobby lining are called *Ophiomorpha* Lundgren, 1891 (Fig. 8.12b). Similar burrows that are unlined, but bear a strong bioglyph of ridges, are referred to *Spongeliomorpha* Saporta, 1887. Other variations such as thickly-lined versions lacking a knobby exterior also occur, but have not been named, and the thickly lined and ornamented form that Kennedy (1967) named *'Spongeliomorpha' annulatum* has not received an ichnogeneric name.

All these forms show a comparable general morphology related to a behaviour pattern at a high level of significance, and therefore could be assembled within a single ichnogenus. The characteristics that at present distinguish them would then continue to be valid as ichnospecific taxobases. Fürsich's (1973) suggestion to synonymize these ichnotaxa

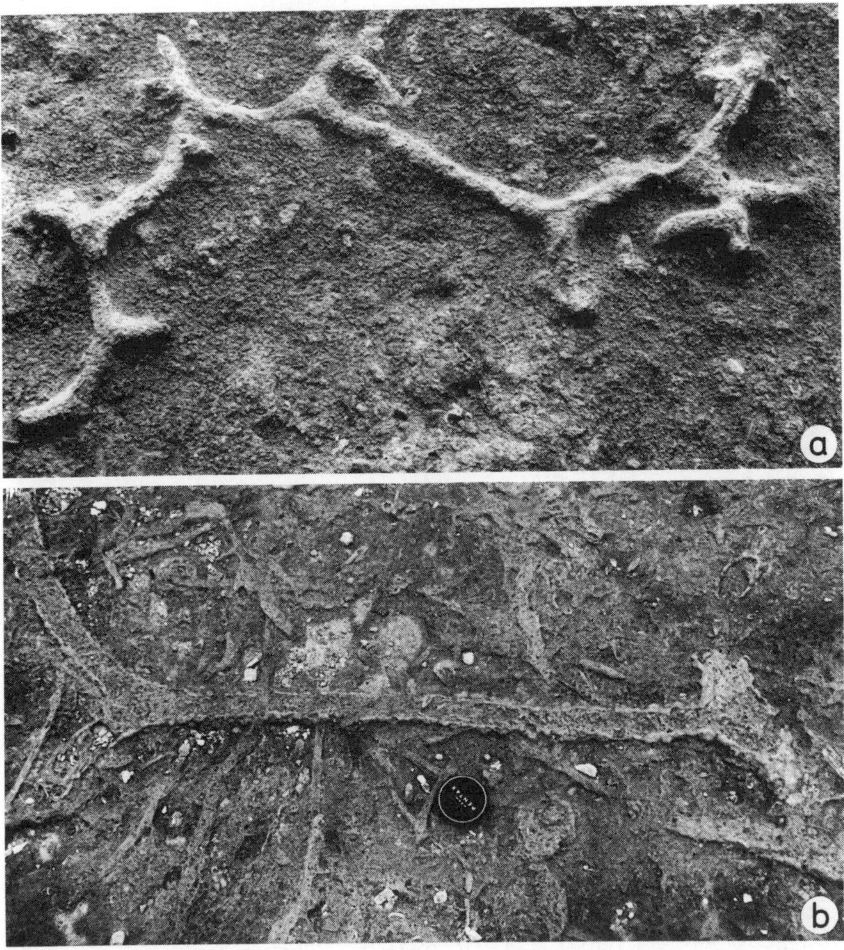

Figure 8.12 Thalassinoid trace fossils. (a) *Thalassinoides suevicus*, base of the Upper Cretaceous chalk at Bruneval, Normandy, France. (b) *Ophiomorpha nodosa*, Lower Globigerina Limestone, Miocene, Gozo, Malta. Lens cap 5 cm, both pictures at same scale.

under the oldest name, *Spongeliomorpha*, however, met with disapproval (Bromley and Frey 1974). This name was a *nomen nudum*, and hardly ever used, whereas *Ophiomorpha* and *Thalassinoides* were ichnological household words.

Subsequently, *Spongeliomorpha* was redescribed on the basis of topotypic material and has become fully available (Calzada 1981). But before this suggested synonymizing is reconsidered, it should be noted that an even older name is available for network structures: *Granularia* Pomel, 1849. This name is normally used in turbidite material for structures that are morphologically close to *Ophiomorpha*.

One of Fürsich's reasons for synonymizing these forms, in addition to their morphology, was that these trace fossils may occur in intimate connection within a single system (e.g. Kern and Warme 1974; section 8.5). However, Kilpper (1962) and Bromley and Frey (1974) demonstrated that spiral parts referable to *Gyrolithes* Saporta, 1884, also occur within such networks, and Hester and Pryor (1972) documented retrusive spreite structures within *O. nodosa* systems that have affinity with the ichnogenus *Teichichnus*.

It is clear that each of these morphological developments represents distinctive behavioural attributes within a complex burrow system, and must be named separately, just as they are when they occur separately. Precisely how they should be named, however, has not yet been determined. Hester and Pryor (1972) applied a varietal name to the spreite sections of the *O. nodosa* systems, calling them *O. nodosa* var. *spatha*.

We have here a form of independence of species- and genus-rank names that is not seen in biotaxa. Thus, it would almost be logical to name the spreiten *Teichichnus nodosus*.

Pemberton and Frey (1982) suggested that in such ichnogeneric intergradations a name should be assigned after the predominant component of the system. This may be feasible in extensive material where predominance can be assessed; but fragmental or otherwise restricted material, such as core, is less easily dealt with.

8.6.2 Cruziana–Rusophycus–Isopodichnus

Cruziana and *Rusophycus* are chiefly Palaeozoic ichnogenera applied to **repichnia** and **cubichnia** (section 9.2.1), respectively. Thus, they were produced by distinctly different behaviour patterns and therefore qualify for separation at ichnogeneric level. Consisting of double ridges in **hyporelief**, they are commonly referred to as 'bilobites'. Most authors consider them to be **endogenic**, having formed beneath a thin layer of sand (Seilacher 1955; Goldring 1985), although evidence for this is not available in every case. Landing and Brett (1987) considered some examples as **exogenic**, having formed in mud firmground. Regardless of which way they become preserved, *Cruziana* and *Rusophycos* have a remarkably sensitive 'fingerprint' in the form of a delicate bioglyph of ridges and grooves.

Seilacher (1955, 1959, 1970) analysed trace fossils belonging to these two ichnogenera and demonstrated convincingly that the tracemakers were trilobites in many cases. His interpretations of trilobite activity on the basis of the fossils were so vivid, in fact, that these ichnogenera are frequently called 'trilobite traces' in the literature.

Seilacher's work (among others, cf. Boucot 1990) leaves no doubt that some trilobites could produce *Cruziana* ispp. and that *some* ichnospecies of *Cruziana* were produced by trilobites. This is one of those familiar half-and-half statements in ichnology. Rare body fossils of trilobites have been preserved *in situ* in their final resting traces (section 8.1). Yet, Seilacher (1960) also showed that other arthropods, and even worms and snails, could produce resting traces that were morphologically like *Rusophycus* (section 7.3).

Most examples of *Cruziana* and *Rusophycus* remain biologically anonymous. If trilobites are bodily preserved in the same unit as the trace fossils, the temptation is too great for many authors to suggest a genetic link between them. Some authors, however, have shown evidence that trilobites did not produce *Cruziana*. Landing and Brett (1987) pointed out that the ornament on *C. reticulata* indicates a leg movement that is not possible in the few (repeat few) trilobites in which the limbs are known. Whittington (1980) found it highly unlikely that the Cambrian trilobite *Olenoides serratus*, in particular, and trilobites, in general, were capable of creating *Cruziana* on the basis of their musculature and articulation (but see Seilacher 1985).

The Burgess Shale (Conway Morris 1979), and comparable units (e.g. Conway Morris 1985; Conway Morris *et al.* 1987; Collins 1987) indicate that Cambrian faunas were dominated by non-calcified arthropods. These forms may well have included many tracemakers.

Those who label bilobite trace fossils as 'trilobite traces' encounter difficulties when these structures are found outside the time and space range of trilobites. At the Precambrian–Cambrian transition, *Cruziana* and *Rusophycus* appear before the first trilobite body fossils (Alpert 1977; Crimes 1994). This trouble can be overcome by citing the uncalcified, 'unpreservable' trilobites that may have preceded the calcified forms (Whittington 1985).

Trilobites died out in the Permian and never colonized the freshwater realm. Thus, post-Permian and non-marine occurrences of *Cruziana* and *Rusophycus* cannot be attributed to trilobites (Fig. 8.13). Small bilobite trace fossils occur in red beds of Devonian and Triassic age; traditionally the ichnogenus *Isopodichnus* was applied to them (Trewin 1976; Hakes 1976; Seilacher 1978, 1985; Pollard 1981, 1985), although this name covers both crawling and resting traces. Bromley and Asgaard (1972a, 1979) placed these bilobites in the ichnogenera *Cruziana* and *Rusophycus* on the basis of their morphology, and presented evidence that the tracemakers were notostracan crustaceans.

However, many authors have preferred to continue the traditional use of *Isopodichnus problematicus* (Trewin 1976, Hakes 1976; Seilacher 1978, 1985; Pollard 1985; Aceñolaza and Buatois 1993; Buatois and Mángano 1993). These authors express uncertainty in separating *Isopodichnus* from *Cruziana* and *Rusophycus*. Pollard (1981, p. 569) wrote 'The size, geological age, facies association, and nature of probable non-trilobite producers suggests that *Isopodichnus* should be retained as a valid ichnogenus' and Seilacher (1985) proposed ichnogeneric separation; for 'geological

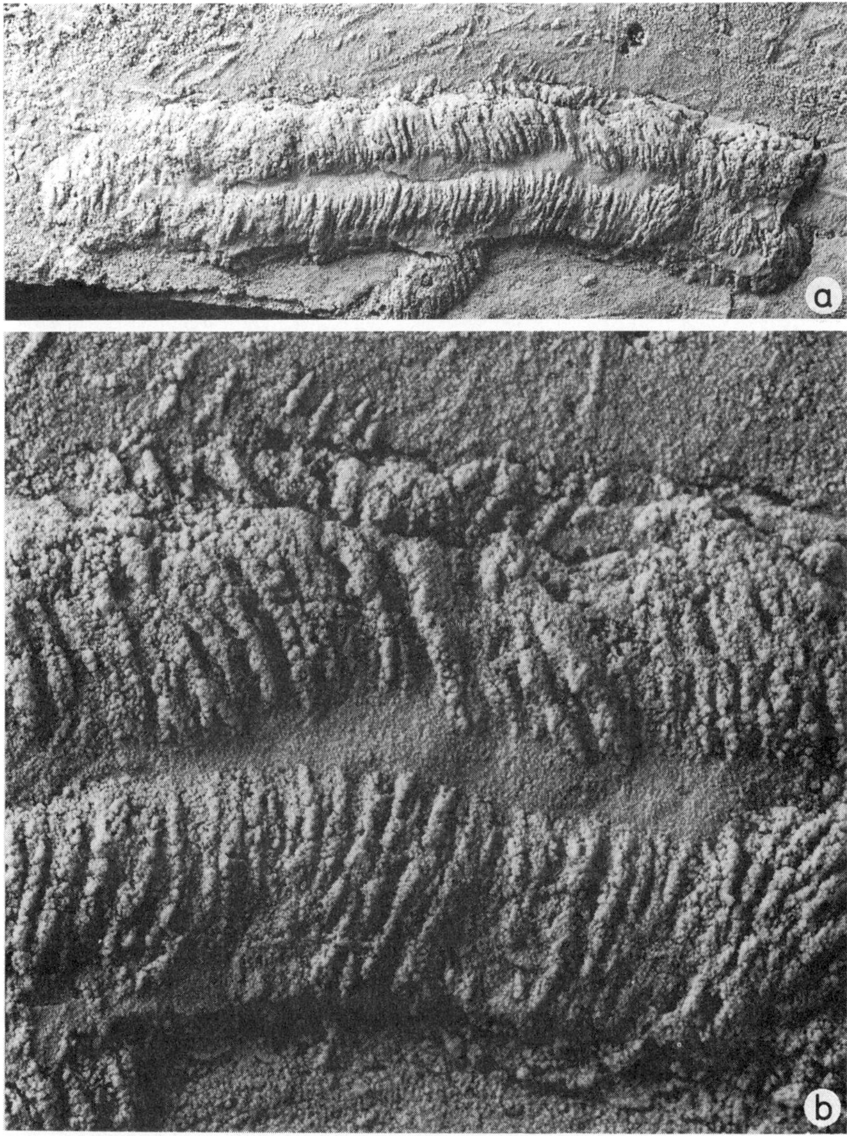

Figure 8.13 A *Cruziana problematica* in convex hyporelief from the fluviatile Ørsteddal Member of the Fleming Fjord Formation (Triassic) of Jameson Land, East Greenland. (a) x2 and (b) x8.

reasons' only. These criteria are not ichnotaxobases, so I cannot see how *Isopodichnus* can be considered available as other than a muddled junior synonym of both *Cruziana* and *Rusophycus* (Romano and Whyte 1987; Pickerill and Peel 1990, 1991; Pickerill 1994; Gradziński and Uchman 1994).

Cruziana and *Rusophycus* have shown themselves to be uniquely suited to stratigraphy, owing to the sensitive 'fingerprint' of the tracemaker. Seilacher (1977, 1992, 1993) has established a *Cruziana* stratigraphy for Gondwanaland sandstones, ranging from Lower Cambrian to Lower Carboniferous. He has done this by recognizing 'species groups' responsible for the rusophyciform cubichnia and cruzianiform repichnia. These groups have been named for convenience under *Cruziana*, although it is the tracemaker that is actually concerned. This nomenclature breaks with ichnological tradition, but it is the biotaxonomic signals that allow a biostratigraphic scheme to be possible using trace fossils. The same tracemaker bias is necessary in naming vertebrate tracks (Lockley 1993) and borings (Bromley 1994).

Crimes *et al.* (1977) have separated some of Seilacher's ichnospecies couplets into *Cruziana* and *Rusophycus* again, and disentangled their diagnoses. To repeat, it is essential that we base ichnotaxa on behaviour, as so elegantly demonstrated by Seilacher (1953a). If we allow the twin nomenclatures to interfere with each other, ichnotaxonomy will collapse and the next revision of the Treatise on Invertebrate Paleontology part W will run into several volumes. We could begin this revision with *Ophiomorpha* and see how many ichnogenera we could create for the work of supposed callianassids, brachyuran crabs, crayfish, real fish, beetles and so on.

8.7 Ichnofamilies

If ichnofamilies are accepted in principle by ICZN (Ride *et al.* 1985), then about a dozen are already valid (Rindsberg 1990). Several were named by Schimper and Schenk (1890) in the belief that they were plant taxa, e.g., groups Alectorurideae and Chondriteae. Fuchs (1895, 1909), well aware that these structures were trace fossils, used major morphological groupings including fucoids for branching forms like *Chondrites*, graphoglypts for meanders like *Cosmorhaphe*, and Alectoruridae for spreite forms like *Zoophycos* (*Alectorurus* is a junior synonym of *Zoophycos*).

Richter (1926) used Rhizocorallidae and Arenicolitidae for U-structures having and lacking a spreite, respectively. More recently, Walter (1983) worked with higher taxa, including Multipodichnia, Pentapodichnia and Tetrapodichnia, based on the number of limbs of the tracemakers. Rindsberg (1994) has used the family Zapfellidae for borings of acrothoracican barnacles, although it was established (Codez and Saint-Seine 1958) to cover the unpreserved tracemakers. Fu (1991) grouped ichnogenera into fucoids and lophocteniids according to their interpreted general function, and Seilacher and Seilacher (1994) have proposed that all trace fossils produced by bivalves be referred to ichnofamily Pelecypodichnia.

Thus, there are three incipient types of ichnofamilies: (1) those based on the work of a single high taxon of tracemaker (Rindsberg 1994; Seilacher and Seilacher 1994); (2) those based on morphological similarity of tracemakers (Walter 1983); and (3) those reflecting morphological and (interpreted) functional similarity of the trace fossil (Richter 1926; Fu 1991). The first group could be called 'zooichnofamilies', the second 'paraichnofamilies' and the last 'euichnofamilies' (I'm sorry).

I admit some doubt over the value of zooichnofamilies. *Protovirgularia* is produced by split-foot bivalves (but may be also by scaphopods). *Solemyatuba* is produced by *Solemya* (but is similar to Y-shaped *Upogebia* burrows). *Lockeia* is the compression cubichnion of wedge-foot bivalves (and hopefully will not be confused with the excavated cubichnion of bivalved conchostracan crustaceans). Let us hope that the shipworm's fingerprint will always distinguish *Teredolites longissimus* from the work of terrestrial wood-boring beetle larvae. Will the palaeontologist find the grouping of these disparate trace fossils as the zooichnofamily Pelecypodichnia useful?

In any case, before we create further ichnofamilies, it would be worth considering the familial suffix. Perhaps -idae should be avoided, as it invites confusion with zoological families (and -acea with the botanical). The suffix -ichnia is already used for ethological categories (Fig. 9.2) and Martinsson's (1965) stratinomic terms (Fig. 9.1). Martinsson (1970) used the vernacular for reference to trace fossil groups, e.g. 'teichichnians'. Other candidates are -ichnidae, -ichna, -ichniae, -ichnida, and, for those who dislike overuse of the Greek (Pemberton *et al.* 1992), -vestigia.

8.8 Confusions and conclusions

Identification of the tracemaker is a risky business beset with traps for the unwary. The most secure method is by the recognition of the architect's 'fingerprints' as Seilacher (1962a) called them. These are features that indicate the mode of trace construction or grain manipulation, thereby revealing the anatomy of the tracemaker. For example, a crustacean cannot 'push and pull' (use double-anchor burrowing technique), it can only excavate. *Lockeia siliquaria* is produced by compression and therefore must be the work of bivalves, not of bivalve-like crustaceans (concostracan or ostracode).

For individual trace fossils, the actual preservation of a tracemaker as a body fossil is a good, if not always trustworthy indication (Figs 4.7 and 9.8). Open burrows have a particular problem, the 'toad-in-the-hole' effect. These burrows offer a microenvironment for high preservation potential for any skeleton, whether architect, squatter or bioclast (section 7.5). Rasmussen (1971) fell into this trap: he described a *Thalassinoides* as a spatangoid burrow on the basis of a reworked spatangoid that had fallen into it.

Whether with or without the presence of a body, attempts at the identification of tracemakers have led to some hair-raising mistakes and confusion. Here are some examples.

Early authors reconstructed the precarious gait of walking pterosaurs, and even *Archaeopteryx* itself, on the basis of trackways preserved in the lithographic limestone at Solnhofen, *Kouphichnium lithographicum*. Later authors have referred this trace fossil to a xiphosuran horseshoe crab, progressing in the opposite direction to the pterosaur (Fig. 8.14). Genuine pterosaur trackways, however, are showing themselves to be surprisingly common (Lockley *et al.* 1995).

Horseshoes? Lockley *et al.* (1994b) reinterpreted supposed fossil tracks of mules and horses (wearing shoes) as the invertebrate trace fossil *Rhizocorallium* and, in some cases, undertracks of dinosaurs. Dinosaur undertracks are dangerous; they can lead creationists to the impression that dinosaurs co-occurred with human beings. Roček and Rage (1994) reinterpreted a supposed amphibian track as the work of a starfish.

A series of cross-cutting sigmoidal scratches in a bedding plane led King (1965) to the interpretation that they were produced by horseshoe crabs. Today, however, these trace fossils are referred to the ichnogenus *Undichnia* and are considered the swimming traces of benthic fish (Higgs 1988).

The burrows of some fish and crustaceans can resemble each other,

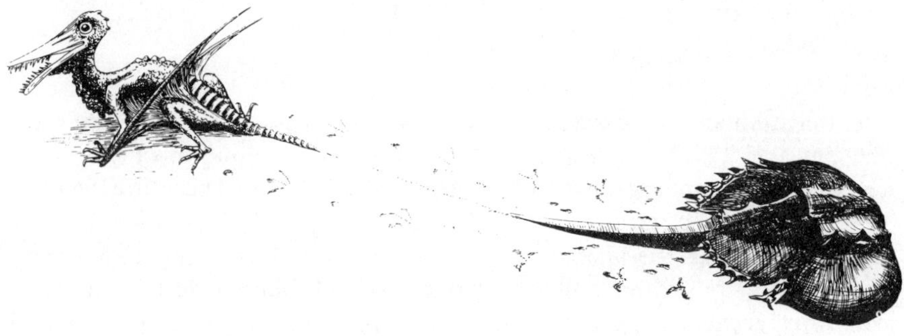

Figure 8.14 The battle of *Kouphichnium lithographicum*. Left half of the figure after Figuier (1866), right after Goldring and Seilacher (1971). Enlightenment came gradually; for a blow-by-blow history, see Abel (1922, 1935) and Caster (1938, 1939, 1941, 1944).

and some freshwater Triassic examples caused much debate. The resulting discussion has taught us a lot about the excavating activity of both lungfish and crayfish (Dubiel *et al.* 1987, 1988, 1989; McAllister 1988; Hasiotis and Mitchell 1989; Hasiotis *et al.* 1993).

Another form of confusion has been caused by Swinbanks' (1981b) revelation that the graphoglyptid ichnogenus *Paleodictyon* morphologically resembles the xenophyophoran *Occultammina profunda.* Both form hexagonal networks, although *Paleodictyon* is normally more extensive than *Occultammina* and much more regularly branched. Is *Paleodictyon* an agrichnian trace fossil or a protistan body fossil? Does it matter? Both have been interpreted in terms of deep-sea deposit feeding, meiofaunal traps and bacterial gardens (Levin 1994). Southward and Dando (1988, fig. 4) illustrated the xenophyophorans lying like fragments of chicken wire on the sea floor rather than just within it.

But *Paleodictyon* occurs as open burrows in the ocean floor as found in box cores (Ekdale 1980; Gaillard 1988), and Pickerill (1990) reported non-marine *Paleodictyon*-like trace fossils; xenophyophores are exclusively marine.

Note added in proof

The International Commission on Zoological Nomenclature has produced a discussion draft of a proposed 4th edition of the Code (O. Kraus and W. D. L. Ride, *in litt.* 1995: Discussion draft of the fourth edition of the Code). The new Code is intended to come into effect 1st January, 1997.

Among many innovations proposed for this new edition, some concern ichnotaxa and, if implemented, will contradict statements in Chapter 8. The reader is therefore advised to refer to the 1997 Code when this appears.

Important are the moves to simplify stabilizing measures for names, so that old, well-established names may continue to be used even though they may be shown to be junior synonyms. The 50-year rule is reinstated, whereby names not used for that period lose validity. The requirement that every genus-group taxon must be designated a type species is extended once more to ichnogenera. Finally, all new species (ichnospecies) must be registered in the *Zoological Record* within 5 years, or they lose validity.

9 Stratinomy, toponomy and ethology of trace fossils

Although there is no ichnotaxonomic superstructure above the rank of ichnogenus, trace fossils may be grouped together in several ways. Two of these groupings are in general use in ichnology: the preservational and the behavioural classifications.

9.1 Preservational classifications

The **toponomy** or morphological expression of an ichnotaxon varies according to whether a trace fossil is preserved within mudstone or sandstone, or at a boundary between the two (section 7.2). A series of terms has arisen to describe these stratinomic possibilities, and several authors have organized these as classifications. These systems in turn have been modified to fit special situations, as reviewed by Ekdale *et al.* (1984a) and Frey and Pemberton (1985).

Seilacher's terms are used here, in which trace fossils are classified as preserved in **full relief** or in **semirelief** (section 6.2). (These terms ultimately derive from the earlier literature, e.g. Maillard 1887; Fuchs 1895.) A somewhat parallel set of terms was devised by Martinsson (1965, 1970; Fig. 9.1). Both of these classifications are based on the relationship of the structure to a **casting medium**, typically a sandstone bed. Spectacular trace fossil preservation occurs on top and, especially, sole surfaces of sandstones in weathered onshore exposures of heterolithic terrigenous units, and it is in relation to this collecting environment that the classifications were developed.

It is important to describe the stratinomic state of different ichnotaxa, as this aids both their identification and their interpretation. *Gyrochorte*, for example, is seen chiefly on top surfaces of the casting medium, i.e. in convex epirelief as a double ridge; more rarely it occurs on sandstone soles in concave hyporelief as a double groove. In contrast, *Cruziana* occurs almost exclusively in convex hyporelief as a double ridge.

Trace fossils have received increasing attention in lithologically monotonous lithologies such as shale and chalk. In such cases, traces normally are preserved exclusively in full relief, although they may be seen clearly on cleavage surfaces of the rock as **cleavage relief**. But, in general, the monotony of the type of preservation in such situations renders stratinomic classification of little value.

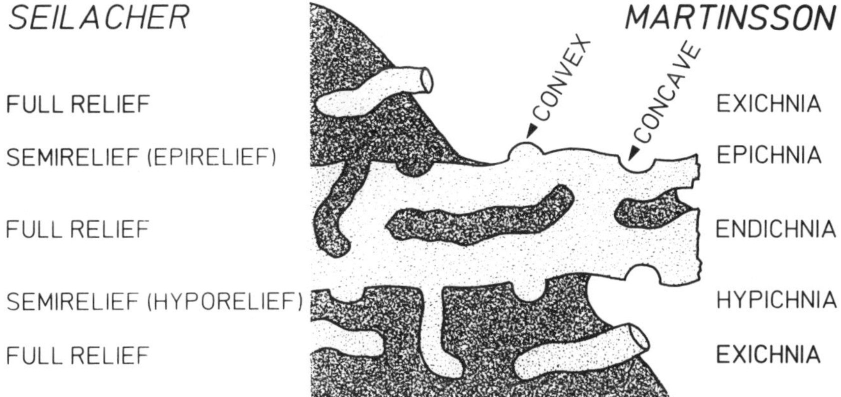

Figure 9.1 Stratinomic classification of trace fossils in relation to the arenaceous casting medium. Comparison of the terminologies of Seilacher (1964) and Martinsson (1965).

These preservational classifications do not accommodate diagenetic features. Although the development of concretions or patchy distribution of cement are of paramount importance in trace fossil preservation, especially where lithology is monotonous, these phenomena have not yet been subject to an attempt at classification. It was the presence of branching flint concretions that first drew attention to the otherwise invisible bioturbation fabric in European white chalk (Bromley 1967), but both the flintified and the non-silicified trace fossils are in Seilacherian full relief.

9.2 An ethological classification

Above all, trace fossils are the tangible evidence of the behaviour of animals, and the most natural way to classify them is according to behavioural patterns. The ethological classification devised by Seilacher (1953a, 1964) serves this purpose and is in general use in trace fossil description and interpretation. This classification is based on the most fundamental aspects of ethology and now comprises 11 compartments (Fig. 9.2). It must be emphasized that there is a natural overlap between compartments, but this merely reflects the intergradations inherent in nature. We must also remember that many structures include different parts that fall into different categories of behaviour.

*9.2.1 Resting traces (**cubichnia**)*

These are structures made by vagile animals digging down for a period and then departing by the same route. Some predators such as starfish do

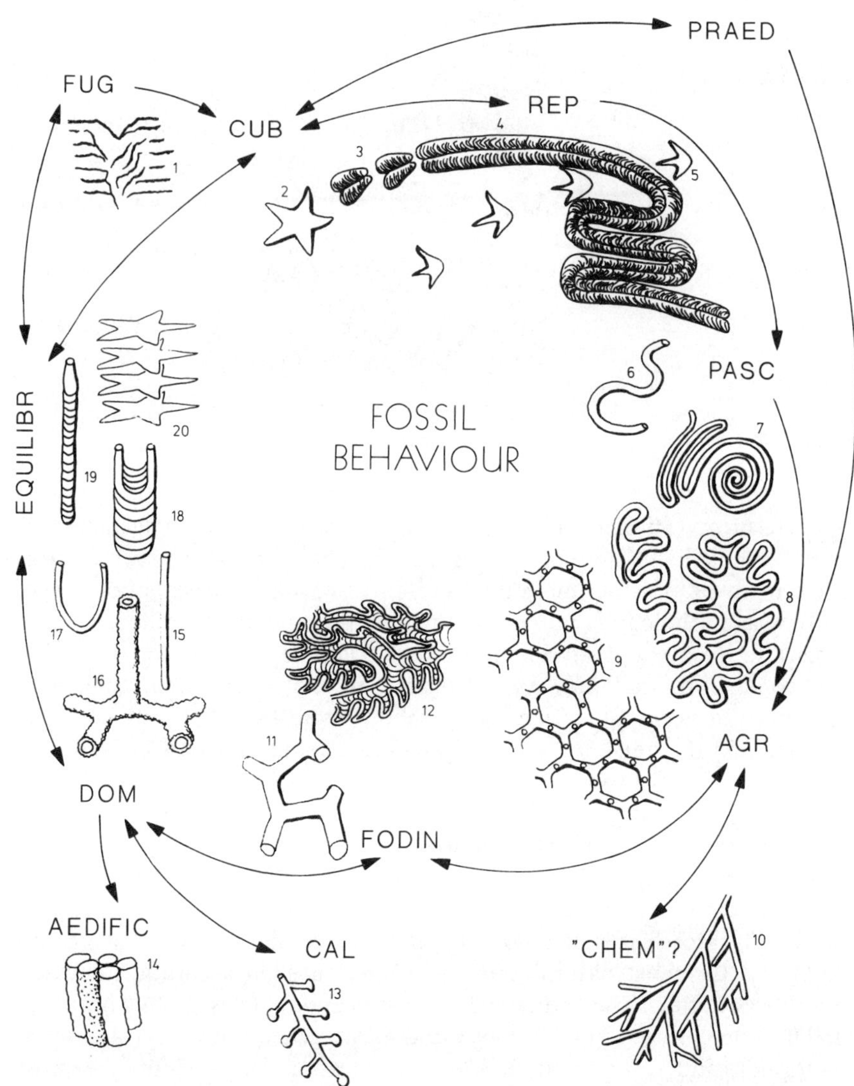

Figure 9.2 The ethological classification of trace fossils. Names of the 11 groups are abbreviated (lacking suffix -ichnia), and interrelationships are suggested with arrows. Should chemosymbiont structures be grouped with agrichnia, or be separated as suggested here (chemichnia)? Ichnogenera and other trace fossils indicated are: 1, escape structure; 2, *Asteriacites*; 3, *Rusophycus*; 4, *Cruziana*; 5, bipedal vertebrate trackway; 6, *Helminthopsis* or *Planolites*; 7, *Helminthoida*; 8, *Cosmorhaphe*; 9, *Paleodictyon*; 10, *Chondrites*; 11, *Thalassinoides*; 12, *Phycosiphon*; 13, beetle brooding burrow; 14, sabellarian sand tubes; 15, *Skolithos*; 16, *Ophiomorpha*; 17, *Arenicolites*; 18, *Diplocraterion*; 19, bivalve adjustment trace; 20, vertically repeated *Asteriacites*.

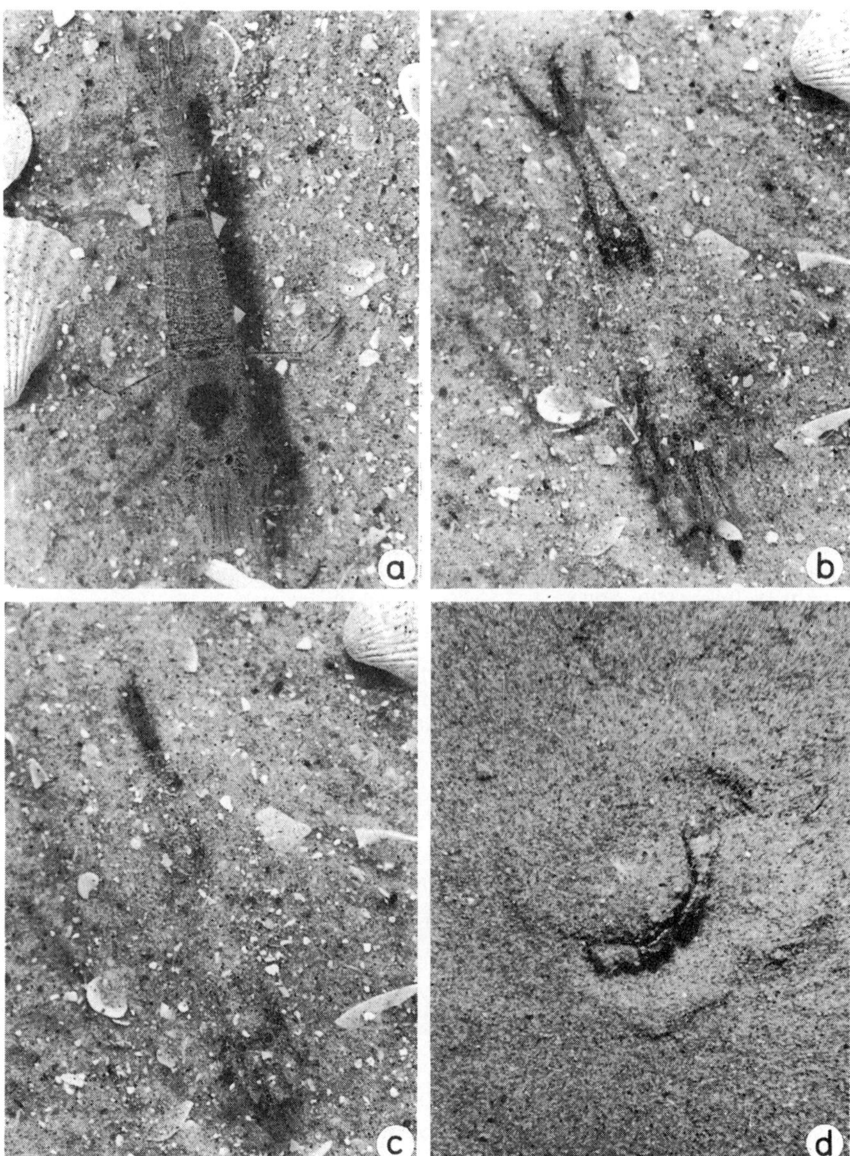

Figure 9.3 Incipient cubichnia. (a–c) A prawn, *Crangon crangon*, buries itself in sand until only its eyes are visible. Danish North Sea coast, slightly enlarged. (d) A shame-face crab, *Calappa* sp., in resting position in sand. Fore-reef slope, Elat, Israel; x0.5.

this in order to capture buried prey. More often, cubichnia are produced endogenically by animals that, living within a superficial sand layer, disturb the top of the underlying substrate each time they dig down to establish themselves (Figs 6.3 and 9.3). Many bivalves show this behaviour (section 4.1.1). Few tracemakers can be considered actually to 'rest', and the production of cubichnia more usually represents concealment, or traces of animals that feed stationarily but shift their position at intervals. Examples are the ichnogenera *Lockeia*, *Rusophycus* and *Asteriacites*.

9.2.2 *Crawling traces (**repichnia**)*

These structures reflect directed locomotion rather than any other activity. The producing animal appears merely to have been wandering from one place to another (Fig. 9.4a). Most typically, these trace fossils follow bedding planes and may be either exogenic or endogenic. Certainly, the animal may have been feeding as it progressed, but this activity is not directly registered in the morphology of the trace fossil. The trackways *Diplichnites* ispp. (Fig. 9.4b), and the simple burrows, *Cruziana* ispp., are typical repichnia.

I would also include here traces produced by swimming and running animals, called 'natichnia' and 'cursichnia' by some workers (Müller 1962; Walter 1980, 1982; cf. Trewin 1994).

9.2.3 *Grazing traces (**pascichnia**)*

Where a trackway or locomotion trace follows a meandering or spiral course, it is clear that an animal has exploited a particular region of the substrate for food. Elegant grazing traces are common on the ocean floor today but most fossil occurrences derive from similar behaviour at a level beneath the sea floor. Grazing traces commonly follow surfaces parallel with the sea floor; examples are the ichnogenera *Phycosiphon* (Fig. 11.11), *Nereites* and *Scolicia*.

9.2.4 *Feeding traces (**fodinichnia**)*

This category is characterized by the combined functions of deposit feeding and dwelling. Thus, the structure has some degree of permanence, and yet its morphology reflects exploitation of the substrate for food. *Thalassinoides suevicus*, *Dactyloidites ottoi* and *Rhizocorallium irregulare* (Figs 6.8, 8.12a and 9.5) are characteristic fodinichnia, and some forms of *Zoophycos* may also belong here.

9.2.5 *Dwelling traces (**domichnia**)*

Structures placed in this group served as semi-permanent domiciles. The tracemaker may be a sessile suspension feeder, an active carnivore wait-

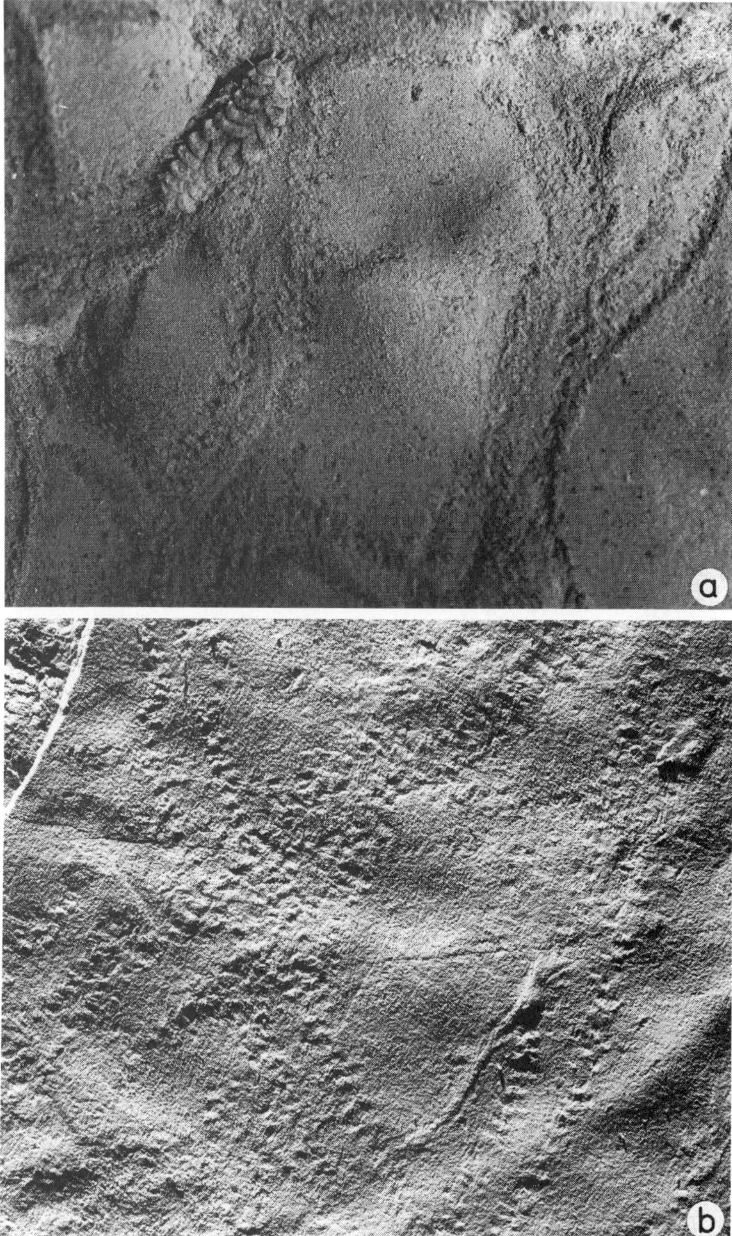

Figure 9.4 Trackways. (a) Incipient repichnia produced by the scalibregmiatid worm, *Gattyana cirrhosa*, in Danish subtidal mud floors. Here in an aquarium, the animal is crawling over the surface, but it also moves similarly beneath it; x2. (b) Comparable trails, but from Triassic non-marine siltstone, Fleming Fjord Formation, East Greenland. Epirelief undertracks of exogene traces; x2.

Figure 9.5 A rather short example of *Rhizocorallium irregulare*, showing slight distal widening as a function of growth. A horizontal deposit-feeder spreite fodinichnion, from Bravaisberget Formation, Middle Triassic, Isfjorden, Svalbard.

ing in ambush, or a worm feeding on the surrounding detritus; but the trace fossil emphasizes the stationary dwelling function and not the trophic group. Examples are the ichnogenera *Skolithos*, *Ophiomorpha* and *Arenicolites* (Figs 8.12b, 10.3, 10.16 and 10.17a).

*9.2.6 Traps and gardening traces (**agrichnia**)*

Regular, patterned structures having various behavioural significance, are grouped together as **graphoglyptids**, and were previously considered to be pascichnia. However, Seilacher (1977) demonstrated that some of these structures are too complicated to be grazing foraging traces. These complex forms show true branching and must therefore have existed as open burrows available for multiple visiting by the tracemaker. Seilacher (1977) suggested two functions for such burrows. The more simple structures may have been traps for migrating meiofauna, similar to the burrows of *Paraonis fulgens* (section 4.4). Those having numerous apertures to the sea floor were clearly vigorously irrigated and meet the paradigm for gardening systems where microbes may be cultured for food.

The classical examples for these structures are *Spirorhaphe* and *Cosmorhaphe* (Fig. 9.6) as traps, and *Paleodictyon* and *Helicolithes* as gardens. Ekdale *et al.* (1984a) erected the category agrichnia to accept these non-foraging graphoglyptids.

It is perhaps also reasonable to place the supposed sulphide pumps

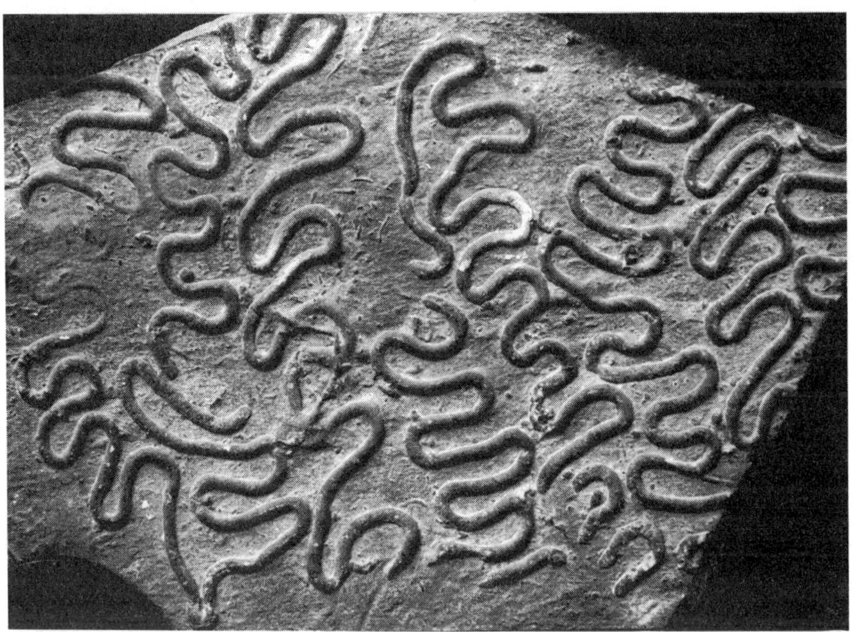

Figure 9.6 The graphoglyptid, *Cosmorhaphe lobata* from Tertiary turbidites, Italy (courtesy of R. Colacicchi) is a two-ordered meander. However, there are short-cut bridges (simultaneous branching) which show that the system was not the faecal string of a deposit feeder (as traditionally thought). Instead, it must have been an open network, allowing repeated revisiting by its occupant, and has been interpreted as a microbe garden (Seilacher 1989). A shallow-tier, **pre-depositional** structure, it was exposed for rapid casting by gentle scour immediately preceding turbidite deposition.

such as *Chondrites* within the agrichnia, rather than erect yet another category for them (Fu 1991; Savrda *et al.* 1991). After all, there is a natural sequence from the trapping of microbes for food, via the culturing of microbes for food, to the culturing of microbes as symbionts. Or should we establish the Chemichnia (Fig. 9.2)?

*9.2.7 Predation traces (**praedichnia**)*

Ekdale (1985) placed trace fossils that arose through predatory behaviour in a group of their own. Such structures are common in hard substrates (Bromley 1981b, 1993), as round drill holes in shells (*Oichnus* ispp.) and as shell damage by predators (durophagy). Soft substrate disturbance resulting from predation (Figs 1.3 and 9.7) is not easily recognizable as such in the fossil record. However, Bergström (1973) and Jensen (1990) documented *Cruziana dispar* that are so related to *Planolites* as clearly to indicate predation, the *Cruziana* animal having captured the *Planolites* animal (cf. Osgood 1970; Osgood and Drennen 1975; Rindsberg 1994).

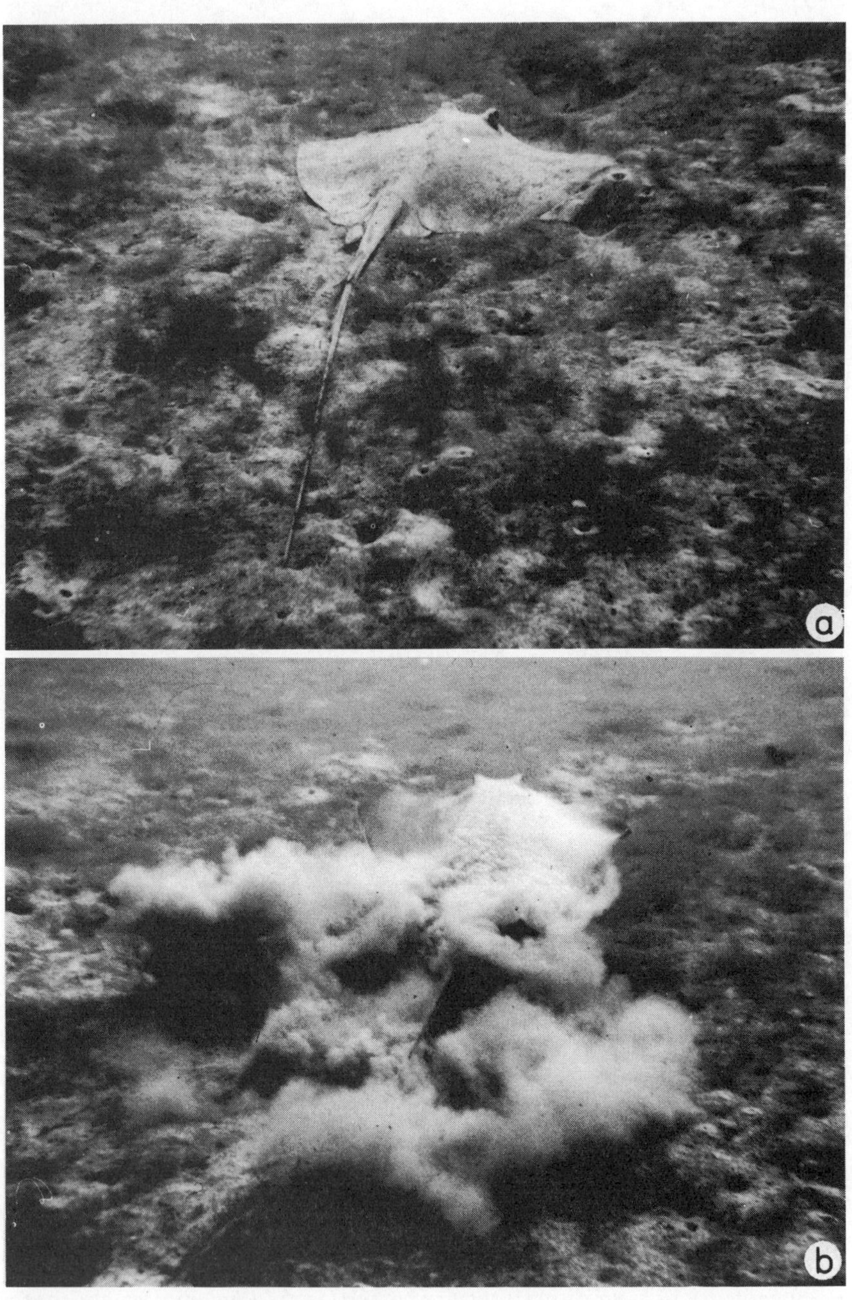
a
b

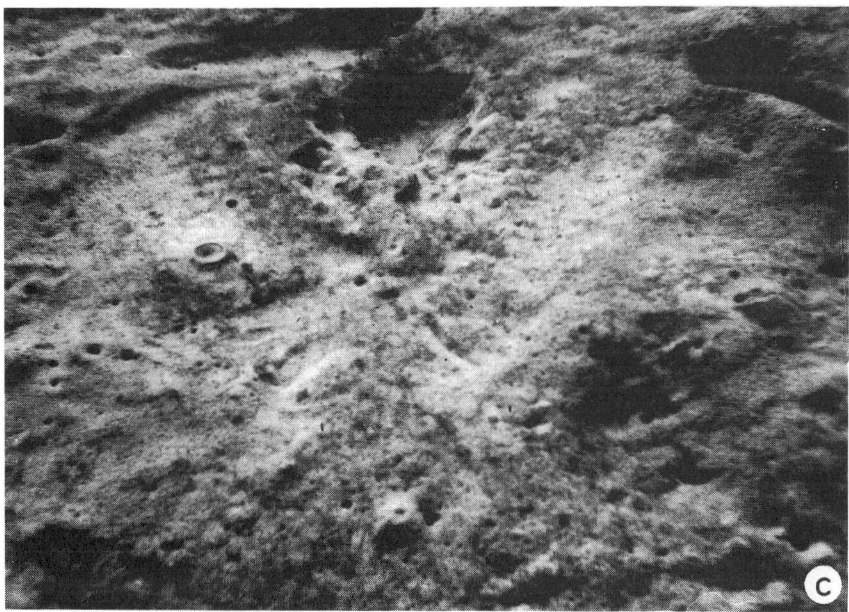

Figure 9.7 A 2 m long sting ray on an algal-bound mud floor, fore-reef slope (–8 m), off Hole Town, Barbados. (a) At rest; note the totally bioturbated substrate. In front of the ray is an old 'ray-hole'. (b) Moving off rapidly causes resuspension of surficial sediment. (c) A recently made 'ray-hole', where a sting ray blew a jet of water into the sediment from the gill openings, producing a pit to expose endobenthic prey animals, thereby resuspending much sediment. The ghostly outline of the predator is detectable but is being obliterated by burrowing activity; only the bottom of the pit has reasonable preservation potential (cf. Gregory *et al.* 1979).

*9.2.8 Equilibrium traces (**equilibrichnia**)*

Beneath gradually agrading and degrading sea floors, the position of infaunal animals and their burrows must constantly be adjusted (section 3.2.2). Depths at which most animals live within the substrate are critical, and contact is usually maintained with the surface. In places where accretion is gradual, the endobenthos must correspondingly shift their burrows upwards. Gradual scour or erosion will cause the reverse effect. Such traces were called 'equilibrium structures' by Frey and Pemberton (1985) in order to distinguish them from escape structures (Figs 3.4c, 3.15c, 4.7, 4.8c and d, and 9.8a).

*9.2.9 Escape traces (**fugichnia**)*

A few endobenthic species can tolerate sudden burial by a package of sediment (sections 3.2.2 and 4.1.3). Under these conditions, panic escape

Figure 9.8 Equilibrichnia and fugichnia. (a) Equilibrium structures produced by *Macoma balthica*. The bivalves climbed in fairly steadily deposited sand in a channel fill. Tidal flat between the rivers Weser and Elbe, the German Bight. Epoxy impregnated box core, photographed through the courtesy of H.-E. Reineck; x0.5. (b) Escape traces in rapidly accumulated sand. Athabasca Oil Sands of the Lower Cretaceous McMurray Formation, Alberta, Canada (cf. Pemberton *et al.* 1982).

reactions are released as the animal flees to the new sea floor, and the sediment is reworked in a manner quite distinct from the calm adjustment of equilibrium traces. Examples are shown in Figs 3.4e, 3.23c, 4.8e, 9.8b and 10.23. Animals avoiding predators may escape laterally (section 4.1.2). I would include here the unsuccessful attempts at escape, which Pemberton *et al.* (1992) called 'taphichnia'.

*9.2.10 Edifices constructed above substrate (**aedificichnia**)*

Structures built of sediment more or less cemented by the architect. The category was first named for mud-dauber wasp nests by Bown and Ratcliffe (1988) and has since been applied to the constructions of termite colonies (Genise and Bown 1994b). In the marine realm the sand 'reefs' constructed by sabellariid polychaetes (Ekdale and Lewis 1993) would fall within this category. Donovan (1994) even suggested spiders' webs and caddis larval tubes as aedificichnia; but perhaps the caddis tubes might fall into the next category.

9.2.11 *Structures made for breeding purposes (**calichnia**)*

Genise and Bown (1994a) proposed this category to cover structures exclusively used for raising larvae or juveniles, e.g. bee cells and scarabeid beetle nests (Retallack 1984; Thackray 1994). Perhaps dinosaur nests and fossil woodpecker nests (Buchholz 1986) may be placed here, but it may prove difficult to place a line of distinction between calichnia and aedificichnia. Presumably, fossil leaf-mining traces of insect larvae (e.g. Müller 1982) belong here, or are they xylic fodinichnial borings? This classification is getting too complicated.

9.3 Evaluation of the behavioural classification

The above classification was founded on the basis of trace fossils and the interpretation of the behaviour that they represented. It was also founded at a time when trophic group analysis in living communities was less advanced than it is today. Nevertheless, the system works well, and it would be difficult to upgrade it without involving excess degrees of subjective interpretation. Overlap between the groupings is unavoidable and some trace fossils are bound to find a place within the classification more easily than others. Certainly, ichnologists are not agreed as to which category should house some of the more 'difficult' ichnogenera. For example, Frey and Pemberton (1985) classified *Zoophycos* and *Planolites* as pascichnia and fodinichnia, respectively, whereas Ekdale (1985) placed these same genera in the fodinichnia and pascichnia, respectively! (I prefer the Ekdale arrangement!)

Feeding activities of various types are inherent in most of the categories. However, the trophic groups recognized by ecologists, i.e. deposit feeders, suspension feeders, carnivores, etc., cut across the compartments of the ethological classification. This is unfortunately a reflection of the difficulty encountered in the pigeon-holeing of ichnotaxa, or individual trace fossils, in precise trophic groups.

9.4 Functional interpretation of trace fossils

Fossil communities are notoriously difficult to analyse ecologically, for three reasons in particular: (1) the degree of transport and physical sorting of skeletons must be evaluated carefully; (2) at best, only a small proportion of the original community is bodily preservable; (3) slow deposition may allow several communities to become mingled at a single horizon; and so on.

Because trace fossils: (1) cannot be transported (without it being blatantly obvious) and (2) are commonly preserved where skeletal material is lost, many authors have looked to ichnology to supply some of the lost palaeoecological data. A third problem – accumulation of more than one community at a horizon – also affects trace fossils (Fig. 10.3).

There are several dangers inherent in 'treating trace fossil assemblages as communities', as is discussed in the next chapter. Let us first, however, briefly examine the trophic analysis of trace fossils.

Early attempts at analysing the function of trace fossils tended to be simplistic. Suspension feeders constructed burrows that were simple, vertical shafts or U-tubes, lacked signs or grain manipulation around them and were passively filled. Deposit feeders, on the other hand, wandered about within the sediment; their burrows were surrounded by grains in random orientation owing to manipulation; and they were actively filled (e.g. Walker and Laporte 1970; Walker 1972).

Ronan (1977), among others, has shown that this series of criteria is excessively generalized and misleading, and the reviews in Chapters 2–4 support Ronan's scepticism. However, many criteria in trace fossil morphology are ecologically informative and should be considered in the evaluation of function.

9.4.1 Shafts and U-burrows

Simple vertical shafts are by no means the unique domain of suspension feeders, as is commonly believed. We have seen, for instance, that carnivorous crustaceans can produce such structures (section 4.4). Furthermore, many vertical conveyor deposit-feeding worms produce transient vertical burrows (section 3.5), as well as detritus-feeding crustaceans (section 3.4.1) and reverse-conveyor worms (section 3.5.4). Nevertheless, there are sabellid, spionid and other polychaetes, anemones and phoronid worms that do produce vertical shafts as domichnia.

Trace fossil material is commonly fragmentary or incompletely exposed, and it is vital to distinguish vertical, isolated burrow structures, ichnogenus *Skolithos*, from vertical shafts that serve to connect deeper structures with the surface. Vertical retrusive equilibrichnia must also be carefully distinguished from shafts (Figs 4.7 and 10.23).

Vertical U-burrows are more typically suspension-feeder structures in that the irrigation current is exploitable. Extant examples are the burrows of *Chaetopterus* (section 3.3.1) and *Urechis* (section 3.3.2). Here, again, however, we must remain sceptical. For while *U. caupo* feeds from suspension in a U-burrow, another echiuran, *Echiurus echiurus*, while producing a similar structure, is classified as a detritus feeder (section 3.3.3). Many U-burrows are produced today by funnel-feeding deposit feeders having various trophic styles but showing little tendency to suspension feed (sections 3.4.2 and 3.4.3).

The trace fossil *Diplocraterion parallelum* is chiefly an equilibrichnion, constructed by shifting a U-burrow (Fig. 8.11). Fürsich (1974b) evaluated the trophic possibilities of the structure and classified it as a suspension-feeder trace. However, in rare cases where the apertures are complete, these may be funnels. This opens the possibility of detritus trapping as seen in funnel-feeding polychaetes, holothurians and hemichordates. Even where funnels can be shown to have been absent, comparison may be drawn with the detritus-feeding mud-flat shrimp, *Corophium volutator* (section 3.4).

Thus, on the basis of comparison with recent structures, the identification of trace fossils as domichnia does not unquestionably lead to an interpretation of the burrowers as suspension feeders. Taghon *et al.* (1980) observed three species of spionid polychaetes, inhabiting incipient *Skolithos*, switch to and fro from detritus feeding to suspension feeding. The change was caused by bottom-current velocity and concomitant changes in suspended seston. The worms did not alter their burrow morphology but merely raised or lowered their tentacles. A comparable case, involving incipient *Diopatrichnus*, is illustrated in Fig. 3.27b.

9.4.2 Burrow boundary

Even during the transient passage of a worm through the sediment, a mucous trace is left behind it. However, where a more substantial lining is present it is reasonable to assume that a burrow was occupied for a longer period of time as an open structure (Figs 11.3 and 11.16).

A massive lining structure may indicate instability of the substrate, as in burrows of callianassid and stomatopod crustaceans in loose sand today (sections 4.3.1 and 4.4) and their fossil equivalent, *Ophiomorpha nodosa* (Fig. 11.5b). Alternatively, a thick, laminated lining (Fig. 8.2) may represent repeated application of material to the walls during lengthy occupation, as in *Amphitrite ornata* (section 3.6).

A. ornata is a detritus feeder, but, in general, such permanent and well established domichnia probably include dominantly suspension-feeding animals. Of course, substantial burrow linings must be distinguished from zoned fills such as in *Ancorichnus ancorichnus* (Fig. 8.3), which is a repichnion, or *Phoebichnus trochoides* (Fig. 11.6), a fodinichnion.

In many trace fossils, a central fill is surrounded by a mantle or disturbance zone that unquestionably arose through deposit feeding. The ichnogenera *Nereites*, *Rosselia* and *Asterosoma* are examples (section 11.3.2).

9.4.3 True branching

If a structure can be shown to have comprised simultaneously open branches, then revisiting and turning round is necessary on the part of the tracemaker. The semi-permanent domichnia of animals feeding in various

Figure 9.9 *Rutichnus rutis* from shallow marine Permian deposits, Scoresby Land, East Greenland. In order to create this branched structure, the animal had to back up and turn around, and beneath the thick mantle there is a meniscate core (D'Alessandro *et al.* 1987). Note the *Chondrites* isp. to the right. Half natural size.

ways can be branched burrows, as can the fodinichnia of stationary deposit feeders. Repichnia and pascichnia cannot truly be branched.

A branched structure can be produced by a probing deposit feeder, backing up or reversing, and backfilling behind it (Fig. 9.9). This is primary successive branching and must be distinguished from the simultaneous form of open burrows. The behavioural patterns and trophic styles are quite different. Primary successive branching is also produced by repair of damaged shaft apertures (Fig. 4.20).

9.4.4 Nature of the fill

Much information is provided on burrow function by the nature of its filling material. Passive fill, of course, with or without a draught fill, betrays an open burrow system or shaft. Collapse without filling may suggest either the same (Figs 11.7b and 11.14b), or compressional passage through softground.

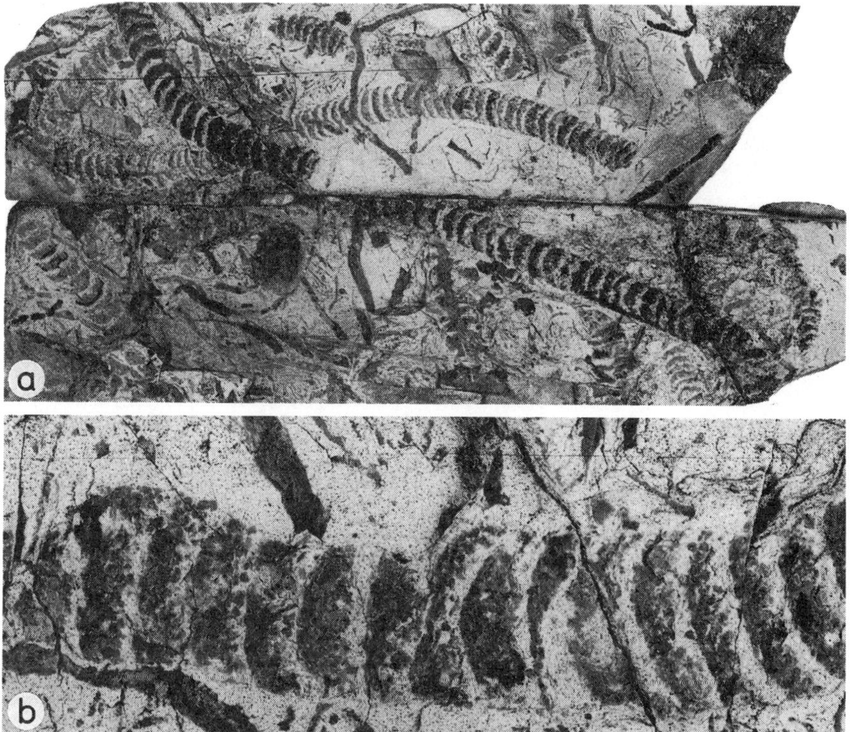

Figure 9.10 The meniscate backfill structure, *Taenidium satanassi*. There is no trace of a wall lining or mantle. The dark menisci are pelleted and probably represent ingested material; the intervening menisci resemble the surrounding sediment (D'Alessandro and Bromley 1987). (a) x0.5 and (b) x5.

An active fill is suggestive of deposit feeding, particularly if it is pelleted. Meniscate backfill, in which menisci are composed of the little-altered substrate sediment, alternating with a modified sediment is highly indicative of deposit feeding, the modified sediment normally being faecal in origin (Fig. 9.10).

Pemberton and Frey (1982) distinguished between the morphologically similar trace fossils *Planolites* and *Palaeophycus* on the bases of fill and walling structure (section 11.3.1). The active fill of *Planolites* is structureless, and differs from the surrounding sediment. There is no lining or mantle and the tracemaker was considered a wandering deposit feeder backfilling behind it. *Planolites* may be classified as a pascichnion as the accent is on mobility.

Palaeophycus, in contrast, has a fill that resembles the surrounding sediment and may be passive. The walling material indicates occupation and stabilization of a permanent burrow. Thus, *Palaeophycus* represents the domichnion of a suspension feeder, detritus feeder or carnivore.

Macaronichnus segregatis, like *Planolites*, is unbranched and has an active fill (section 11.3.1; Clifton and Thompson 1978). It differs from *Planolites* in having a distinct mantle of different grains selected out during the feeding process, and usually darker than both fill and surrounding sediment (Fig. 11.9). This structure is clearly the work of a vagile deposit feeder.

Curran (1985, pl. 1c) identified a similar trace fossil as *M. segregatis*, but illustrated a Y-shaped simultaneous branching point. Thus, this structure cannot be considered trophically or ichnotaxonomically equivalent to *M. segregatis*. In vertical section (Curran 1985, pl. 1d), and therefore in core, however, these structures would be indistinguishable. Branching is vitally important, but is not easy to see in vertical section (section 11.2.3).

9.4.5 Spreite

Spreiten arise in several ways. Most important are:

1. active strip-mining by a deposit feeder moving part of its burrow laterally (Seilacher 1967a), as in some *Zoophycos* and some *Rhizocorallium* ispp.;
2. equilibrium movement in a vertical plane to keep pace with sea-floor fluctuation (Goldring 1964), as in *Diplocraterion parallelum*; and
3. gradual, passive, gravitational collapse of roof material, which collects as laminated floor sediment and displaces the burrow upwards over a retrusive spreite, as in some *Thalassinoides* ispp. (Frey and Seilacher 1980).

Horizontal spreiten belong to the first category. Vertical spreiten are less easy to evaluate, as, for example, *Teichichnus rectus*, which is nearly always retrusive and probably represents deposit feeding in most cases, but may be an equilibrichnion in less stable settings. Regular incorporation of pellets into the spreite would prompt an interpretation as deposit feeding rather than suspension feeding.

It should be remembered that the collapse of roof onto floor after the burrow has been abandoned may cause drastic changes. An internal deposit will accumulate that resembles a biogenic spreite, and an upward migration of the cavity may extend this pseudospreite vertically. Wallace (1987) modelled this process for stromatactis cavities, some of which may have originated as burrow systems.

9.4.6 Chemosymbiosis

Identification of trace fossils associated with sulphide or methane symbiosis would seem to be particularly hazardous. We must seek structures that might resemble deep sulphide or methane wells, as in *Solemyatuba upsilon* (section 4.1.4b).

Seilacher (1990) has made a convincing case for chemisymbiosis for *Chondrites*. Previously considered to represent deposit-feeding activity, he pointed out that the *Chondrites* animal deliberately did not rework the sediment entirely, but left spaces between successive probes. This organization does not correspond to the deposit-feeding paradigm. Instead, *Chondrites* corresponds nicely to a sulphide well that is continually relocated by successive probes, the previous probe being backfilled (Kotake 1989; Fu 1991) and, indeed, the structure resembles the well constructed by some lucinid bivalves (Fig. 4.10).

Chondrites is normally situated in anoxic sediment, not only in the deepest tier (Bromley and Ekdale 1986) beneath richly oxygenated sea floors (Leszczyński and Uchman 1993), but also in dysoxic sea floors (Bromley and Ekdale 1984a). Locally, *Chondrites* shows strong affinity to organic-rich sites; it is commonly associated with shallower-tier burrow fills, linings or the sediment immediately surrounding these (Fig. 7.2; Bromley and Frey 1974; Frey and Bromley 1985). *Chondrites* may also be found in close association with body fossils; Underwood (1993) reported *Chondrites* entering the rhabdosomes of graptolites.

Exaggerated association of a trace fossil with sites of organic richness may indeed be important in recognition of chemosymbiont activity. Such

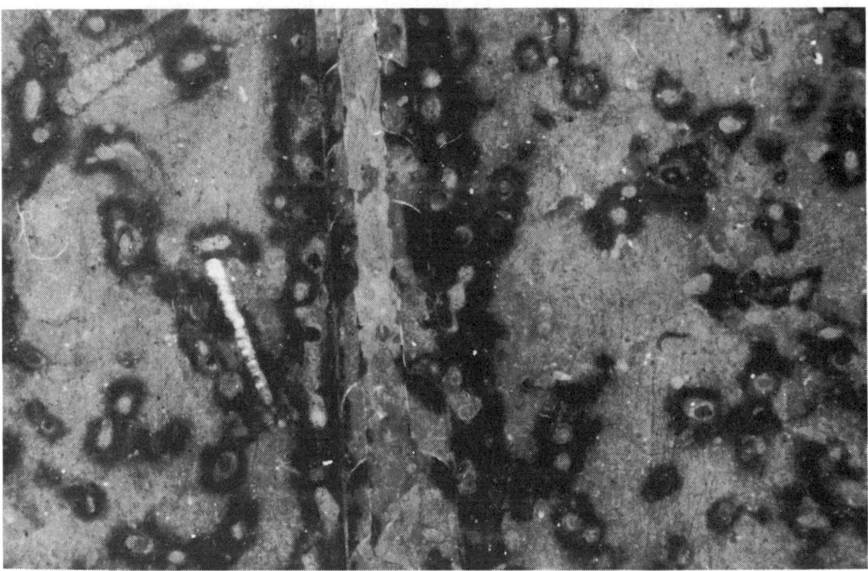

Figure 9.11 Polished horizontal section (floor stone) of Orthoceratite Limestone, Ordovician, Öland, Sweden. The cephalopod shell has been very attractive to an animal constructing a *Thalassinoides*-like trace fossil. At the level of section, numerous shafts to apertures have been cut, having a strong wall mineralization. At a level 2–3 cm lower down, the branched network is developed. Natural size.

an association is reported by Bromley and Ekdale (in press) for *Thalassinoides*-like systems in Ordovician limestone. These trace fossils show strong affinity to shells of orthocone cephalopods (Fig. 9.11); the burrow walls are heavily mineralized, suggesting long-term biological activity; and the systems have an excessive number of apertures to the sea floor. The size and network morphology resemble the burrow systems of the chemosymbiont cerianthid anemone *Cerianthus vogti* (section 3.2.2).

9.5 Functional interpretations: conclusions

Trace fossils have the reputation of being the key to fossil animal behaviour. There has therefore been a tendency to append functional labels too readily to individual trace fossils and to morphological traits. However, whereas earlier, optimistic and simplistic criteria for interpretation cannot be applied uncritically, there are nevertheless clear indications of function to be drawn from several morphological aspects of trace fossils. Of course, a single burrow system may represent more than one type of activity.

A thick burrow lining, especially where associated with passive fill, indicates a long-term domichnion. Whether around a shaft, a U-structure or a boxwork, such lining generally suggests the dominant trophic activity to be suspension feeding. On the other hand, well-developed funnels at the apertures of a U-burrow or shaft generally indicate detritus feeding.

Panic escape and quiet equilibrium structures are normally easily identified, even though physical escape structures (gas and water) can closely resemble biological escape.

Extensive horizontal trace fossils having no lining, and a fill that contrasts with the surrounding sediment, suggest deposit feeding (pascichnia). This becomes conclusive where meniscate structure and pelleted fill are preserved.

Horizontal spreiten normally represent deposit feeding of the subsurface strip-mining type. Truly branched, horizontal mazes, however, represent a more normal form of mining to produce open galleries; this combines dwelling and deposit feeding (fodinichnia).

Gardening has been envisaged as the most reasonable explanation for the excessively complicated networks (agrichnia), while chemosymbiosis has been suggested for some of the deep-tier trace fossils that fit neither the deposit-feeding nor the gardening paradigm. Gardening and chemosymbiosis may well be secondary activities in many other types of burrow systems.

All these trophic and other activities must be measured on the demanding budget of supply and survival. Ecologists operate with the 'optimal foraging theory', balancing benefits against costs. The theory assumes that within the constraints of biologically possible feeding behaviours, one behaviour will maximize the difference between benefits and costs. 'There is no free lunch, yet some types of foods are better bargains than others' (Taghon 1989, p. 224). Trace fossils represent this trophic precision, if only we can extract the information.

10 Trace fossil assemblages, diversity and facies

Palaeoichnologists possess a rich vocabulary of terms with which to describe occurrence and distribution of trace fossils. These terms are mainly borrowed from body fossil palaeoecology, but as bodies and traces are not biologically equivalent concepts, problems emerge with the equivalent terminologies. In fact, these terms have been used in many different ways and there is some disagreement over their definitions. In particular, there is difficulty over scale, or hierarchical ranking, of the terms.

10.1 Terminology of trace fossil associations

10.1.1 Trace fossil assemblage

This is the basic collective term, embracing all the trace fossils occurring within a single unit of rock. The term is equivalent to the assemblage of body fossils (e.g. Kidwell and Bosence 1991), and is non-committal as to the origin of the collection of trace fossils. Thus, the trace fossils of an assemblage may have been emplaced simultaneously as a single ecologically-related group, or they may represent several overprinted events of bioturbation. As physical transport of trace fossils is uncommon, the trace fossil equivalent of the taphocoenosis is correspondingly rare (but see Belt *et al.* 1983; Baird and Brett 1986; Donselaar 1989; Fürsich *et al.* 1992; Hunt *et al.* 1994). Goldring (1991) called reworked trace fossils **ichnoclasts**.

10.1.2 Ichnocoenosis

This term has given many problems. Dörjes and Hertweck (1975) clearly identified **ichnocoenosis** as covering 'an association of [traces] that can be related to one definite biocoenose'. These authors thus confined the term to the neoichnological realm.

Many geologists have used the word to define fossil equivalents, i.e. assemblages of trace fossils that appear to be the work of a single community (Bromley and Asgaard 1979; Bottjer *et al.* 1986). Strictly speaking, the term **palaeoichnocoenosis** would be more appropriate here, to pair with palaeobiocoenosis, and its equivalent, life assemblage. As in the body fossil life assemblage, identification of a palaeoichnocoenosis is difficult. Despite the rarity of physical redeposition, there are other factors that readily cause trace fossil assemblages to be ecologically impure groupings.

Perhaps the most threatening factor is the rapid succession of communities that can develop on an aquatic floor. Time-averaging by relatively slow depositional rates ensures that the structures produced by each successive community are superimposed in the same rock unit (Fig. 10.1). Cross-cutting relationships might demonstrate which trace fossil was emplaced last and which first. But this may be confused with tiering within a single community, where deeper structures come to transect shallower ones as the floor accretes. This is quite distinct from a temporal evolution of successive communities in the same substratum (Fig. 10.2; section 5.3.4).

The opposite problem is equally valid. Ecological tiering may spread the work of different members of an endobenthic community over several units more than a metre apart (Figs 10.3 and 11.4), so that structures belonging to different communities come to lie together in the same sediment. This has been called **'palimpsesting'** (e.g. Rollins *et al.* 1990).

There is another understanding of the term ichnocoenosis, the difference being chiefly one of scale. Dörjes and Hertweck (1975) suggested the fossil equivalent of ichnocoenosis to be ichnofacies, and this thought has led Frey and Pemberton (1987) to treat ichnocoenosis as the unfossilized equivalent of the recurring, global, Seilacherian ichnofacies (section 10.1.4).

This is a quite different definition from the usual one and ultimately derives from the unsatisfactory status of the term ichnofacies. A term may be necessary for the collective 'communities' of traces that may give rise by fossilization to a given Seilacherian ichnofacies; however, it should not be confused with the work of the individual endobenthic communities that patchily and temporarily colonize the sea floor. I follow Ekdale *et al.* (1984a), and define ichnocoenosis on the small scale.

10.1.3 Suite

In many cases, an assemblage of trace fossils can be seen to comprise different subunits that appear to represent the work of separate endobenthic communities. These ecologically coherent subunits were called **suites** by Bromley (1975), who studied such compound assemblages at non-depositional omission surfaces. In such cases, successive benthic environments occur where a change in depositional regime involves a hiatus. Each of these environments may be represented by a distinct trace fossil 'community' that is emplaced within the same sediment, subjacent to the omission surface (Fig. 10.4). The compound assemblage may thus be subdivided into a **pre-omission suite** of softground structures; an **omission suite** of trace fossils emplaced during the period of non-deposition, often

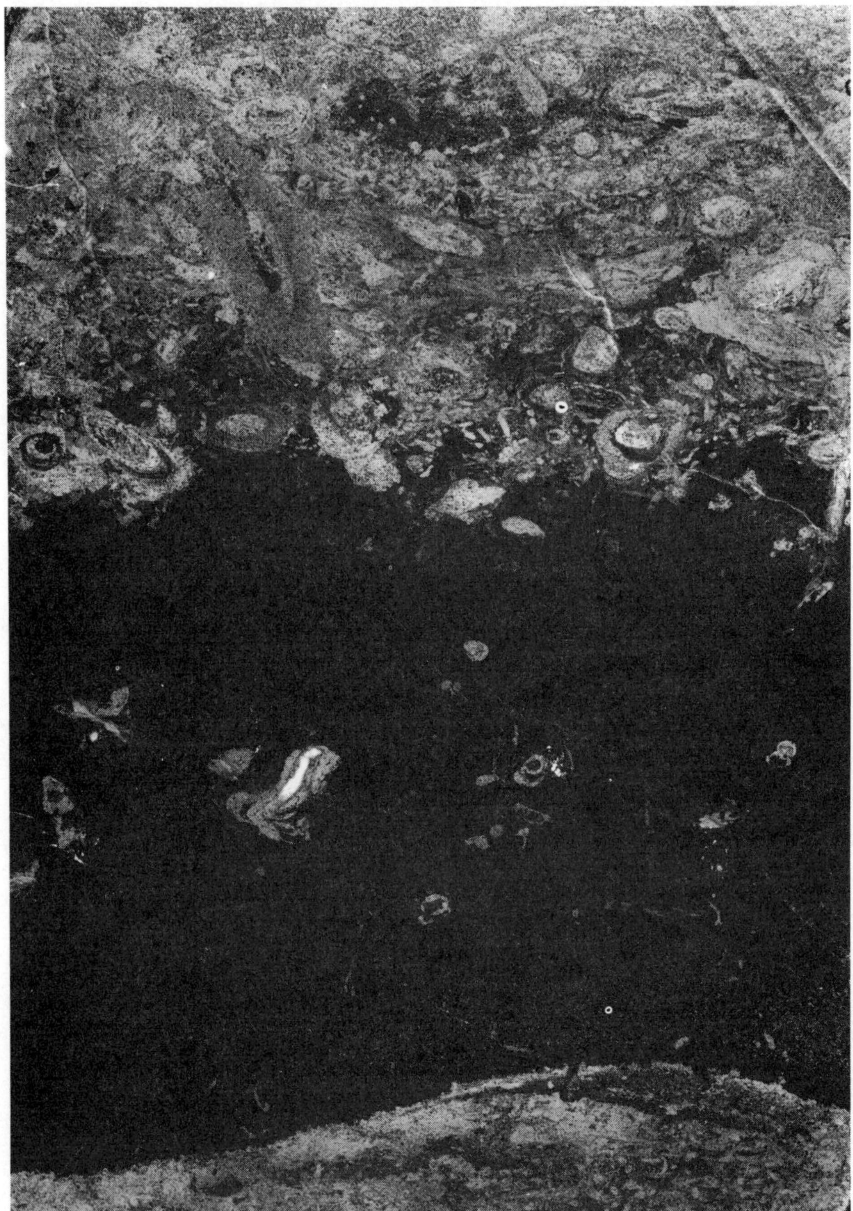

Figure 10.1 A complex ichnofabric from a series of argillaceous diatomite and volcanic ash beds, Harwich Member, London Clay Formation (Palaeocene), Essex, England, x2. Tiering profile and community succession are not clearly defined. The top part of the ash bed is mixed with the overlying diatomite and the original thickness of the tephrite is uncertain. The small trace fossils within the ash may represent not the earliest, pioneer community that recolonized the sea floor, but rather the deeper tiers penetrating from later successional stages (Fig. 10.2c). The thickly-walled and laminated-walled *Palaeophycus* ispp. that dominate above are intersected by the smallest burrow fills, which represent deeper tiers.

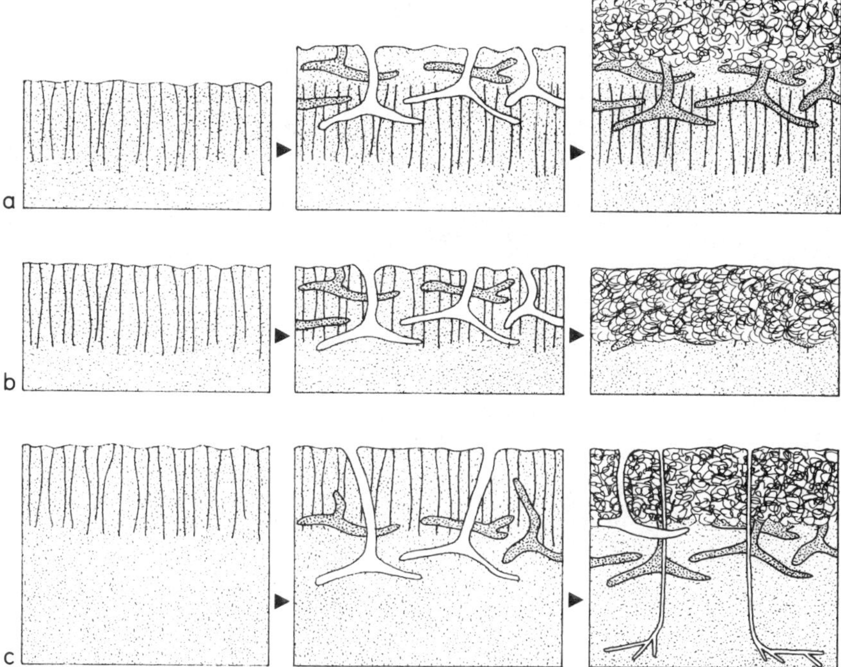

Figure 10.2 Community succession and ichnofabric. (a) With gradual accretion of the sea floor, the work of successive communities may be preserved one above another. (b) More usually, rapid succession and slower deposition lead to obliteration of earlier structures by later ones. (c) As tiering develops in later successional stages, a profile may develop that is superficially similar to that of (a), but is radically different in origin.

in a firmground; and a **post-omission suite** extending down from the later sediment that ultimately buried the omission surface.

Sea-floor cementation during the hiatus, producing a hardground, introduces a further omission suite into the palimpsest assemblage: a firmground **pre-lithification suite** then is followed by a **post-lithification suite** of borings.

Another example of suites was given by Bromley and Asgaard (1979). In fluviatile Triassic environments a community of arthropods produced a variety of trace fossils in a subaqueous softground. However, periods of drying out produced desiccation cracks before the substrate was resubmerged. In the interval during which the exposed sediment was a damp firmground, the sediment was invaded by other arthropods, perhaps insects, which introduced a series of structures that cross-cut those of the aquatic suite (Fig. 10.5).

A suite does not necessarily represent an individual ichnocoenosis, but

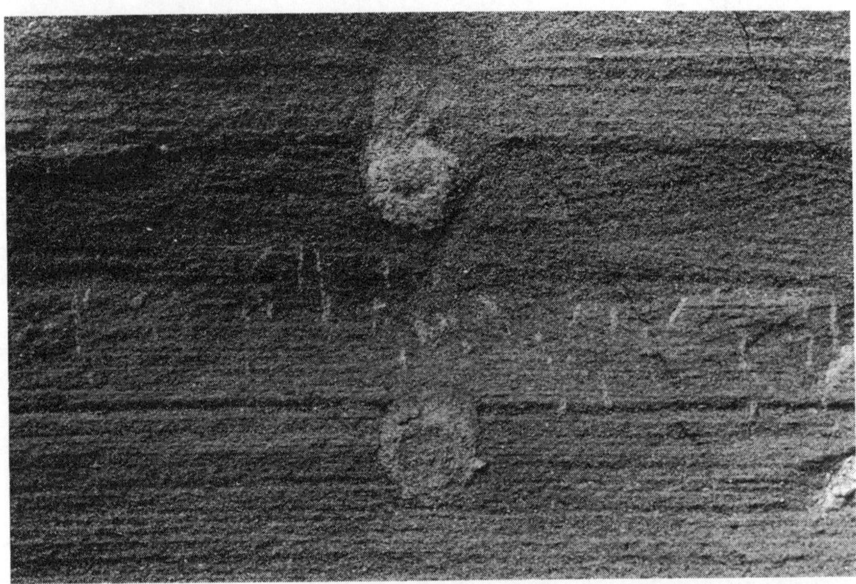

Figure 10.3 Heterogeneous trace fossil assemblage comprising small *Skolithos* isp. and large *Ophiomorpha nodosa*, representing different suites connected with different colonization surfaces. The *O. nodosa* descend from a bedding plane 1 m higher than the *Skolithos* isp. Pleistocene, Ladiko, Rhodes, half natural size.

is a wider grouping in most cases, owing to time-averaging. However, the concept is useful as it enables assemblages to be subdivided in a way that aids environmental and ecological analysis.

10.1.4 Ichnofacies

This term has been used at three scales by geologists having different requirements. Seilacher (1964, 1967a) introduced the term to cover recurring associations of trace fossils through the Phanerozoic on a global scale. In Seilacher's synthesis, the associations were related to both sedimentary facies and depositional environment (section 10.7).

Like ichnocoenosis, however, the term is also used at the opposite end of the scale, for individual rock units (e.g. Hayward 1976; D'Alessandro *et al.* 1986). In these cases the term is useful in defining locally-recurring rock facies on the basis of their contained trace fossil assemblages. Such ichnofacies are normally named after the dominant ichnotaxon.

Both of these usages are valuable and a new term should be coined for one of them. Herein I refer to the large-scale ichnofacies as 'Seilacherian'. Frey and Pemberton (1987) used the epithet 'archetypical'.

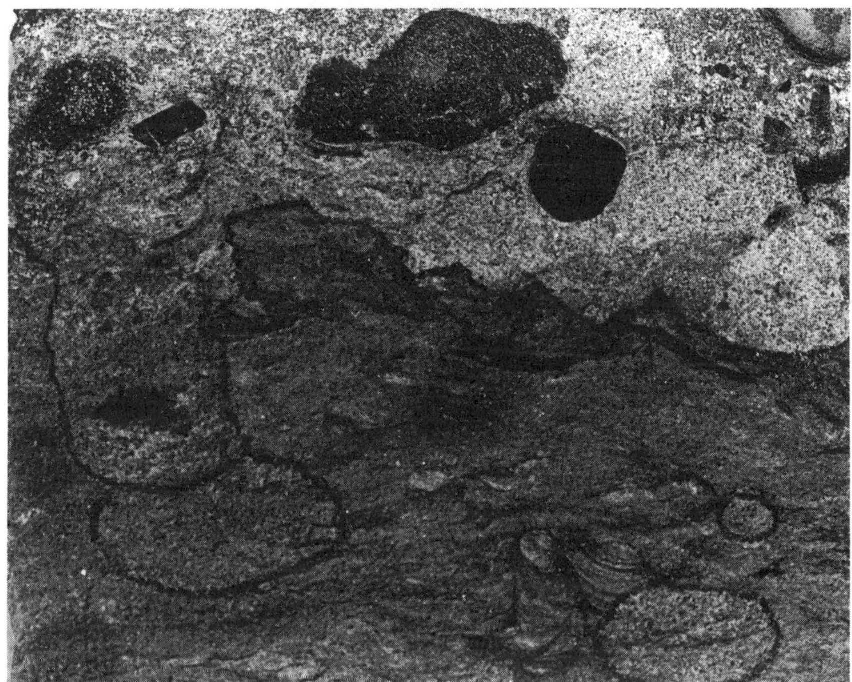

Figure 10.4 Bounding surface between two sequences in the North Sea Jurassic, offshore Norway, seen in core. The pre-omission suite is dominated by *Teichichnus rectus* in fine-grained sediments of the inner shelf. This fabric is truncated by an erosion surface and is overlain by a conglomeratic sandstone of the shallow subtidal, having a *Thalassinoides* ichnofabric. Post-omission *Thalassinoides* isp. are cut into the subjacent *Teichichnus* ichnofabric. The bounding surface itself (enhanced here by a pencil line) has a cusped, thalassinoid sculpture, and is very sharp, suggesting a firmground consistency. Natural size.

A medium scale between these two extremes also has been used. Lockley *et al.* (1987) defined a *Curvolithus* ichnofacies for a recurring association of trace fossils, worldwide and throughout the Phanerozoic. It was emphasized, nevertheless, that this association was a subset of Seilacher's *Cruziana* ichnofacies.

10.2 Diversity and ichnodiversity

As emphasized in Chapter 6, trace fossils relate to actions and are conceptually different from organisms. Thus, just as ichnotaxonomy is held distinct from biological taxonomy (section 8.2.2), then we must not confuse trace fossil diversity with species diversity. The number of trace fossil

Figure 10.5 A bedding plane from the fluviatile Triassic of Jameson Land, East Greenland. It is covered with groups of scratches (convex hyporelief) that are related to an aquatic suite of trace fossils including *Cruziana problematica* (Figs 8.13 and 9.4b). Cut through this fabric is the trace fossil *Spongeliomorpha carlsbergi*; its powerful transverse bioglyph and vuggy, incomplete fill (leading to compactional cracking and flattening) clearly show this to belong to a terrestrial firmground suite. *S. carlsbergi* is accompanied by other ornamented trace fossils and desiccation cracks, and may be the work of an insect (Bromley and Asgaard 1979), possibly a mole cricket (Metz 1990). x3.5.

taxa present in an assemblage depends on factors that are quite different from those that determine the diversity of a body fossil assemblage.

10.2.1 Fossilization potential

The fossilization barrier provides an effective taphonomic filter between the activity of the endobenthic community and the ichnofabric that documents that activity in the fossil record. Some sorts of behaviour leave more durable structures than others. The carefully constructed domichnial tube of a polychaete and the precisely backfilled repichnion of an excavating echinoid stand a far better chance of producing discrete trace fossils than the eddy diffusion caused by, say, an intruding predaceous snail. Furthermore, the deeper the structure is emplaced beneath the depositional interface, the less likelihood there is of destruction by physical reworking or biological disturbance (section 6.2).

These taphonomic factors are similar to those affecting body fossils, and the disparity in preservation potential likewise distorts the ecological information available in the ichnofabric. However, there are other factors that are solely the problem of trace fossils.

10.2.2 Cross-cutting tiers

Mutual destruction of biogenic structures provides a further twist to the quality of the preserved assemblage. As Seilacher (e.g. 1974) has often pointed out, surface trails and other exogenic structures have a poor potential for survival owing to the likelihood of their obliteration by burrowing activity. This process is equally valid in the case of the individual tiers of an endobenthic community (section 5.5.3). More deep-seated structures will cut through and tend to obliterate shallower structures (Fig. 10.6).

In the case of incomplete bioturbation of a sediment, complete obliteration of shallow structures by deeper ones cannot occur. Even here, however, the deeper structures, being more completely preserved, are likely to be the eye-catching elements of the ichnofabric. On the other hand, where a mature endobenthic community has become established (section 10.5.2), complete bioturbation is normal and the quality of the preserved ichnofabric will then depend on the activity levels within each tier (Fig. 10.7).

Tiering taphonomy, then, decreases the preservation potential of shallowly-emplaced structures, and thereby emphasizes deeper structures. This emphasis is the opposite of the biomass distribution and activity patterns of the original endobenthic community, in which 'everything happens at the top', within the mixed layer (section 5.4.4). Thus, it is clear that ichnodiversity and biological diversity are very distinct concepts and their trends are virtually unrelated. We must take the ichno- prefix seriously!

10.3 Tiering and ichnofabric

Bromley and Ekdale (1986) showed the effects that tiering has on the taphonomy of trace fossils and by which the deepest structures come to be the most conspicuous (Fig. 6.6).

1. Deepest structures are the most complete. Only their upper parts are cross-cut by shallower structures. Activity within each tier cross-cuts structures produced in shallower tiers and tends to obliterate them.
2. The substrate of the deeper tiers contains less water and is firmer than that of the higher ones. Thus, the deeper structures tend to have clearly defined, sharp boundaries and are taphonomically well suited for fossilization.
3. The uppermost structures suffer more compactional deformation than the deeper ones, which are emplaced in already somewhat compacted sediment.
4. The deeper structures produced by deposit feeders are commonly filled with material that contrasts in colour and texture with the surround-

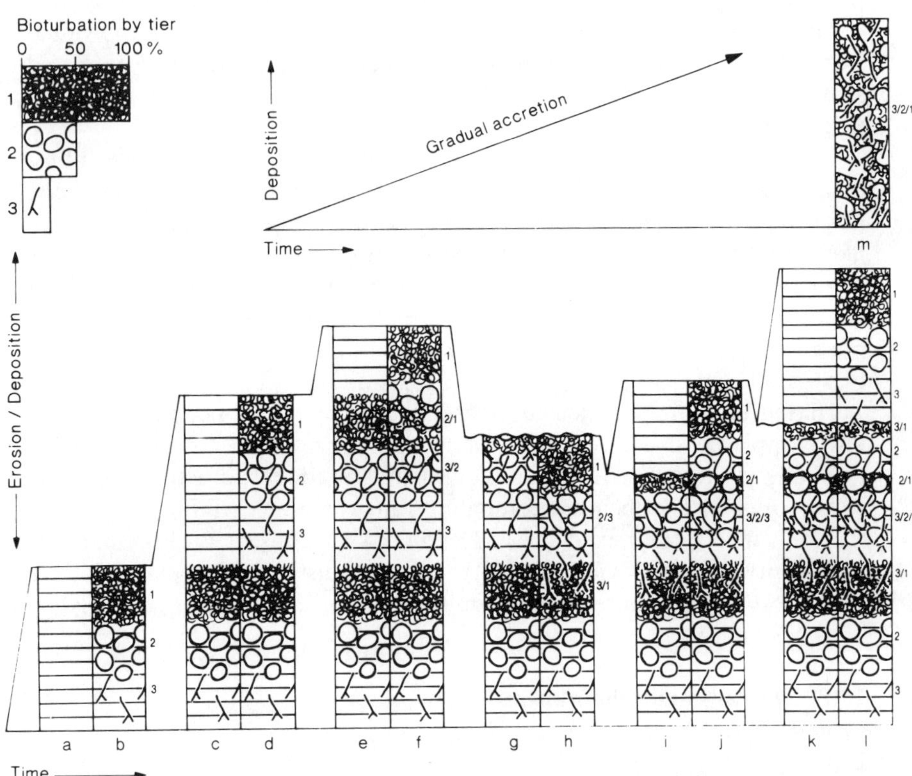

Figure 10.6 Effect of tiered bioturbation in different depositional regimes, based on the system shown top left. Below, event deposition produces many different effects. (a) Rapid deposition followed by non-deposition (omission) allows colonization (b). (c) Rapid burial with a package thicker than the bioturbation zone preserved the first bioturbated bed as a frozen tiering profile. (d) New colonization. (e) Deposition of a thinner package leads to overprinting (f) and produces a palimpsest fabric. (g) Erosion followed by non-deposition followed by colonization (h) of the erosion surface. (i) Slight erosion followed by deposition. The erosion surface is not a colonization surface and is largely obliterated by later bioturbation (j). Erosion followed by deposition of a thicker package of sediment (k, l). Tier overprinting indicated by numbers. Above, (m) the result of gradual accretion is the development of an ichnofabric having characteristic cross-cutting relationships.

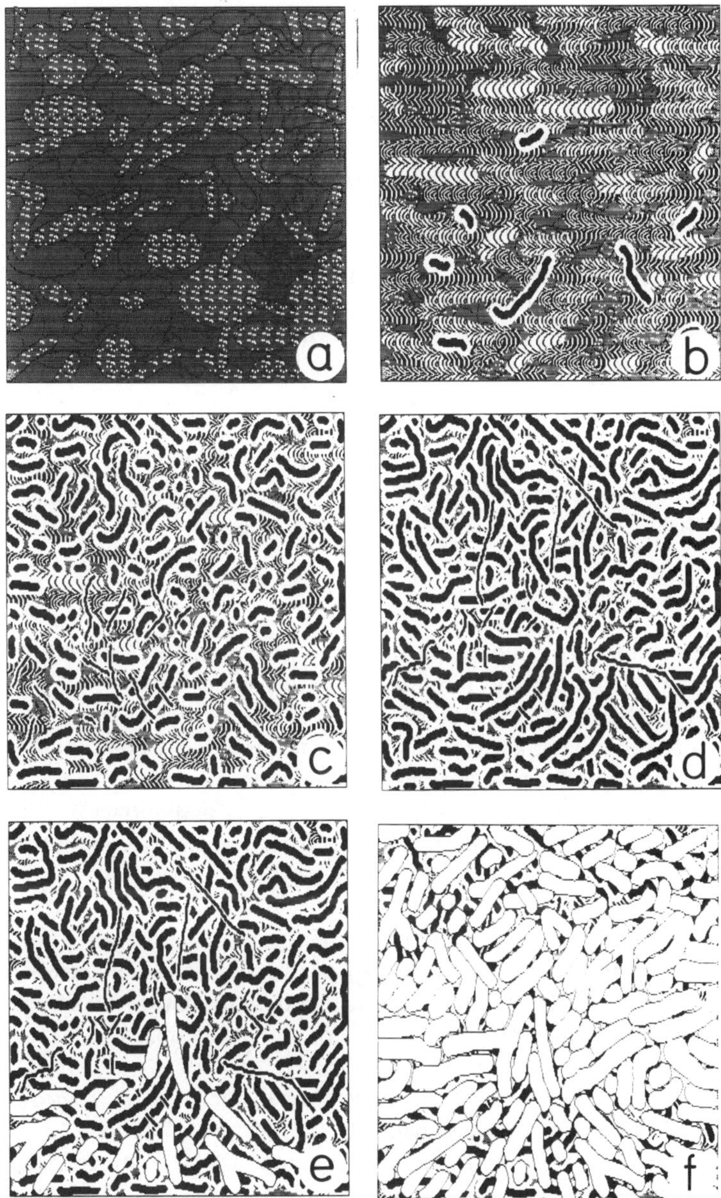

Figure 10.7 Computer model showing successively increasing activity downwards, tier by tier, and the resulting effect on the ichnofabric. A steady-state situation is envisaged, involving even and gradual sea floor accretion. (a) Complete bioturbation by the *Thalassinoides* tier. (b) Appearance of a deeper tier of *Taenidium* activity has largely obliterated the *Thalassinoides* fabric. In turn, the *Taenidium* fabric becomes partially (c) or almost completely masked (d) by increasing activity in the *Phycosiphon* tier. Small *Phycosiphon* cuts deeper than large. Finally, (e) a deep tier of *Chondrites* is superimposed on the fabric. (f) Total bioturbation within this tier in places reduces ichnodiversity to 1.

ing sediment. This may be because the relatively low nutrient value of the deeply buried sediment demands rigorous particle selection by the deposit feeder. Mixed with faecal material, the backfill or spreite has a special constitution.

5. Reverse conveyor feeders in the deep tier introduce material from high levels into the deeper sediment, which contrasts in colour and texture with the matrix.
6. The contained faecal sediment, surface material and metabolic products in the special fill of the deep structures are chemically dynamic, being emplaced within reducing, organic-poor sediment. This creates a special diagenetic microenvironment that may initiate mineralization and further enhancement of the visibility of the structure.

There are thus many reasons why the deeper structures dominate the ichnofabric. In this respect they resemble elite structures (section 5.1.2) on the other side of the fossilization barrier and many become **elite trace fossils** (section 6.4).

10.3.1 Modelling ichnofabric

Depending upon the degree of activity within each tier of a community of sediment-manipulating animals, the resulting ichnofabric may take on a wide spectrum of textures. Figures 10.7 and 10.8 illustrate this by means of computer graphics, based on the same basic community. If the community works from a stationary floor, there being no deposition until the substrate is suddenly buried, a fabric such as that shown in Fig. 10.6b might be preserved. This is known as a 'frozen tiered profile' (Savrda and Bottjer 1986).

In Figs 10.6m, 10.7 and 10.8, however, a steady accretion of the floor is envisaged. Depositional rate is slower than the rate of biogenic reworking. During the gradual upward movement of the successive tiers, deeper structures cross-cut shallower ones. Consequently, the nature of the resulting ichnofabric varies according to the rate of activity in different tiers.

In most endobenthic communities today, the rate of activity tends to fall off with distance beneath the depositional interface. This accounts for the mixed layer immediately beneath the surface, overlying the transitional layer containing few burrows (section 5.4.4). However, the fossil record shows that, not uncommonly, a key bioturbator in a deep tier can cause complete bioturbation and thus leave its indelible impression on the fabric as an elite trace fossil. Such a key bioturbator is *Echinocardium cordatum* (Reineck *et al.* 1967; Reineck and Singh 1971; Bromley and Asgaard 1975; Bromley and Ekdale 1986; Bromley *et al.* 1995).

The presence of such a key bioturbator in a deep tier greatly reduces the ichnodiversity of the assemblage. If the deepest tier is highly active, then ichnodiversity may be reduced to one (Fig. 10.9).

The unwary investigator, rashly comparing ichnodiversity with biological diversity, might seek a stress factor in the environment to account for such an 'impoverished ichnofauna'. Indeed, very minor fluctuations in the environment may cause changes in the activity of the deepest tiers, but these changes will be registered as profound differences in the assemblage. The appearance and disappearance, bed-by-bed, of mid-tier ichnotaxa alternating with an elite deep-tier ichnotaxon may have a taphonomic explanation. The mid-tier tracemaker may have been present all the time, but its work was obliterated in certain periods by temporary appearance of a deeper key bioturbator. In this case, it is the taphonomic environment caused by the deepest burrower's activity that is the controlling factor and external environmental controls need not be invoked to account for these changes.

Even the deepest burrower has to maintain contact with oxygenated water (Fig. 5.14). Excessive interference from near-surface tiers, or a key bioturbator in shallow or middle tiers, might temporarily render the deepest niche unavailable. On the other hand, substrate conditioning might be involved, brought about by the activity of other species in the community (section 5.2). This would be a case of 'behavioural amensalism', having its roots in the ecology of community succession, not in the physical environment and community replacement (section 5.3.3).

10.3.2 Tiering and oxygen

One of the factors controlling the spatial distribution of animals within the substrate is the availability of oxygen. All metazoans are dependent on a supply of oxygen. Yet animals can penetrate far below the RPD as long as they have access to oxygenated water. Some animals pump oxygenated water through open burrows well down into the anoxic zone. The oxygenated halo around such burrows creates a niche for other endobenthos (sections 5.1.2 and 5.4.3). Some animals reside within the anoxic sediment, but periodically move into oxygenated water to respire (section 3.5.2).

Those animals that come into contact with anoxic sediment develop physiological adaptations to cope with that environment, as R. Thompson and Pritchard (1969) showed in the case of two callianassids having contrasting lifestyles (sections 4.3.2 and 4.3.4), and Pals and Pauptit (1979) for *Heteromastus filiformis* (section 3.5.2).

The floors of densely-stratified basins commonly range from fully-oxygenated regions in shallow water into anoxic environments in deeper parts, via a dysaerobic middle ground of low oxygen pressures. The oxygen minimum zones of oceans show similar features. In such oxygenation profiles, several trends are seen in the benthic faunas (Rhoads and Morse 1971; Byers 1977; Savrda *et al.* 1984). As dissolved oxygen

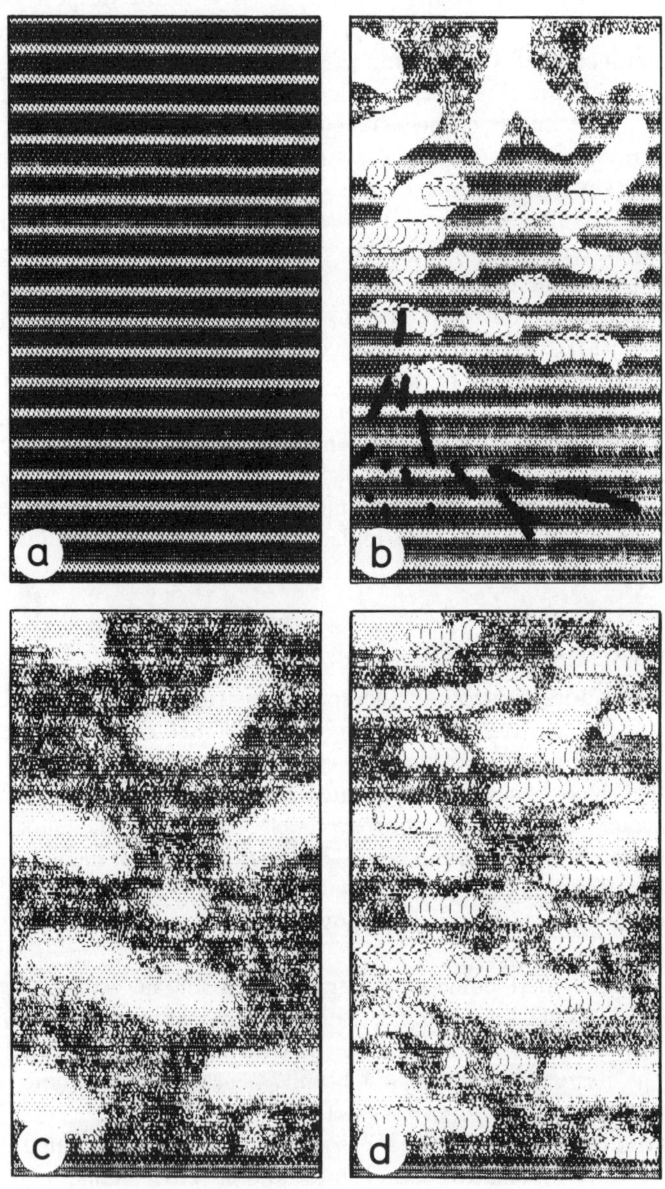
a
b
c
d

Figure 10.8 Computer model of ichnofabric variation involving smaller degrees of bioturbation than in Fig. 10.7. (a) Primary lamination, zero bioturbation. (b) During non-deposition, a frozen tiering profile may develop; ichnodiversity 3, as well as a meiofaunal mixed zone at the top. (c–f) Gradual deposition. (c) Activity in the *Thalassinoides* tier; outlines are unsharp as structures were emplaced shallowly in a soupground. (d) The deeper *Taenidium* structures cross-cut the shallower, and are sharper, the substrate having a firmer consistency at that level. (e) Introduction of freely ranging *Chondrites* isp. (f) Commonly the deepest tier activity is restricted to *Thalassinoides* fills.

decreases, the benthos becomes less diverse, less abundant, smaller in size, less heavily calcified and dominated by endobenthos.

In beds bioturbated during dysoxic intervals in generally anoxic basins, it is common that the ichnogenera *Chondrites* and *Zoophycos* are reported (Sandberg and Gutschick 1984; Bromley and Ekdale 1984a; Baird and Brett 1986; Savrda and Bottjer 1987a). The originating burrow in these structures has a small to minute diameter.

Similar trends are recognizable in the vertical profile of the endobenthic community in a well-oxygenated floor. Greatest activity normally occurs in the uppermost tiers, where species vary widely in size from minute to large. Lower tiers commonly exhibit less activity (low bioturbation density), lower ichnodiversity and a progressive downward reduction in the maximum diameters of burrows. Thus, somewhat as predicted by Rhoads (1975), the deepest-burrowing animals generally are smallest, penetrating furthest downwards into the anoxic zones of the substrate. Significantly, these deepest tiers are commonly represented by ichnospecies of *Chondrites* and *Zoophycos* (Fig. 6.6; Reineck and Singh 1971, pl. 13; Ekdale 1977; Swinbanks and Shirayama 1984).

Figure 10.9 Biogenic lamination, an ichnofabric produced by 100% *Teichichnus rectus* bioturbation. Any work there may have been in shallower tiers will have been obliterated, reducing ichnodiversity to 1. Ichnofabric index 5. Vardekløft Formation, Jurassic, Jameson Land, East Greenland. About natural size.

With gradual lowering of the oxygen content of the bottom waters, the RPD rises toward the depositional interface and a corresponding shift in endobenthic tiering also will be expected (Bromley and Ekdale 1984a). The species occupying the oxic uppermost tiers, which are dependent on abundant oxygen, will be excluded and their place taken by the less demanding animals of the mid-level tiers. Ultimately, after further lowering of oxygen pressure, the inhabitants of the deepest tiers will find a niche immediately below the surface, before anoxic conditions are reached and they too are eliminated (Fig. 10.10).

The bioturbation densities characteristic of the respective tiers may be expected to be maintained, so bioturbation will normally be incomplete where upper tiers are missing. It follows that if *Chondrites* is found alone, cutting primary fabric and representing a low per cent of bioturbation, then dysoxic bottom waters are likely. If, on the other hand, *Chondrites*, although alone, comprises total bioturbation (Fig. 10.11), then interpretation should be more cautious. It may be hiding something!

Savrda (1986), Savrda and Bottjer (1986) and Bottjer *et al.* (1986) proposed and applied a model for the reconstruction of palaeo-oxygen levels of bottom waters on the basis of progressively impoverished ichnocoenoses. These authors presented a series of ichnofabrics that represented

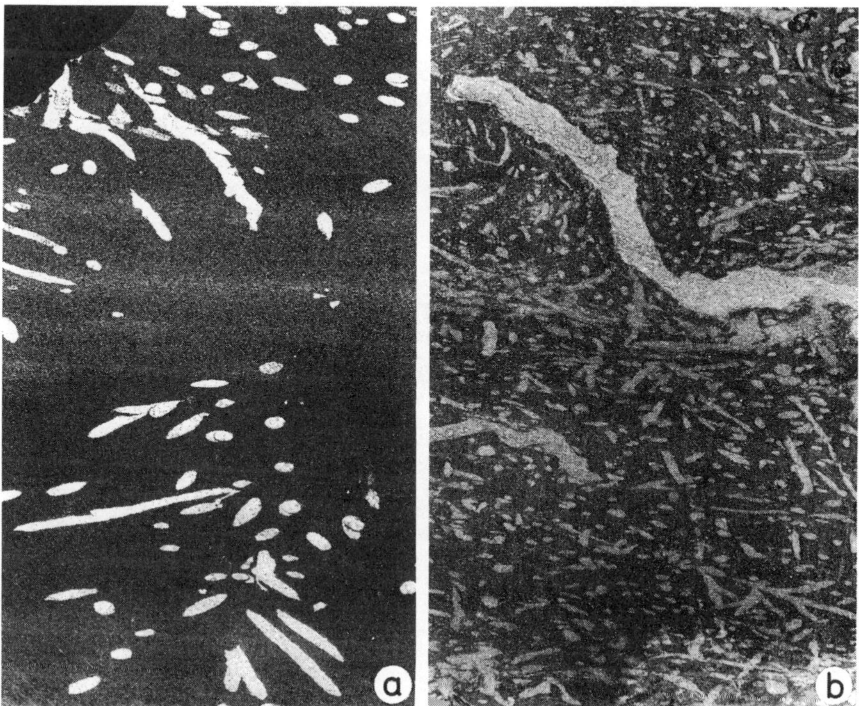

Figure 10.10 *Chondrites* in oxygen-poor environments. (a) In a laminated channel fill, large *Chondrites* isp. alone has disturbed the sediment. Ichnofabric index 2. Cretaceous Austin Chalk, near Austin, Texas. (b) Small, irregular *Zoophycos* isp. accompany small *Chondrites* isp. in vaguely laminated marly chalk. Ichnofabric index 3. Plenus Marls, Cretaceous, Hannover, Germany (specimen courtesy of Heinz Hilbrecht). Both x2.

different degrees of bottom-water oxygenation (Fig. 10.12). However, application of this model requires careful distinction between historical (Fig. 10.12) and steady-state models (Fig. 10.7). Does the profile show an upward trend during steady sedimentation, the different trace fossils representing community replacement; or does it reveal frozen tiered profiles beneath periodic colonization surfaces, the trace fossils belonging to a single tiered community? In other words, are there four successive colonization surfaces in Fig. 10.12, or a single one at the top? A study of cross-cutting relationships will resolve this problem. Other applications of palaeoxygen interpretations are discussed in section 12.1.1.

10.4 Quantity of bioturbation

The overall density of degree of bioturbation is a sediment property that has received much attention. One hundred per cent bioturbation of the

Figure 10.11 Completely bioturbated fabric dominated by *Chondrites* isp. Bravaisberget Formation (Triassic), Svalbard, x2.

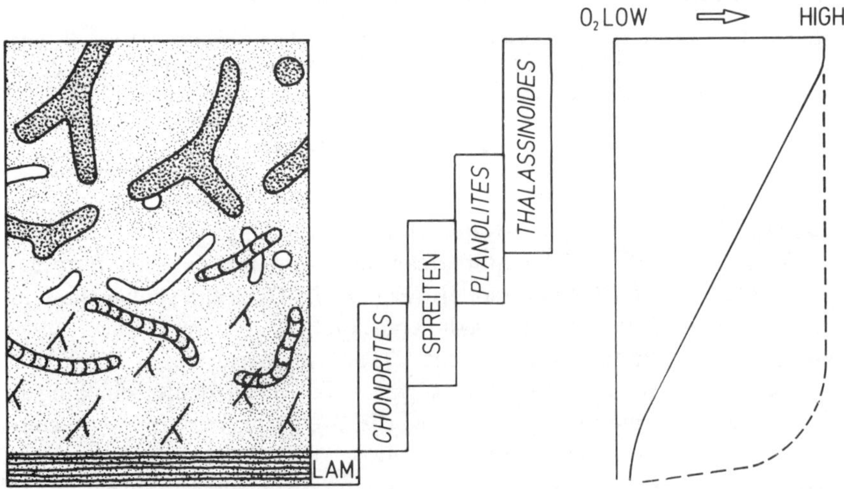

Figure 10.12 Rhythmic unit from the Cretaceous Niobrara Chalk, Kansas, and an interpretation of the fluctuating oxygen sequence revealed by the trace fossils (solid curve). Data from Bottjer *et al.* (1986). If, however, the unit were interpreted as a frozen tiering profile (Fig. 10.2c), a different oxygen sequence would be indicated (broken line); in fact, possibly no oxygen fluctuation is indicated.

substrate is the natural end-product of the activity of the endobenthos. Failure to reach 100 per cent, or failure of that state to be preserved in the rock record, are conditions that require explanation. But, first, the degree of bioturbation has to be judged and evaluated.

10.4.1 Judging the degree of bioturbation

The detailed studies of recent shallow-marine and estuarine sediments by use of box core and vibracore, by Howard and Reineck (1972, 1981), Reineck (1976), Howard and Frey (1975), Frey and Howard (1986) and

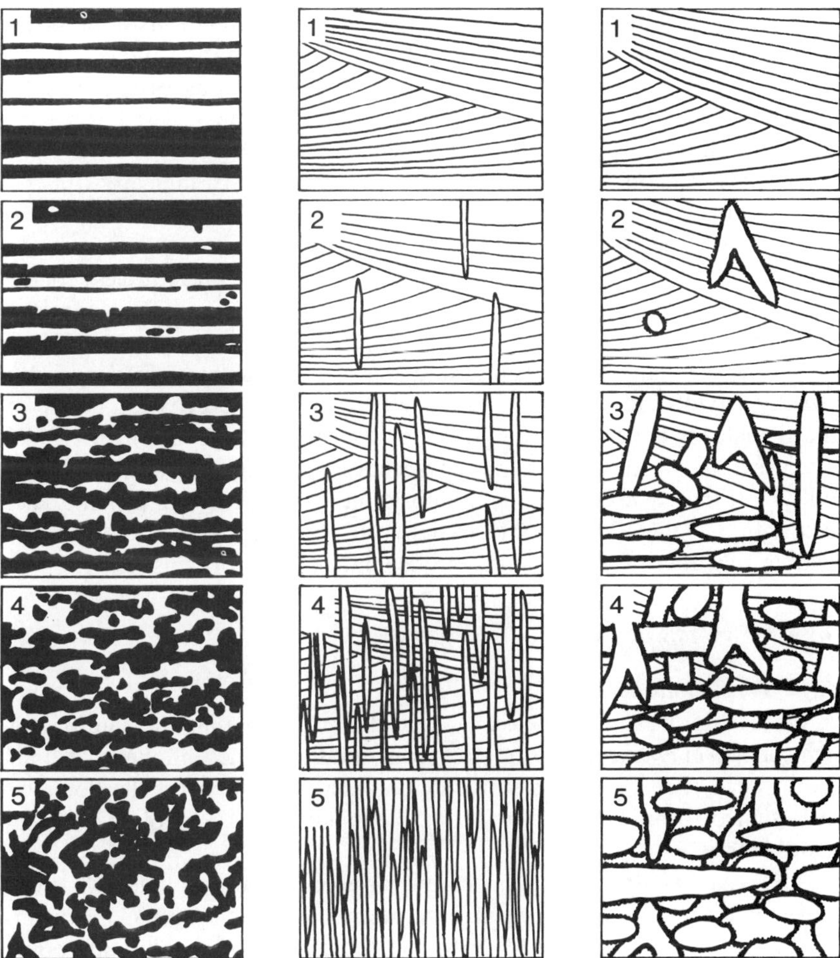

Figure 10.13 Bioturbation density and the recognition of ichnofabric indices. Left: schematic diagrams of ii 1–5 for sediments deposited in shelf environments. Centre: ii for high energy nearshore sands dominated by *Skolithos*. Right: ii for nearshore sands dominated by *Ophiomorpha*. Modified after Droser and Bottjer (1986, 1989).

Hongguang *et al.* (in press) have carefully defined the percentage of bioturbation at each site. These workers succeeded in distinguishing by eye several grades: 0, trace, 30, 30–60, 60–90, 90–99 and 100 per cent. Howard and Reineck (1972) illustrated examples of these degrees of bioturbation using X-radiographs.

Estimation of the degree of bioturbation is not easy. Problems include variable visibility of structures; range of quality of the fabric, i.e. different types of trace fossils; and different scales of disturbance, from cryptobioturbation by meiobenthos to vertebrate diggings. Nevertheless, as a general guide, the work of these authors bears out the great success of their scheme.

Droser and Bottjer (1986, 1987) tried to ease evaluation of bioturbation by producing 'flash cards' that depict five ichnofabric indices (**ii**) for use in the field or core lab (Fig. 10.13). However, this system has the same shortcomings as the first. Clean, equivalently-weathered rock surfaces or comparable quality of core must be available.

In some versions of the ii scheme, an index of six indicates 100 per cent bioturbation and is represented by structureless fabric. However, although homogenization may be the result of mixing in the uppermost tiers (section 5.2.2), deeper tiers tend to heterogenize. Structureless rock is not the end-product of all bioturbation processes (Figs 6.6, 8.12, 10.1, 10.11, 10.17b, 10.21, 11.1, 11.3, 11.4, etc.).

Another system was introduced under the name 'bioturbation index' (**BI**) by Taylor and Goldring (1993). Whereas zero bioturbation is one on the ii scale, it is zero on the BI scale. These scales are useful for quick overviews of variously bioturbated successions (Figs 10.9 and 10.10). In fully-bioturbated composite ichnofabrics it is tempting to provide each tier with a value (Fig. 12.11; Ekdale and Bromley 1991), though it is seldom that there are sufficiently precise data to support this.

10.4.2 Evaluating variation in quantity

(a) Total bioturbation

A totally bioturbated rock clearly provides evidence that the rate of biogenic reworking exceeded that of sedimentation. The endobenthic community had time to mix its substrate completely. Moreover, the floor can be assumed to have offered a favourable environment for colonization by endobenthos.

These statements contain too many variable factors to offer any precise measurements of environmental parameters. Rate of biogenic reworking of the sediment varies according to: (1) density of the endobenthic population; (2) temperature of the water, which may vary seasonally; (3) presence or absence of key bioturbator species; (4) type of activity; and (5) the tier. Moreover, Thayer (1979), among others, suggested that the general rate of bioturbation has greatly increased during the Phanerozoic.

Thayer (1983) provided an exhaustive compilation of rates of reworking by different endobenthic species. Such rates nevertheless should be treated with caution when extrapolating to palaeoichnocoenoses. A dense population of *Donax variabilis* (section 6.1.2) is capable of totally reworking the upper tier of its substrate within hours and the meiofauna in the tropics can homogenize sediment within days. However, equivalent denizens of upper tiers of bathyal environments work far more slowly.

Warme (1967) estimated that a population of *Callianassa californiensis* producing incipient *Thalassinoides suevicus* can rework the topmost 75 cm of the intertidal zone in under a year. But it would not be wise to apply this figure uncritically to, say, the work of animals producing *T. suevicus* in the middle tier of a Cretaceous shelf chalk community.

Some trace fossils themselves indicate such intrinsic differences. For example, the diameter of the marginal tube of *Zoophycos* increases in size distalward, reflecting growth of the creating animal. One *Zoophycos*, therefore, would relate to a lifetime, perhaps years, of work. *Chondrites*, on the contrary, has a constant diameter within each system and would seem, therefore, to have been emplaced more rapidly. Thus, total *Zoophycos* bioturbation cannot be equated in terms of rate with total *Chondrites* bioturbation.

It is not even clear how much sediment disturbance an individual animal may cause, by merely identifying its lifestyle. A suspension-feeding polychaete may reside, immobile, in a single domichnial shaft in an aquarium for months. In the busy sea floor, however, such worms suffer interference from other animals and may shift their position frequently (sections 3.2.2, 5.2.1 and 6.3).

Total bioturbation, therefore, is more than just a simple statement of the state of a rock fabric. Different types of trace fossil, belonging to different tiers or different lifestyles, may account for the total mixing, each having a different significance (Fig. 10.7).

(b) Incomplete bioturbation

This implies a stress factor that has prevented the completion of the work of the endobenthic community (section 10.5.1). Middlemiss (1962) and Shourd and Levin (1976) interpreted the varying density of bioturbation in a succession as reflecting variations in the rate of sedimentation. Savrda (1986) modelled the effect of varying depositional rate pictorially.

Clearly, however, other stress factors may inhibit the complete bioturbation of the substrate. In particular, fluctuation in salinity may severely reduce the biological diversity of a fauna and this will be reflected in its work. Another important limiting factor is the availability of oxygen. In order to distinguish between the factors negatively affecting the community, it is necessary to take the quality of the assemblage into account, and identify individual ichnotaxa. Wightman *et al.* (1987) and

Savrda and Bottjer (1986) investigated these characteristics of brackish and oxygen-starved ichnocoenoses, respectively.

(c) Zero bioturbation

The lack of bioturbation may be caused by either of two contingencies. It may result from an original lack of endobenthic activity, or it may be due to failure to preserve biogenic structures. In the former case, the perfect preservation of primary lamination in the absence of evidence for physical reworking may convince the observer that 'no trace fossils means no animals'. In the second case, if physical erosion cuts deeper than biogenic reworking, no bioturbation will be preserved. In shallow-marine set-

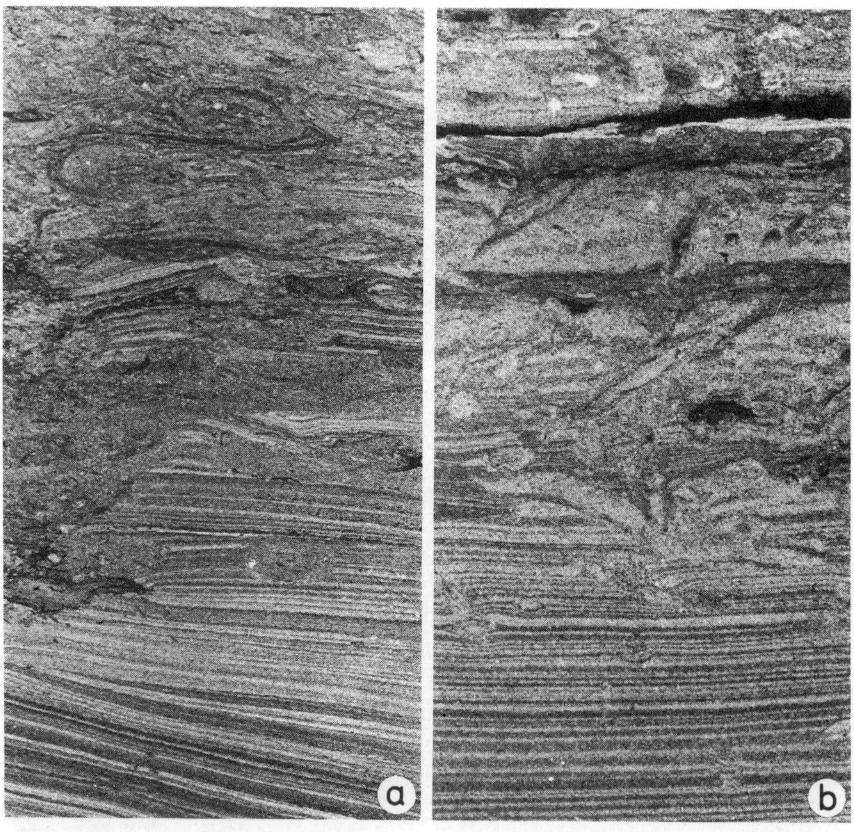

Figure 10.14 Two examples of laminated sand as seen in core. Tops of the units are bioturbated, producing a 'laminated-to-scrambled' (or 'lam–scram') unit. As in Fig. 10.1, it is not easy to identify a tiering profile in the ichnofabric owing to overlap by time-averaging of several successive communities (Fig. 10.2b). Three amalgamated units are probably represented in (b). Jurassic of the Norwegian offshore, North Sea, natural size.

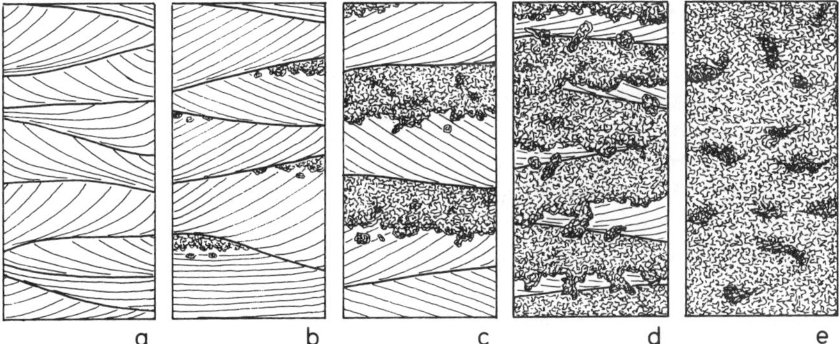

Figure 10.15 A series of models to illustrate the balance between amount of erosion and thickness of deposited units, assuming constant rate and depth of bioturbation. In (c) a well-balanced rhythm is shown. During pauses in deposition, tops of units are colonized and bioturbated. In (b) erosion cuts deeper and obliterates the colonization surfaces and much of the bioturbated levels (Fig. 11.14c). In (a) no bioturbation is preserved, but it would be unwise to assume that the depositional environment was devoid of life. In (d) the units are thinner and only small islands of laminated structure survive (top of Fig. 10.14a and b). Finally, in (e) the depositional history has been obliterated by bioturbation.

tings, sediments are commonly totally bioturbated in summer but are completely reworked by winter storms.

Where a sufficiently thick unit of sediment is deposited rapidly, such as a storm layer, its lower regions may be beyond the burrowing depth of the animals that colonize and rework the top of the bed (Fig. 10.14). Repeated storm events may remove the bioturbated top of the previous bed and obliterate the biological record, giving a spurious appearance of lifelessness. The thickness of each package, and the amount of scour, will determine the amount, if any, and style of bioturbation preserved (Fig. 10.15).

10.5 Opportunistic and equilibrium ecology

Miller and Johnson (1981) and Ekdale (1985) were first to apply the principle of **opportunistic** (r-selected) versus **equilibrium strategies** (K-selected) to trace fossils. **Opportunistic species** include those organisms that have high reproductive rates, rapid growth rates, broad environmental tolerances and generalized feeding habits (Pianka 1970). Such species are commonly the **pioneers** that speedily colonize a habitat after a major and abrupt environmental change (e.g. storm deposition) or are the animals that flourish in high-stress (e.g. beach) or low-resource (e.g. starved basin) environments. They are typified by high abundances of individuals in low diversity, relatively short-lived communities.

In contrast, **equilibrium species** are slow to enter new environments but, in the long run, are adaptively superior to the more rapid colonizers (Pianka 1970). Relative to opportunists, the equilibrium species generally show lower reproductive and growth rates, and their environmental tolerances are narrow. Most equilibrium species are rather specialized feeders that have adapted to occupy specialized niches. In the marine realm, they tend to be stenobathic, stenohaline and stenothermic. They are typically members of high-diversity, persistent, **climax communities**. Species diversities may be high but population densities are generally low and no single taxon dominates the assemblage.

It should be realized that opportunist and equilibrium strategies are end-members of a spectrum. No organism is completely opportunistic or completely equilibrium selected, but all must reach some compromise between the two extremes. Transferring this classification to the ichnological world involves few problems (Ekdale 1985). Nevertheless, owing to possible confusion with equilibrichnia, I refer to the work of equilibrium species of the climax communities as **climax trace fossils**.

10.5.1 Opportunist trace fossils

Ichnocoenoses of opportunistic ichnotaxa commonly have a low diversity and high density of trace fossils. Sedimentological data may indicate that the burrows were produced over a short period of time and also that the depositional environment was inhospitable to most life forms. This may be attributed to oxygen depletion, variable salinity, uneven rates of sediment accumulation or merely a newly-deposited, biologically-unconditioned substrate.

Cuomo and Rhoads (1987) described a capitellid polychaete whose larvae are attracted to, and actively metamorphose on, substrates rich in hydrogen sulphide: a specialist in hostile environments.

Vossler and Pemberton (1988b) gave an excellent review of opportunist ichnocoenoses in connection with storm-deposition sequences. These authors brought together from the literature many examples in which two contrasted ichnoassemblages had been documented in a single sequence: a fair-weather assemblage and a storm assemblage. The assemblages occurring in the tops of storm layers tend to have a very low diversity and a marked dominance by domichnia (Fig. 10.16), whereas the background assemblages represent climax communities of higher ichnodiversity.

These results compare favourably with recent examples of substrates where the endobenthos had been abruptly removed or killed off (Vossler and Pemberton 1988b). In most cases, the pioneer species are small worms that produce vertical domichnia equivalent to *Skolithos*. Many pioneer structures have intrinsically high preservation potential, involving well built tubes to effectively isolate the burrow from the hostile pore water of the sterile sediment (cf. Aller 1980; Rhoads and Boyer 1982).

The speed with which these opportunists colonize the empty substrate

Figure 10.16 Y-shaped *Polykladichnus* isp., I-shaped *Skolithos* isp. and U-shaped *Arenicolites* isp. Three suspension-feeder structures comprising a pioneer community of opportunists, seen here at two horizons in a shallow marine, rapidly deposited Pleistocene sand unit at Ladiko, Rhodes. Natural size.

is impressive. McCall (1977) experimentally sterilized an area of shallow subtidal sea floor and found that within ten days the sediment was densely populated by one species of polychaete, two others occurring in small numbers. None of these species was among those of the climax background community that had been removed from the area, and which later became re-established. Rhoads *et al.* (1978) obtained similar results following disturbance caused by dredging operations.

Ekdale (1985) pointed out that turbidite assemblages could be considered in the same light. Trace fossil assemblages of turbidites consist of two suites (Seilacher 1962b; Kern 1980). The pre-event suite has a high diversity, dominated by agrichnia (e.g. *Paleodictyon*, *Desmograpton*) and regularly meandering pascichnia (e.g. *Nereites*, *Scolicia*), and represents the equilibrium community of narrow trophic specialists inhabiting the muddy pelagic background environment (section 6.2.1). This ichnocoenosis is cast in semirelief on the soles of the turbidite beds, which represent sudden input of sand into the pelagic realm. This sandy substrate is colonized by a community or communities of opportunists that produce the post-event suite (*Granularia*, *Phycosiphon* and irregular *Scolicia*).

In a succession of turbidites containing very rare pre-depositional trace fossils and a low diversity of post-depositional opportunists comprising *Chondrites, Planolites* and *Helminthoida*, Uchman (1992b) identified low sediment oxygen levels as the limiting factor. However, Wignall (1990) pointed out that steady dysoxia (as in upwelling systems) will encourage K-strategists, whereas unstable, pulsating oxygen levels will favour r-strategists (opportunists).

The identification of assemblages representing opportunistic communities is fairly straightforward, resting as it does on sedimentological setting and ichnodiversity levels. However, labelling individual ichnotaxa as opportunistic is less reliable. Opportunistic ichnocoenoses are commonly heavily dominated by *Skolithos linearis* (Vossler and Pemberton 1988b), but ichnospecies of *Skolithos* also occur in fully established ichnocoenoses. Many suspension feeders tend to be r-strategists because their food supply is unpredictable (Wignall 1990).

Ekdale (1985) considered *Chondrites* an opportunistic ichnogenus; its strategy reveals opportunism in severely oxygen-depleted environments, where it may occur alone (Bromley and Ekdale 1984a; Vossler and Pemberton 1988a), and Wightman *et al.* (1987) have found *Chondrites* in brackish-water situations. However, the new interpretation of *Chondrites*

Figure 10.17 Two contrasting lacustrine ichnocoenoses from the Triassic Fleming Fjord Formation, Jameson Land, East Greenland. (a) In rapidly deposited sand units, tops of beds contain a low diversity ichnocoenosis of *Skolithos*, *Arenicolites* and *Polykladichnus* ispp. indicating suspension-feeding opportunists of a 'storm assemblage'. (b) Finer-grained units, deposited slowly, are totally bioturbated. Ichnodiversity is low here also (as is usual for non-marine ichnocoenoses), but a 'fair-weather assemblage' is indicated. *Lockeia siliquaria* (from suspension-feeding bivalves) is locally obliterated by deeper-tier deposit-feeding activity represented by *Fuersichnus communis* (Bromley and Asgaard 1979). (a) Vertical section, (b) sole surface; both x2.

as a chemosymbiont (section 9.4.6), specially adapted to sulphide-rich sediment, gives the ichnogenus the narrow trophic character of a climax form (Fu and Werner 1994).

Miller and Johnson (1981) interpreted an occurrence of crowded *Spirophyton* in a Devonian marginal marine setting as opportunistic. Although *Spirophyton* superficially resembles *Zoophycos*, in that it has a spiral spreite, *Spirophyton* is probably not a once-in-a-lifetime structure. *Zoophycos*, on the other hand, can appear together with *Chondrites* in oxygen-depleted environments, and, like *Chondrites*, may be considered a climax specialist.

Ichnocoenoses described by Bromley and Asgaard (1979) from Triassic lacustrine environments show degrees of opportunism. Total bioturbation with *Fuersichnus communis* (Fig. 10.17b) clearly represents a shallow tier and the trace fossil is a relatively simple, rapidly produced fodinichnion. Ichnodiversity is two, *Lockeia* occurring at some horizons where *F. communis* has not obliterated it.

At intercalated horizons, slender ichnospecies of *Arenicolites* and *Polykladichnus* co-occur at colonization surfaces in amalgamated storm beds and represent rapid, short-term colonization events (Fig. 10.17a).

10.5.2 Climax trace fossils

Communities that succeed the pioneering opportunists, as a substrate becomes stabilized, tend to penetrate increasingly deeply (Fig. 10.2c; Rhoads *et al.* 1977; Rhoads and Boyer 1982). The work of later communities will obliterate that of earlier ones, at least at rates of sedimentation such as 1–5 mm per year (McCall and Tevesz 1983). So it is unlikely that the complete succession of communities, developed perhaps over a decade or two, will be documented in the final ichnofabric (Figs 10.18 and 10.19). If rate of deposition is increased, so as to physically separate the horizons of different communities (Fig. 10.2a), this sedimentation speed may itself constitute a stress factor that will modify community succession. Savrda and Bottjer (1986) were able to recognize community replacement where the rate of succession was controlled by gradually changing oxygen concentrations in bottom waters (Fig. 10.12).

Climax forms occur in stable habitats where conditions tend to change gradually and predictably. Ekdale (1985) characterized these as producing very diverse trace fossil associations, in which any single ethological category is represented by many ichnogenera. Structures produced by the endobenthic climax species are exemplified by complex, ornate burrow systems that reflect long-term occupation or highly specialized feeding behaviour.

Certain individual ichnogenera may be labelled as climax forms. Most of the ornate graphoglyptids, whether agrichnia or pascichnia, are safe candidates (Ekdale 1985). I would also include *Chondrites* and *Zoophycos*

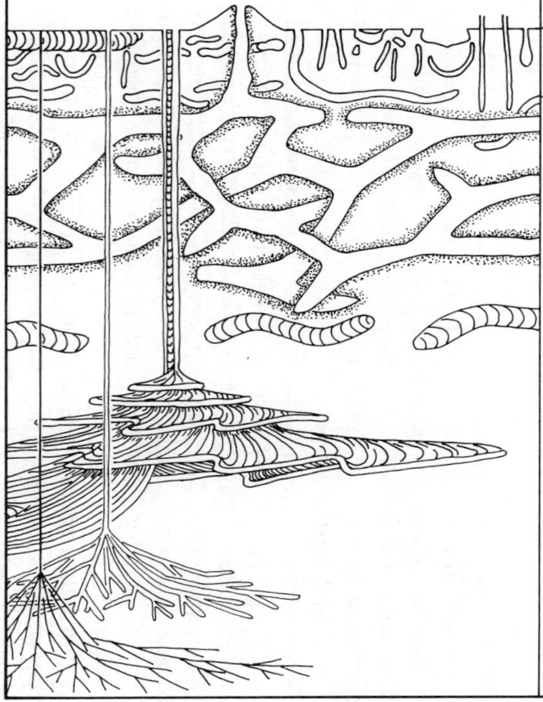

ZONE	TIER	AMOUNT OF BIOTURB'N. IN %	% OF WORK OF COM-MUNITY	% OF PRESERVED BIOTURB'N.	% OF WORK OF TIER PRESERVED	CHARACTERISTIC TRACE FOSSILS	SUGGESTIVE BODY FOSSILS
MIXED	1	100	70	0	0	NONE	SPATANGOIDS, ASTEROIDS, OPHIUROIDS, GASTROPODS, BIVALVES, SCAPHOPODS, ANNELIDS, ETC., ETC.
MIXED	2	100	10	0	0	NONE	
MIXED	3	100	10	15	5	*Planolites*	
TRANSITION	4	90	5	80	60	*Thalassinoides*	RARE CRUSTACEANS
TRANSITION	5	10	2	2	80	*Taenidium*	NONE
TRANSITION	6	1	2	2	95	*Zoophycos*	NONE
TRANSITION	7	0.5	0.5	1	98	*Chondrites* (large)	NONE
TRANSITION	8	0.05	0.5	0.5	100	*Chondrites* (small)	NONE

here (section 10.5.1). Such relatively rapidly produced structures as *Planolites*, *Thalassinoides* and *Scolicia* may belong to an intermediate category.

10.6 Ichnoguilds

Analysis of ichnofabrics produced by tiered communities reveals a number of recurring patterns involving such parameters as trace fossil type, activity level within the substrate and feeding style. Trace fossils are grouped in this way into **ichnoguilds**.

10.6.1 Ecological guilds and functional groups

The term **guild** was introduced into the ornithological literature by Root (1967, p. 335) for 'a group of species that exploit the same class of environmental resources in a similar way. This term groups together species, without regard to taxonomic position, that overlap significantly in their niche requirements.' Guilds have been used in connection with invertebrate ecology with emphasis on trophic grouping (e.g. Fauchald and Jumars 1979).

A different term, 'functional group', was introduced by Woodin and Jackson (1979, p. 1030), to include 'all organisms [that] use and affect their environment in approximately similar ways'. Their term was 'different from the concept of a guild ... which was defined solely on the basis of modes of exploitation of resources'.

However, Bambach (1983) extended the guild concept to more or less cover the functional group. He based the guild on three broad aspects of species groups: (1) structural plan of the body (bauplan); this also includes physiology and growth, and restricts the species taxonomically at class level; (2) food source; (3) use of space, whether pelagic, epibenthic or endobenthic. The purpose of guild analysis is to examine the habitat structure of a community as it functions in some place at some time.

Figure 10.18 A generalized tiering diagram for the Maastrichtian chalk of Denmark. Eight tiers are indicated, the two uppermost are hypothetical, but indicated by body fossils. The statistical columns show, from left: the amount of reworking of the substrate within each tier; the per cent of the work of the whole community represented by each tier; the per cent of the preserved ichnofabric represented by each tier (*Thalassinoides* dominates the fabric); and the amount of the work of each tier that is preserved. The numbers represent 'a reasonable guess' based on innumerable observations.

10.6.2 Guilds in ichnology

The guild concept is well suited for characterizing the ecological complexity of ichnocoenoses. In transferring the term to the ichnological realm, however, and thereby emphasizing the biological origin of trace fossils, it must be remembered that biogenic sediment structures are not organisms and, consequently, ichnoguilds are not guilds. Bambach's three factors are adapted as follows.

1. **Bauplan**. By removing taxonomic restrictions from this factor, this becomes the least important of the three. Here we categorize whether structures are produced by: (a) stationary burrows, semi-permanent or possibly branching, such as fodinichnia, domichnia and agrichnia; or (b) by vagile animals in transitory burrows, producing pascichnia, repichnia and cubichnia.
2. **Food source**. The trophic type is basic to trace fossil analysis. Here we have categories such as deposit feeding, suspension feeding, gardening and chemosymbiosis.
3. **Use of space**. This corresponds to the tier.

Labelling of individual ichnoguilds is necessary for their identification as recurring entities. Ideally, the name should incorporate elements from all three factors, which produces lengthy labels like 'mobile deep-tier deposit-feeder structures'. Acronyms could be based on these (MDDF), but these are unfriendly and ugly. Usually, having defined the ichnoguild with a name as above, it is referred to within the individual case study by a reference number, IG-I, IG-II, etc. (Ekdale and Bromley 1991; Bromley and Asgaard 1993).

Ichnoguild analysis is still in its infancy, however, and labelling of local ichnoguilds by number is a transient phase. Ultimately, it may be hoped that universal ichnoguilds will be recognized, and these may receive the names of their characteristic trace fossils. Possible examples of these are given below.

10.6.3 Examples of ichnoguilds

The structures of several Cretaceous chalk ichnocoenoses are known in some detail (Ekdale *et al.* 1984a; Frey and Bromley 1985; Ekdale and Bromley 1991). These assemblages have provided a starting-point for the definition of distinctive ichnoguilds.

(a) ***Chondrites–Zoophycos*** *ichnoguild*

Non-vagile, deep-tier deposit-feeder or chemosymbiont structures. This ichnoguild comprises the deepest tier normally present in chalk assemblages (Fig. 10.19). Ichnotaxa represented are several size-classes of *Chondrites* and *Zoophycos* and, in some chalks, ichnospecies of *Teichichnus*.

Figure 10.19 Close-up of Fig. 6.6 showing the deep-tier ichnoguild, dark and pale *Zoophycos* spreiten and dark and pale minute *Chondrites* isp. Near the bottom, a dark *Zoophycos* spreite has been reworked by dark *Chondrites* activity. x3.

Savrda (1992) suggested the inclusion of *Trichichnus* in this ichnoguild. McBride and Picard (1991) recorded *Trichichnus* in Miocene turbidites that were two to four times as deep as the *Chondrites* they cross-cut. (Dare I mention the deep-tier, slender, chemosymbiontic pogonophores?)

The *Chondrites–Zoophycos* ichnoguild is by no means ubiquitous in chalk, and the significance of its presence or absence is not known. However, it is an ichnoguild of narrow specialists, precareously colonizing inhospitable zones at the very fringe of habitability. Possibly neither of the major ichnogenera represent deposit feeding; both are reverse-conveyor structures, and chemosymbiosis and several other trophic styles have been suggested (Fu 1991; Bromley 1991; Savrda 1992). Minor fluctuations in the general (depositional) environment or ecological (in-community) environment may affect the availability of such niches to the endobenthos.

Outside the chalk environment, this ichnoguild is widespread in deeper basinal settings, and from the continental shelf break into deep water (Ekdale 1977, pl. 3d; Scholle *et al.* 1983, figs 57 and 58; Tyszka 1994, fig. 3c and e; Savrda in press, fig. 2A).

(b) ***Thalassinoides*** *ichnoguild*

Semi-vagile and vagile, mid-tier deposit-feeder structures. This ichnoguild is present in most chalks in oxygenated situations, commonly leading to high degrees of bioturbation (Fig. 10.20). The quality of this ichnoguild varies, containing normally one or several forms of *Thalassinoides suevicus*, accompanied locally by *Teichichnus rectus* and *Teichichnus zigzag*.

Thalassinoides-dominated ichnoguilds in other settings should be considered as separate from the chalk example. In sands the *T. suevicus* probably belongs in a shallower tier than in chalks. Heinberg and Birkelund (1984) found *T. suevicus* to occur in two quite distinct ichnocoenoses in the Jurassic of Greenland. The significance of this may become apparent when the tiering structure of these ichnocoenoses has been worked out.

(c) ***Planolites*** *ichnoguild*

Vagile, shallow-tier deposit-feeder structures. This ichnoguild is normally poorly visible but clearly ubiquitously present in chalks from well-oxygenated sea floors. Overall size and consistency of fill of the *Planolites* vary. Shallow-tier *Planolites* ichnoguilds similar to that of the chalk occur in many multi-tiered or simpler ichnocoenoses (Figs 11.7a, 11.8, 11.10, 11.13 and 11.17; Enos 1977, fig. 19a; Hattin 1981, fig. 22).

(d) ***Phycosiphon*** *ichnoguild*

Vagile, middle-to-deep-tier deposit-feeder structures. In many situations the deeper tiers of an ichnocoenosis are dominated by *Phycosiphon incertum*, cutting deeper than *Thalassinoides* but less deeply than *Zoophycos*

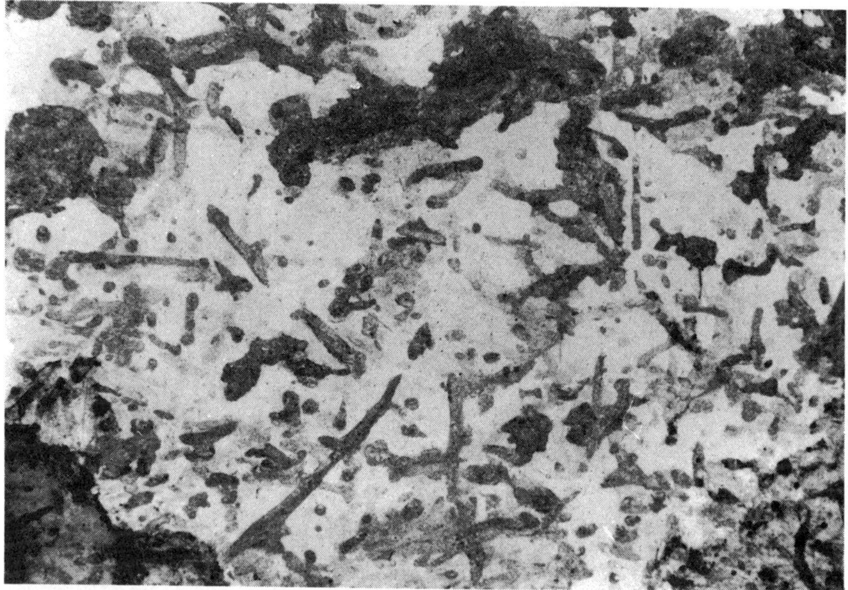

Figure 10.20 Over 2 m beneath a channel filled with brown phosphatic pellet chalk, the white chalk contains conspicuous trace fossils. The omission and post-omission suites are filled with the brown sediment advected down at least 3 m, locally to 5 m (Jarvis 1992). Recognizable are vertical *Skolithos* shafts, oblique *Teichichnus rectus* and two size-classes of *Thalassinoides suevicus*. Cretaceous, Beauval, Piccardie, France, x0.25. Smoothed vertical section.

(Fig. 10.21; Seilacher 1978). This ichnoguild commonly includes more than one size-class of *P. incertum* together with other vagile deposit-feeder structures (Figs 11.1 and 11.10). Goldring *et al.* (1991) provided valuable details and excellent pictures; these authors considered the ichnoguild to occupy a shallow tier, especially in opportunistic situations.

(e) ***Skolithos–Ophiomorpha*** *ichnoguild*

Stationary, deep-tier suspension-feeder structures. This is the ichnoguild of unstable sand substrates in hydrodynamically energetic environments (Fig. 11.5b). *Skolithos linearis* and/or *Ophiomorpha nodosa* dominate, the *O. nodosa* comprising mainly shafts. This is an ichnoguild of opportunist trace fossils and must not be confused with *Skolithos* occurrences in chalk or deep-sea sediments (Figs 5.15 and 10.20); nor identified with the horizontal networks of *Ophiomorpha* in chalks (Fig. 8.12b) or in bathyal fans (Armentrout 1980).

(f) Deep ***Scolicia*** *ichnoguild*

Vagile, deep-tier chemosymbiont structures. In Tertiary and Quaternary muddy sands and sands, the deepest tier is commonly occupied by *Scoli-*

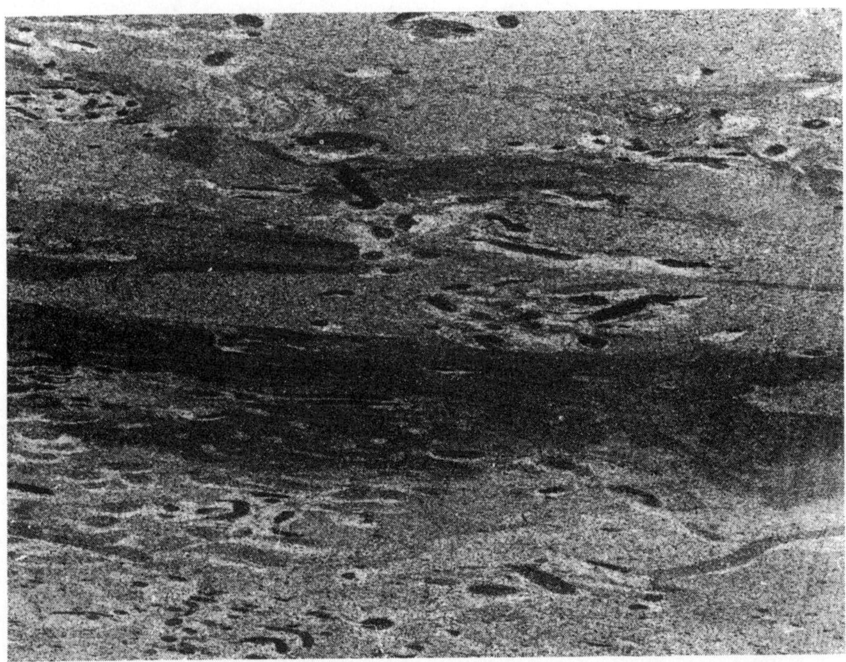

Figure 10.21 *Phycosiphon incertum* dominates this ichnofabric, in places reworking the otherwise poorly visible fills of *Planolites* isp. Carboniferous, Barents Sea, offshore Norway, x2.

cia isp. Bromley *et al.* (1995) recently argued for a chemosymbiotic explanation for the deep burrowing of some species of spatangoid echinoids. The resulting elite trace fossils require further attention from this aspect.

The identification of ichnoguilds is only possible when the tiering structure of an ichnocoenosis is known. The concept therefore encourages careful analysis of individual trace fossil communities and it is hoped it will prove a useful tool in developing a synthesis of the structure of ichnocoenoses in space and time.

10.7 Seilacherian or archetypal ichnofacies

Seilacher (1964, 1967a) noted that recurrent assemblages of certain ichnotaxa must have palaeoenvironmental implications, and established the concept of ichnofacies on this basis. An ichnofacies is an association of trace fossils that is recurrent in time and space, and that directly reflects environmental conditions such as bathymetry, salinity and substrate character.

Initially, Seilacher (1967a) established six ichnofacies, named after characteristic ichnogenera. (Incidentally, I do not italicize the eponymous ichnogenera because it is a facies and not an ichnotaxon that is under discussion. The ichnotaxon concerned is not necessarily present in a given assemblage of that ichnofacies. This is in line with usage of taxa in biostratigraphical zones, e.g. Bifrons zone.)

Four of these ichnofacies were explicitly based on bathymetry: the Skolithos, Cruziana, Zoophycos and Nereites ichnofacies. A fifth, the Glossifungites ichnofacies, was characteristic of omission surfaces and reflected firm to hard surfaces. A sixth, the Scoyenia ichnofacies, characterized non-marine red beds.

Subsequently, two further ichnofacies were defined on the basis of substrate consistency: the Trypanites ichnofacies for lithic substrates (hardgrounds and rockgrounds) (Frey and Seilacher 1980) and the Teredolithes ichnofacies for xylic substrates (wood) (Bromley *et al.* 1984).

In the last decade, further ichnofacies have been proposed at an accelerating rate. Lockley *et al.* (1987) proposed a Curvolithus ichnofacies as a subset of the Cruziana ichnofacies (section 10.1.4); Bromley and Asgaard (1991) introduced an Arenicolites ichnofacies for opportunistically colonized sand event beds; Buatois and Mángano (1993) suggested a Mermia ichnofacies for freshwater turbidites; Bromley and Asgaard (1993b) divided the Trypanites ichnofacies into Entobia and Gnathichnus ichnofacies; Lockley *et al.* (1994a) indicated the need for five vertebrate ichnofacies; and Hunt *et al.* (1994) even suggested basing ichnofacies on fossil excrement as 'coprofacies'.

The vertebrate and coprolite candidates have no sedimentological connotation and should be retained as ichnocoenoses or associations (Goldring 1995b). Nevertheless, the remaining plethora comes close to chaos (Table 10.1).

Ichnofacies are basically sedimentary facies defined on the basis of trace fossils (Fig. 10.22). They all have the two important components of biological input and taphonomic loss. The relative importance of these two components vary considerably in different ichnofacies which, in fact, fall into two groups: those that are characterized dominantly by ecology of the tracemakers (biofacies) and those that are distinguished chiefly on the basis of taphonomic bias (taphofacies) (Bromley and Asgaard 1991).

On this basis, the following survey does not closely resemble the traditional summaries of recent years (Bromley *et al.* 1984; Ekdale *et al.* 1984a; Frey and Pemberton 1984, 1985; Ekdale 1985; Frey *et al.* 1990; Pemberton *et al.* 1990, 1992). One should be aware (with Metz 1995) that our understanding of non-marine assemblages is still very scanty. Again, although terrestrial (palaeosol) trace fossils have been recognized back to the Ordovician (Retallack and Feakes 1987; Johnson *et al.* 1994), details are lacking and no attempt at defining a palaeosol ichnofacies has been made.

Table 10.1 Scheme indicating relationships of ichnofacies with environment. Names followed by a question mark are tentatively suggested herein

Woodground	Rockground	Firmground		Loose- and softground		Sedimentology/environment		
		Marine	Freshwater	Freshwater	Marine	Energy	Bathymetry	Grainsize
Teredolites	Trypanites; Entobia	Glossifungites	Scoyenia		Psilonichnus		Backshore	Sand
				Rusophycos?	Skolithos	High	Beach	Sand
	Gnathichnus			Arenicolites?	Arenicolites	Event	Shelf	Sand, silt
				Fuersichnus?	Cruziana	Medium	Lagoon/shelf	Sand, silt
				Mermia	Nereites	Event	Slope to abyssal	Sand, mud
					Zoophycos	Low		Mud

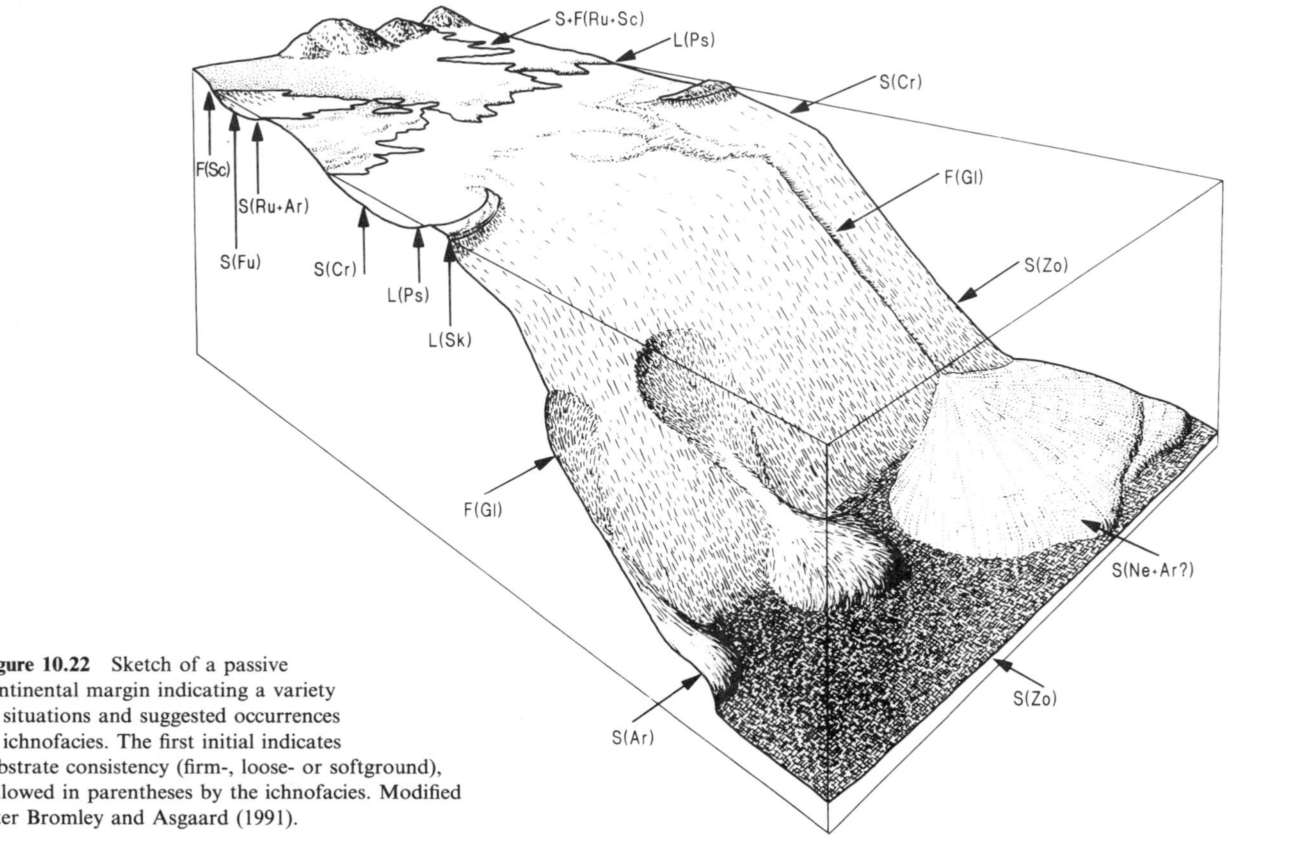

Figure 10.22 Sketch of a passive continental margin indicating a variety of situations and suggested occurrences of ichnofacies. The first initial indicates substrate consistency (firm-, loose- or softground), followed in parentheses by the ichnofacies. Modified after Bromley and Asgaard (1991).

10.7.1 The salinity barrier

Seilacher (1967) introduced the Scoyenia ichnofacies on the basis of Triassic red-bed assemblages, but later (Seilacher 1978) admitted to the non-distinctiveness of the association. Frey and Pemberton (1984) came to a similar conclusion.

Bromley and Asgaard (1991) suggested abandoning the Scoyenia ichnofacies and replacing it by extending the marine ichnofacies into the non-marine realm. Trace fossils are very useful for documenting salinity levels and changes (section 12.1.2), but palaeosalinity is interpreted on the basis of trace fossil size, diversity and ichnospecies presence/absence, and not on ichnofacies.

However, Bromley and Asgaard's (1991) untraditional solution was found unacceptable (Pemberton *et al.* 1992; Pemberton and Wightman 1992; Pickerill 1992; MacNaughton and Pickerill 1995). Instead, therefore, I arrange the ichnofacies here with partial equivalents on either side of the salinity bottleneck. I have had to upgrade some associations to the rank of ichnofacies in order to fulfill this role.

10.7.2 Scoyenia ichnofacies

This ichnofacies covers freshwater firmground associations, in sands and muds connected with shallow lacustrine and fluviatile settings commonly subject to periodic subaerial exposure. The trace fossils have wall sculpture developed in varying degrees, and are typified by *Scoyenia gracilis, Skolithos (Cylindricum) antiquum, Spongeliomorpha carlsbergi* and U-shaped spreite structures, both scratched (*Glossifungites*) and unscratched (*Rhizocorallium jenense*) (Seilacher 1967; Fürsich and Mayr 1981).This is a narrower definition than the earlier broad, indefinite descriptions (e.g. Frey and Pemberton 1984) and is based on the firmground character of the eponymous trace fossil and non-overlap with the other ichnofacies proposed herein. This is principally a biofacies, based on communities restricted by substrate consistency.

10.7.3 Glossifungites ichnofacies

Originally, this ichnofacies had no marine connotation (Seilacher 1967). *Glossifungites saxicava* was originally described by Lomnicki (1886) from a non-marine succession, although it is possible that the bed containing it represents a marine incursion (A. Radwanski, pers. comm. 1990). Anyway, more recent commentators have considered the ichnofacies as marine (Pemberton and Frey 1985).

Commonly, the Glossifungites ichnofacies is recognized as a transient phase of benthic community succession as an omission surface passes from softground through firmground to a hardground consistency

(Gruszczyński 1979, 1986; Goldring and Kazmierczak 1974; Fürsich 1978; Savrda and Bottjer 1994). Furthermore, local erosion to expose compacted sediment as firmground is also a major setting for this ichnofacies (Pemberton and Frey 1985). The long periods of non-deposition represented by firm omission surfaces allow the development of the trace fossils that represent growth of the tracemaker rather than its movement (Bromley and Hanken 1991). There is no bathymetric restriction: Bromley and Allouc (1992) reported a well-developed Glossifungites ichnofacies phase in the development of bathyal hardgrounds to over –3000 m.

Characteristic ichnotaxa are *G. saxicava, Spongeliomorpha* ispp. and *Thalassinoides paradoxa*. Various other scratch-ornamented forms such as *Strophichnus xystus* and *Glyphichnus harefieldensis* may also occur (Fürsich *et al.* 1981; Bromley and Goldring 1992). This ichnofacies is a marine parallel to the Scoyenia ichnofacies as emended here, and likewise has a strong biofacies tone.

10.7.4 Psilonichnus ichnofacies

This ichnofacies is developed in the backshore, between the foreshore zone and the terrestrial realm. The ichnofacies is a biofacies, having been largely based on unbiased biological data. However, equivalent assemblages attributable to this ichnofacies occur in the Pleistocene of Georgia (Frey and Pemberton 1987) and Bahamas (Curran 1994), and also have been identified in the Jurassic of Portugal (Fürsich 1981). The ghost crab ichnogenus *Psilonichnus* is present in each of these few examples, and vertebrate trackways will also be expected (Pemberton *et al.* 1992).

10.7.5 Skolithos ichnofacies

The Skolithos ichnofacies is 'indicative of relatively high levels of wave or current energy' (Pemberton *et al.* 1992, p. 53). It represents 'lower littoral to infralittoral, moderate to relatively high-energy conditions ... associated with muddy to clean, well sorted, shifting sediments subject to abrupt erosion or deposition' (Frey *et al.* 1990, p. 57).

Modern communities in these shifting sands do not lack tracemakers in the upper tiers. Enormous bioturbation is caused by suspension-feeding bivalves (section 6.2; Fig. 6.3), trapping worms (section 4.6; Fig. 4.39), and vagile deposit feeders also are present in all clean intertidal sands (Clifton and Thompson 1978; Ronan *et al.* 1981). However, almost all the structures these animals produce are destroyed at once or seasonally by physical processes and stand negligible chance of preservation (Fig. 11.5; Howard and Reineck 1981).

Slow suspension feeders that inhabit the shifting sands environment seek security by burrowing deeply and remaining stationary for long periods. These deeper tiers of endobenthic activity have a good chance of preservation, though they may be less accessible for examination in the recent environment (section 4.3.1).

Thus, the structures typical of the Skolithos ichnofacies have the low ichnodiversity, low density and vertical orientation of the deep suspension-feeder ichnoguild, and are dominated by *Skolithos*, *Ophiomorpha* and *Diplocraterion*; domichnia and equilibrichnia. If middle-tier levels are preserved, the deposit feeders are represented, commonly by *Macaronichnus* (Fig. 11.9; Saunders and Pemberton 1990).

The upper parts of these structures, like the upper tiers of the

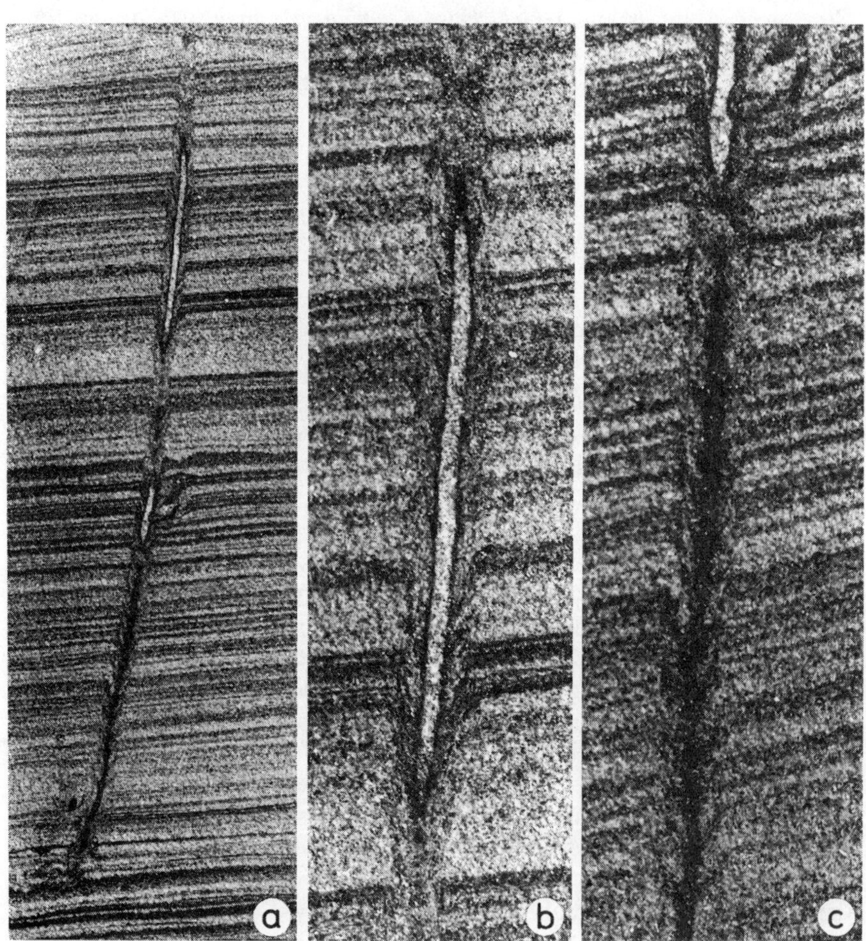

Figure 10.23 A remarkable equilibrium structure. In steadily accreting sand, a lined shaft has been extended upwards in pace with deposition. The cone-in-cone structure around it is interpreted as a small conical funnel at the aperture, repeated upwards. Thus, the trace fossil may be called *Skolithos (Monocraterion) tentaculatum*. Jurassic, Norwegian sector, North Sea. (a) x0.5, (b) and (c) x2.

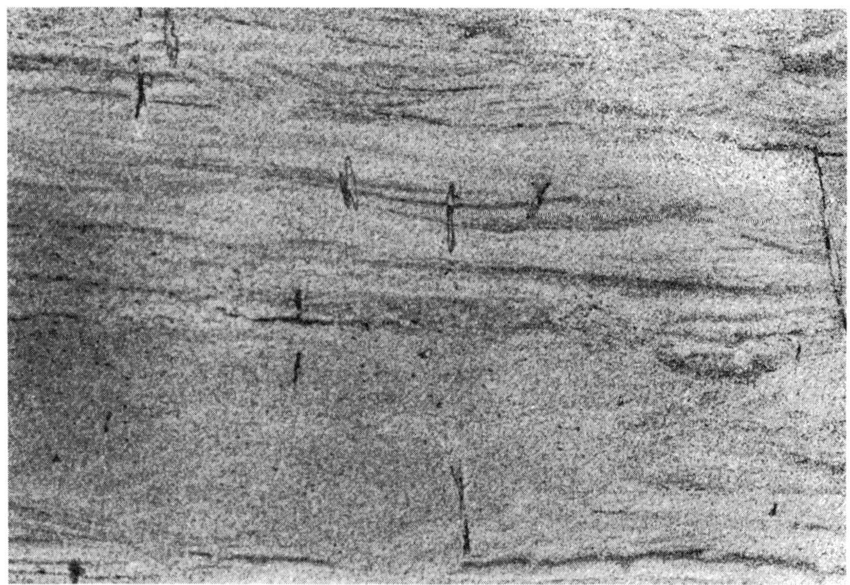

Figure 10.24 Low-density assemblage of *Skolithos linearis*; thinly lined with organic carbon, they are easily mistaken for plant roots. Jurassic, Norwegian sector, North Sea, natural size.

palaeoichnocoenosis, only exceptionally become preserved (Fig. 10.23); but, in characterizing ichnofacies, we are concerned with the norm not the exception. The ichnodiversity certainly does not convey much idea of the high species diversity and large biomass of the endobenthic community that may inhabit the shifting sands. Thus, the Skolithos ichnofacies is a taphofacies.

In neighbouring environments, quite a different taphonomic regime may dominate. Thus, in rapidly accreting sands, such as the foresets of migrating sandwaves, there may be no erosional loss of upper tiers. In such cases, however, rapidity of burial may inhibit maturation of communities, allowing only temporary colonization by a stressed pioneer community (Figs 10.24 and 12.5) and we approach the Arenicolites ichnofacies (section 10.7.8).

Ichnofacies show evolution through the Phanerozoic, and the Skolithos ichnofacies is no exception. The spectacular piperocks of the Cambrian, dominated by elite *Skolithos*, were a short-lived phenomenon (Droser 1991). In the absence of disturbance by predators or deposit feeders the populations of *Skolithos*-makers bloomed. However, piperocks dwindle in quality and quantity through the Palaeozoic and are replaced in the post-Palaeozoic by *Ophiomorpha* fabrics (Bottjer and Droser 1994).

10.7.6 Cruziana ichnofacies

This ichnofacies characterizes the region between daily wavebase and storm wavebase (Frey and Pemberton 1985). From a taphonomic point of view, this situation profoundly increases the preservation potential of the upper tiers of the ichnocoenosis.

In regions of rapid deposition, bioturbation may be far from complete (Fig. 10.25). Deeper-tier activity then will not have time to obliterate shallower structures, and ichnodiversity may be correspondingly high. Rapid rate of sedimentation may also prevent the maturing of communities, deep tiers may not be occupied, and the *Zoophycos–Chondrites* ichnoguild is generally not represented.

A high biological diversity is reflected by a high behavioural diversity, including suspension feeders and deposit feeders. Cubichnia in the uppermost tiers are commonly preserved (e.g. ichnogenera *Asteriacites, Rusophycus, Lockeia*). However, repichnia, fodinichnia and domichnia dominate the ichnocoenoses (ichnogenera *Cruziana, Protovirgularia, Phycodes, Rhizocorallium, Skolithos, Arenicolites*). Rare in the Palaeozoic, the *Thalassinoides* ichnoguild comes to play an important role in this ichnofacies in the Mesozoic and Cainozoic.

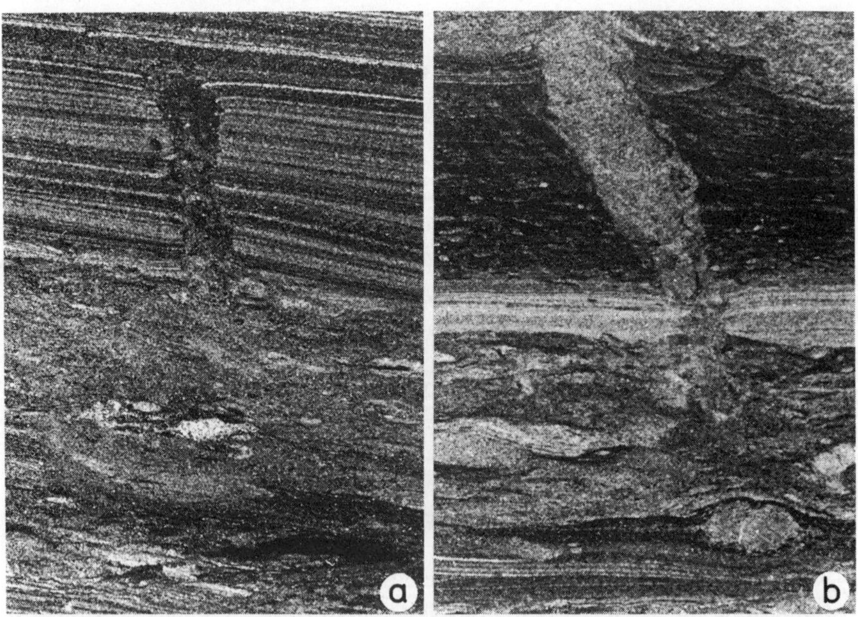

Figure 10.25 Escape traces rising from the tops of laminated-to-scrambled units. Jurassic, Norwegian sector, North Sea, x2.

10.7.7 Rusophycus ichnofacies?

In the non-marine realm, similar assemblages of trace fossils to those of Cruziana ichnofacies have been reported in fluvial and shallow lacustrine, mud, silt and fine-sand softgrounds. Such an assemblage was described by Bromley and Asgaard (1979) as the aquatic suite of the 'Rusophycus ichnocoenosis', and similar occurrences are not uncommon (Walter 1983; Aceñolaza and Buatois 1991, 1993; Pickerill 1992; Buatois and Mángano 1993; Miller and Collinson 1994). The ichnofacies is dominated by repichnia (*Cruziana* ispp. and many walking arthropod traces like *Diplichnites* and *Multipodichnus*) and cubichnia (*Rusophycus* ispp.).

10.7.8 Arenicolites ichnofacies

Recurring assemblages of vertical trace fossils characterized by a low diversity, opportunistic suspension-feeding ichnoguild were referred to an Arenicolites ichnofacies by Bromley and Asgaard (1991). The ichnofacies applies to short-term colonization of storm beds and other sandbodies occurring in incongruous environments. Ichnogenera are usually small *Skolithos* and *Arenicolites*, sometimes accompanying *Polykladichnus* and *Diopatrichnus*.

Vossler and Pemberton (1988, 1989) documented fair-weather and storm trace fossil assemblages occurring in the same sequence. The fair-weather, climax communities produced trace fossil assemblages referrable to the Cruziana and Zoophycos ichnofacies. Storm sand beds intercalated within the succession of mudrocks contained an entirely different assemblage that was interpreted as resulting from rapid colonization by opportunists, and heavily dominated by diminutive *Skolithos* isp. This situation is not unusual, e.g. Dam's (1990b) *Arenicolites* isp.1 ichnocoenosis (Fig. 12.9a and b) and Droser *et al.* (1994).

These occurrences of *Skolithos* were included in the Skolithos ichnofacies, the distribution of which thereby became extended into the deep sea (Frey *et al.* 1990; Pemberton *et al.* 1990). Of course, in unusual cases of deep-water strong contour current activity (e.g. Colella and D'Alessandro 1988), a case for the Skolithos ichnofacies can be argued. However, the occurrence of *Skolithos* in tempestite and turbidite tops is entirely different.

The Skolithos ichnofacies reflects communities adapted to ongoing high hydrodynamic energy and shifting sands, conditions which allow the preservation of only deeper-tier biogenic structures. The Arenicolites ichnofacies, although normally occurring in sand deposited by a raised-energy event, represents post-event colonization, and reflects tranquil conditions allowing preservation of all tiers including the shallowest levels. The **colonization window** (section 12.2.3) is but briefly open for this sand community prior to smothering by background depositional mud.

Proximally, the Arenicolites ichnofacies grades with the Skolithos ichnofacies where storm energy levels are frequently or temporally raised, providing colonized hummocky cross-stratified units (Frey 1990; Frey and Howard 1990; Frey and Goldring 1992; Buckman 1992). Distally, the ichnofacies grades into the post-turbidite sand assemblages where, under quiet conditions, a sandy sea floor may be colonized for some time to produce an approach to a climax community, especially containing *Ophiomorpha* isp. (Crimes 1977; Uchman 1991a, b, 1992a). Storm assemblages in lakes are closely similar to marine occurrences and, although oligochaete rather than polychaete tracemakers are probably involved, the behavioural response to the sedimentological event is identical (Fig. 10.17).

10.7.9 Zoophycos ichnofacies

This is the black sheep of the family of marine softground ichnofacies. Some authors have doubted its validity (Osgood 1970). In most accounts, ichnologists have difficulty in ascribing any characteristic ichnogenera to it apart from *Zoophycos* itself (Seilacher 1978, fig. 6; Frey and Pemberton 1985, fig. 28); and many are eager to point out that *Zoophycos* ispp. can occur in many different associations and settings. (But none of the eponymous ichnogenera are restricted to their ichnofacies.)

Frey and Pemberton (1985) and Pemberton *et al.* (1992) characterized the environment as circalittoral to bathyal, quiet water conditions, more or less deficient in oxygen; offshore sites are below storm wavebase to fairly deep water.

Sediments of Zoophycos ichnofacies usually show total bioturbation. The quiet accumulation of mud allows climax communities to develop, spread over many levels, the *Zoophycos–Chondrites* ichnoguild occupying the deepest tier.

In higher tiers, *Thalassinoides* and *Phycosiphon* ichnoguilds are normally present (Figs 10.21 and 11.10). Owing to the dense bioturbation, upper levels are homogenized mixed layers and commonly only the lowest tiers are conspicuous. Hence, the ichnofacies is characterized by low ichnodiversity; and since *Chondrites* is labelled as a facies-breaking form (Seilacher 1978), whereas *Zoophycos* surprisingly is not, *Zoophycos* is stated to be the only characteristic ichnotaxon!

Furthermore, because this deepest tier occupies the oxygen-depleted zone of the substrate, the ichnofacies has been characterized as representing oxygen-depleted sea floors, which is quite another matter. Nevertheless, where sea-floor oxygen concentrations do become a limiting factor, then the upper mixing zone tiers disappear, leaving the *Zoophycos–Chondrites* ichnoguild in an otherwise unbioturbated substrate (Fig. 10.10; section 10.3.2).

Wetzel (1983b, fig. 4) noticed in a series of bathyal to abyssal cores, a gradual telescoping of the tiering structure. Passing from –1000 to –4000 m, the tiers all became less deeply emplaced. Thus, while *Zoophycos* perhaps is produced about 1 m below sea floor in the shallow end of the range, it displaces upwards to 20–30 cm by –4000 m depths.

As we pass into the deeper water of the abyssal regions, the lowest tiers appear to become sparsely occupied or disappear altogether, again judging by evidence from recent sediments. Ekdale and Berger (1978) and Ekdale *et al.* (1984a) wrote of an **abyssal** or **deep-sea ichnofacies** to cover these variants.

The deepest documented occurrence of *Thalassinoides* may be that of Hayward (1976) at possibly –3000 m. If *Thalassinoides*, *Zoophycos* and *Chondrites* all disappear at lower abyssal and hadal depths, leaving only cryptobioturbation and *Planolites* to obliterate the surface trails, there remains little foundation on which to base an ichnofacies!

The ichnogenus *Zoophycos* occurs in shallow-water settings in Palaeozoic strata (e.g. Osgood and Szmuc 1972; Yurewicz 1977; Marintsch and Finks 1982; Brett *et al.* 1990). It also appears to occupy a shallow tier, and these facts suggest that the Zoophycos ichnofacies is not a useful concept in the Palaeozoic (Miller 1991).

In the Mesozoic, the ichnogenus moved into deeper water (Ekdale 1978; Bottjer *et al.* 1987, 1988). It is tempting to speculate if the explosive increase in the *Thalassinoides* ichnoguilds at the beginning of Mesozoic time did not displace *Zoophycos* into deeper environments.

10.7.10 Nereites ichnofacies

Seilacher defined this ichnofacies in terms of turbidite deposition and its characteristics derive from the special depositional and taphonomic framework. The characteristic ichnotaxa are ornate and complicated pascichnia and agrichnia, such as the ichnogenera *Paleodictyon*, *Helminthoida*, *Spirorhaphe* and *Cosmorhaphe*. These are all pre-depositional in the sense that they are preserved by slight erosion and sudden burial as semirelief casts on the sole of the overlying turbidite (Fig. 9.5). This special depositional process preserves with great emphasis the ephemeral structures of the top tier of the mud ichnocoenosis (section 6.2.1; Leszczyński and Seilacher 1991).

For some time, the sandy turbidite provides an arenaceous substrate for a post-event community. After a while, however, this gradually becomes buried in pelagic sediments and a mud environment returns. Where they are visible, the poorly preserved, deeper, pre-depositional tiers are seen to resemble those of the Zoophycos ichnofacies (Seilacher 1987; Uchman 1991c).

The turbidite ichnological assemblage can be envisaged as a tripartite taphonomic system based on a two-community palimpsest: (1) the background community, inhabiting the pelagic sediment, is that which supports the Zoophycos ichnofacies; (2) the turbidite erosional/obrusion event preserves the topmost tiers of that community, producing the predepositional suite that supports the Nereites ichnofacies; (3) the sandy substrate brought in by the turbidite current is colonized by a short-term sand community, the post-depositional suite which normally (but not logically) is included in the Nereites ichnofacies. This suite is more closely allied to the Arenicolites ichnofacies, but has been confused with the other sand-based ichnofacies, i.e. the Skolithos ichnofacies.

Wetzel (1983a, 1984) found that deep-sea deposits in the Sulu Sea are incompletely bioturbated, and no cryptobioturbation has occurred. Thus, the deep burrowers have not obliterated the work of those in the upper tiers, and Wetzel was able to demonstrate the Nereites ichnofacies agrichnion *Protopaleodictyon* and the deep-tier *Zoophycos* co-existing in the same box core.

Thus, it would seem that the extreme contrast in quality between the Zoophycos and Nereites ichnofacies in terms of ichnotaxa is the result of differing taphonomic histories. The original endobenthic mud communities were one and the same, whereas the preserved trace fossils are widely different. The one suite demonstrates its well-oxygenated upper tiers, the other only its oxygen-starved lower tiers. On that basis, there are no grounds for considering the Nereites ichnofacies to represent deeper water than the Zoophycos ichnofacies. The evolutionary trend from shallow to deeper water has been documented in detail for graphoglyptid trace fossils (Crimes *et al.* 1992; Crimes and Fedonkin 1994; Crimes and McCall 1995). During Cambrian time the Nereites ichnofacies was restricted to shallow water, and deeper water was colonized during the Ordovician.

10.7.11 Fuersichnus ichnofacies?

Bromley and Asgaard (1979) distinguished a *Fuersichnus* ichnocoenosis for a lacustrine trace fossil assemblage below a fair-weather wavebase. It was dominated by *Fuersichnus communis*, a deposit-feeding structure that superficially resembles a spreite (Fig. 10.17b). This trace fossil has since been found in similar settings elsewhere (e.g. Gierlowski-Kordesch 1991; MacNaughton and Pickerill 1995).

Having a 'deeper-water aspect' than the proposed Rusophycus ichnofacies, the Fuersichnus ichnofacies might approach a freshwater pair for the marine Zoophycos ichnofacies. Most larger lakes have a stratified water column, and their deeper regions, where one might look for a freshwater equivalent of the Zoophycos ichnofacies, are normally anoxic.

10.7.12 Mermia ichnofacies?

Buatois and Mángano (1990, 1993) described a remarkable occurrence of trace fossils in a Carboniferous lake succession characterized by turbidites. The '*Mermia* association' was dominated by pascichnia, mostly non-specialized grazing patterns, but was marked by an absence of meniscate backfills and by the small size of the trace fossils. Possibly we have here a candidate for a non-marine equivalent for the Nereites ichnofacies. While awaiting further details of this assemblage, it may be noted that Archer and Maples (1984) and Pickerill (1990, 1992) have reported non-marine examples of the graphoglyptid ichnogenus *Paleodictyon.*

10.8 Do we need archetypal ichnofacies?

For hierarchical reasons it is presumably necessary that a major rank of ichnofacies exists to cover more minor and local groupings. However, recent efforts at improving this archetypal system have not shown much success or produced a concensus. The marine ichnofacies fall into two camps, having taphonomic and biological bias, respectively. And the non-marine ichnofacies, through the diverse nature of non-marine environments, have long been considered so variable as to be virtually impossible to classify (Asgaard and Bromley 1983).

Goldring (1993, 1995a) has questioned whether ichnofacies have reached their limits of resolution. We can recognize far more sedimentary facies than ichnofacies by combining all aspects of sedimentology, ichnology and palaeontology in an integrated study (section 12.5).

11 Ichnofabric and trace fossils in core

Ichnology is finding major application in the descriptive logging of cores. Some of this material derives from the exploration of the ocean floor in the form of piston cores, vibracores, box cores and can cores, including the copious material of the Deep-Sea Drilling Project and Ocean Drilling Project. Still more demanding is the material provided by cored wells of oil companies. In the often hasty and impatient atmosphere of the core laboratory, kilometres of costly core are at this moment receiving inadequate attention from far too few, overworked, specialist geologists. All but two of the photographs illustrating this chapter were taken by me under such conditions, while working on Jurassic cores from the Norwegian sector of the North Sea.

11.1 Outcrop versus core

The nature of the information derivable from outcrop exposure and core rock, respectively, is so different that direct cross-reference is made difficult. We must begin by examining the two sources of data.

11.1.1 Advantages of core

The advantages of working with drill core material are many. (1) Cores are usually long and provide a vertical continuity. This allows a far more complete and detailed overview of the succession of facies than is normally available in outcrop. (2) There is usually a complete lack of weathering in core material; the rock is fresh, and has sometimes been protected from excessive diagenesis by the passage through it of hydrocarbons. In onshore outcrops, fissile weathering of mudstone to shale can obscure features of bioturbation, but in core material such an argillaceous unit may offer much ichnological detail (Fig. 11.1). (3) The location of the drill site is strategically chosen in order to provide optimum subsurface information for correlation of facies and for basin analysis. (4) A fourth advantage is the large amount of geological and geophysical data that are generated from cores. Ichnological data are supported by a wealth of supplementary information from the same core. (5) Finally, in

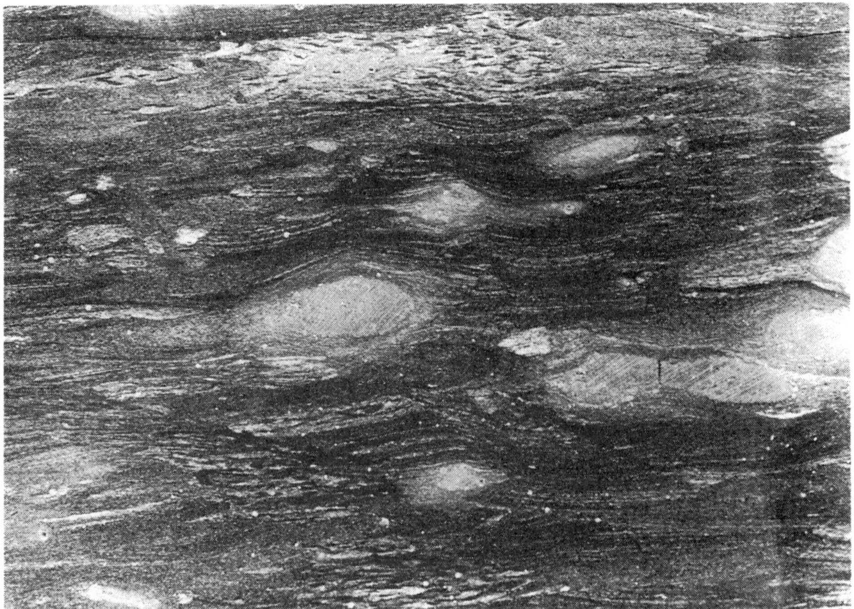

Figure 11.1 Highly argillaceous unit from outer shelf setting. Cretaceous, Norwegian sector, North Sea. Possible compacted *Thalassinoides* fills at top have been reworked by *Phycosiphon incertum*. The lamination is at least in part biogenic: compacted *Teichichnus* spreiten locally reworked by *P. incertum*. Natural size.

many situations, both onshore and in the ocean basins, outcrops are lacking and cores offer the only opportunity to see the rocks.

11.1.2 Disadvantages of core

The disadvantages of working with core material are also numerous, but it is here that the challenge of the long, thin format becomes exciting. The narrowness of the sample is a major problem. Each horizon is represented by rock having but a few centimetres of lateral extent. Local lateral continuity of a rock unit often has to be estimated on the basis of intuition. The likelihood of large trace fossils characteristic of that bed being represented in the sample is small indeed.

Furthermore, the cored rock is generally offered to the logger in the form of sawn slabs. Chances are small that characteristic structures be recognizably displayed on the surface of such samples; the use of hammers is rarely encouraged! Horizontal or bedding-plane views of trace fossils are rarely available in core material, and then, of course, only of very limited areal extent. Almost all information has to be derived from often a single plane of section that is perpendicular or oblique to bedding.

11.2 Trace fossils in core

The difference in format between outcrop exposures and those of core material require that we rethink our attitudes to ichnological data when dealing with core. After all, our basic principles, ichnotaxonomy and combined experience are largely based on wide bedding-plane exposures in outcrop. As far as is possible, this corpus of knowledge must be translated to core format.

11.2.1 Techniques for studying core

Material is prepared for logging in different ways by different laboratories. Sometimes, a half or quarter core is available. In this case, more than one surface can be studied, which allows a three-dimensional image of the structures to be reconstructed. In most cases, however, the core is slabbed and the B-cut slabs, 2 or 3 cm thick, are permanently mounted on backing boards or metal trays. This treatment secures and displays the core nicely, but it is unfortunate for the ichnologist in that one side of the slab is obscured, and the vital information it would have provided is lost. Because three-dimensional reconstructions of the morphology of structures are largely dependent on serial sections or sections lying in two planes, it is often necessary to refer to other cuts of the core.

The smoother the surface of the rock the more detail that can be obtained. The rough outer surface of the core may be suitable, but often this is fluted with grooves caused by the drilling bit or rotation of the core within the barrel. Nevertheless, this surface should be cleaned of drilling mud in water and inspected.

Sawn surfaces provide the best results. Wetting the surfaces may improve viewing, though not always. However, repeated wetting and drying causes some clay-rich rocks to crack up. If geochemical analyses are not to be made, the surface may benefit from application of a light oil to enhance visibility of structures (Bromley 1981a). Detail can further be improved photographically, using a high-contrast monochrome film and hard printing grade.

Fluorescence by ultra-violet radiation also may be tried where structures are poorly visible. This procedure may also be effective where there are hydrocarbon residua in the rock. The use of X-radiography is recommended where the slabs are not fixed to backing (Fig. 6.5; Wetzel 1983a, 1984). Stereopairs of radiographs can produce excellent information without damaging the core. Wherever possible, the two-dimensional straitjacket should be relieved by examining any horizontal cuts or fractures that are available (Fig. 11.2).

For those with access to computed tomography, 'serial section' images can be made of core and rock samples without damage (Fu *et al.* 1994),

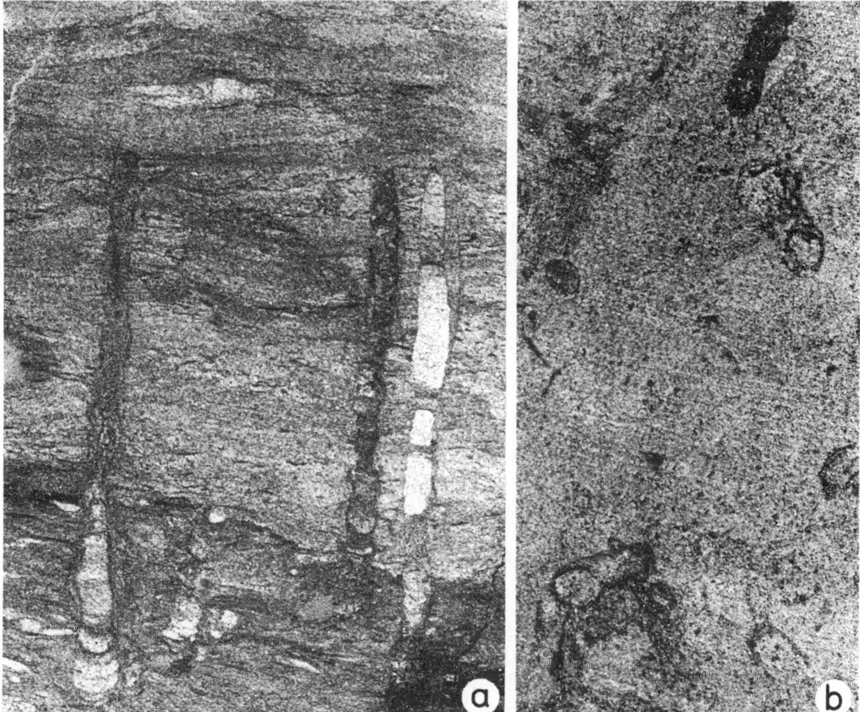

Figure 11.2 *Diplocraterion habichi* seen in core (Fig. 8.10): (a) in vertical section, natural size, (b) on a horizontal cut, x2. Note the narrow spreite.

but resolution of the images is normally disappointing. However, there is likely to be improvement of this technique.

Finally, having obtained a reasonable photographic or other image of an ichnofabric, this can be manipulated by computer analysis. Digital enhancement, involving point-by-point modification and spatial convolution of computer-scanned images can improve the visualization of chosen features of the ichnofabric (Magwood and Ekdale 1994).

11.2.2 Seeing in two dimensions, thinking in three

Working with cores exercises the ability to reconstruct three-dimensional structures on the basis of their expression at a transecting plane. Most geologists are good at this anyway, having training in constructing tectonic models on the basis of geological maps, working with thin sections, and recognizing sediment structures and fossils from random cross-sections in outcrop surfaces.

Chamberlain (1978, fig. 1) demonstrated the many interpretations that

Figure 11.3 The faintly visible spreite at top is *Rhizocorallium* isp. Below, the fabric is dominated by thick-walled *Palaeophycus heberti*, and thin, white walled, largely crushed *Terebellina* isp. Small, black *Phycosiphon incertum* are visible locally right of centre, together with several other indeterminate forms. In weathered onshore exposure it would probably be the *Rhizocorallium* that dominated the fabric (Fig. 9.5). High diversity of suspension- and deposit-feeder structures: fully marine, probably shallow subtidal. Natural size.

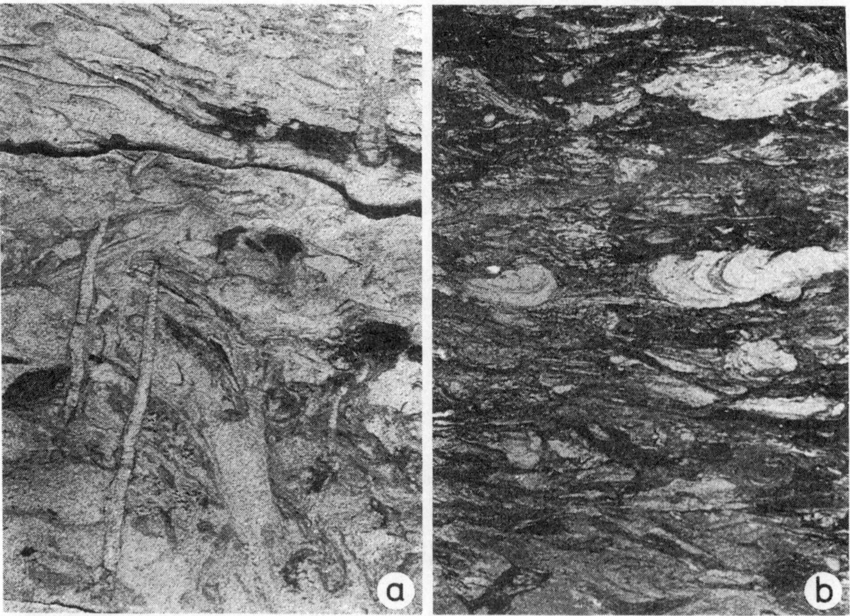

Figure 11.4 Two contrasting fabrics, x0.5. (a) Bioturbation of this sandstone is not complete. A *Palaeophycus* ichnofabric is cross-cut by *Skolithos linearis*. Diversity is high, including deep-tier, tiny black *Phycosiphon incertum* in patches. Fully marine, around a fair-weather wavebase. (b) Muddy sandstone containing a high diversity dominated by *Rhizocorallium* isp. (centre), from an inner-shelf environment.

are possible of circular or elliptical outlines in cross-sections, and Berger *et al.* (1979, fig. 14) took this conceptual model further. The eternal trap for the hasty logger is the bleb of contrasting sediment: is it a cross-section of a cylindrical burrow fill, or of a round clast?

Similarly, the longitudinal section of the cylindrical meniscus backfill of *Taenidium* ispp. can easily be confused with sections of *Zoophycos* or *Rhizocorallium* spreiten. These structures are fundamentally different in form, and could not be confused in outcrop, but in the confines of a sectioned core they can closely resemble each other (Figs 11.3 and 11.4b).

11.2.3 Recognition of ichnotaxa

Soundly established ichnotaxa are normally based on a whole range of taxobases such as three-dimensional morphology, wall structure, branching characteristics and fill material (section 8.3). These characteristics are very incompletely portrayed in the format of a single vertical section through the structure. Consequently, the positive identification of ichnotaxa in core encounters many problems.

In two dimensions, the overall morphology of a structure can be difficult to ascertain, where only fragments are visible in random section. Shafts may be *Skolithos*, but may equally well comprise limbs of *Arenicolites* or vertical parts of other structures. Branching characteristics are almost impossible to determine unless the structure is very small and its branching nodes are closely spaced. Yet the manner (or absence) of branching is a vital ichnotaxonomic criterion.

On the other hand, the nature of the burrow boundary or walling material, and the details of the fill, are particularly well displayed in core. In particular, the wall lining is commonly degraded or lost through weathering in outcrop, and while this preservation may render the structures conspicuous and spectaclar in the field, it hinders closer comparison with unweathered material.

Thus, the emphasis on various features of trace fossils in outcrop and core are radically unalike, and this leads to discrepancies when attempting to correlate and compare data from the two sources. Core emphasizes full-relief structures whereas outcrop emphasizes structures in semi-relief, or at bedding planes and in cleavage relief. Core draws attention to small and minute structures that are easily overlooked in weathered outcrop; large structures like *Thalassinoides* or *Phoebichnus*, which are eye-catching in the field, are generally too large to make much impact in core (Figs 11.5a, 11.6 and 11.10).

This discrepancy in the quality of the data is a major problem. Ekdale (1977) suggested identifying trace fossils in core to ichnogenus only. This is untenable, however, owing to the uneven application of ichnotaxobases. For example, ichnospecies of *Ophiomorpha* can be identified readily in core because they are based on wall structure, whereas those of

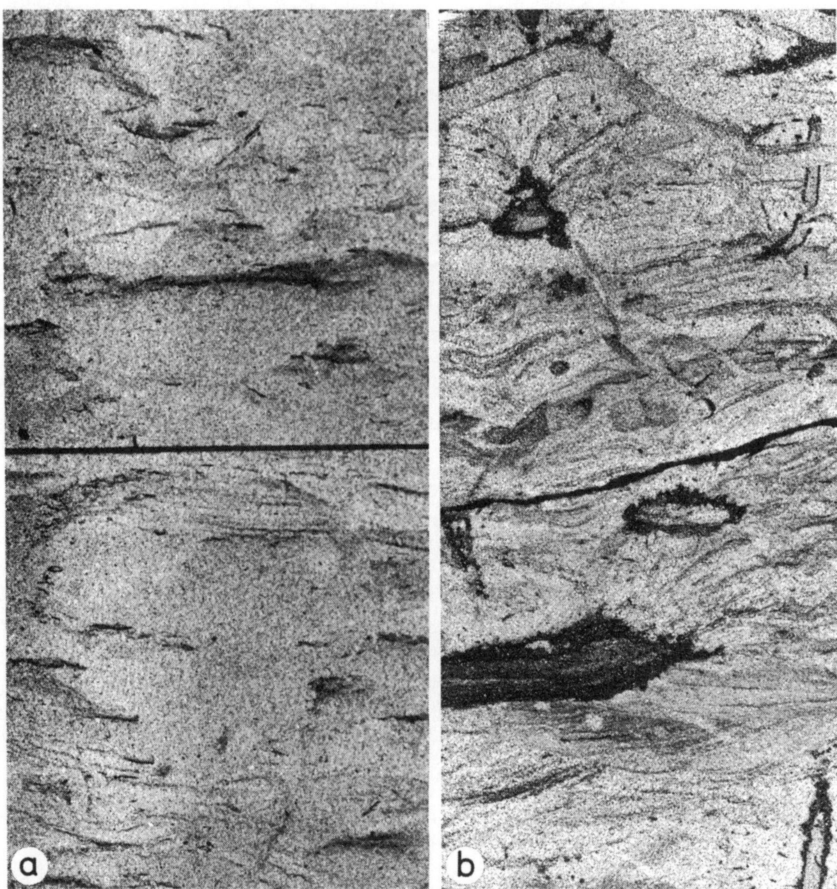

Figure 11.5 The ichnogenera *Thalassinoides* and *Ophiomorpha* are morphologically similar (Fig. 8.12) but have completely different expressions in core. (a) An incompletely bioturbated sandstone containing large burrow fills that suggest *Thalassinoides* isp. (b) The dark mud pellet lining of *Ophiomorpha nodosa* is strikingly visible in a sandstone containing, in addition, smaller structures referable to the ichnogenera *Planolites* and *Palaeophycus*. x0.5.

Rhizocorallium are less easy because they are distinguished on general form. However, with experience, a great deal can be achieved, and the tendency to relegate all structures to '?*Planolites*' isp., '?*Palaeophycus*' isp., '?*Teichichnus*' isp. and '?*Zoophycos*' isp. can be avoided!

11.3 Some ichnotaxa as seen in core

Rigorous description of outcrop trace fossils should routinely include internal features seen in cross-section as far as these are preserved. In

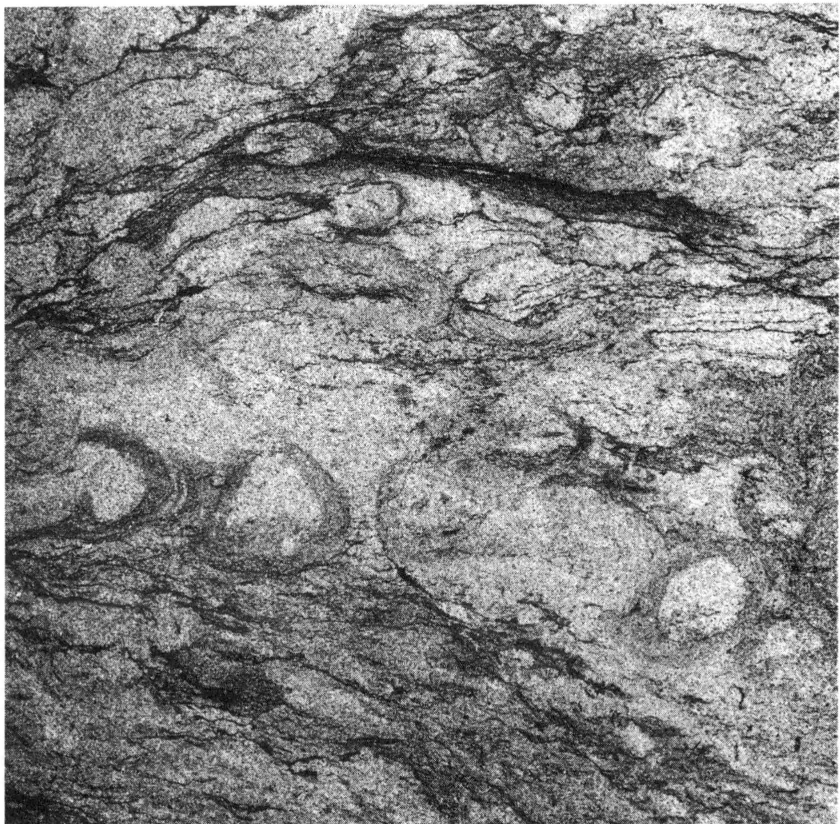

Figure 11.6 A muddy sandstone dominated by middle-tier stationary deposit-feeder ichnoguild, here represented by unusually clear sections of *Phoebichnus trochoides* (Figs 8.5 and 8.6). Inner shelf. Natural size.

unbedded lithofacies or in limestones having only diagenetic bedding, trace fossils commonly can only be observed in cross-section, so this problem of identification is not restricted to core. In such cases, the use of a hammer and serial sectioning on extensive material may demonstrate the three-dimensional form of the structure (e.g. Bromley and Ekdale 1984b).

Such studies are particularly valuable to the core ichnologist. The first attempt to bring this approach to bear on core material was that of Chamberlain (1975, 1978), in which trace fossils were illustrated in block and compared with their appearance in core. Chamberlain's initiative has been followed by other workers. Some of the more debatable groups are treated here.

11.3.1 Planolites, Palaeophycus *and* Macaronichnus

The ichnogenera *Planolites* and *Palaeophycus* are distinguished by the details of wall and fill (section 9.4.4), and thus are names much used for core (Figs 11.7a and 11.16). Characteristically, *Chondrites* is patchily distributed and shows such dense branching that it is usually distinguishable from *Planolites* (Fig. 10.10).

Chamberlain (1978) used the name *Terebellina* for very pale tubes, seen especially in offshore muddy settings. These structures are commonly crushed, having been incompletely filled (Figs 11.3, 11.7b, 11.8 and 11.14b). *Terebellina* is supposedly a short, horizontal or oblique tube, acuminate and closed distally (Frey and Howard 1981). According to A. K. Rindsberg (pers. comm. 1986), the holotype may be an agglutinating foraminiferan, *Bathysiphon* sp. Certainly these large foraminifera closely resemble the crushed white *Terebellina* tubes (Fig. 2.2; cf. Miller 1988; Gooday *et al.* 1992; Crimes and Uchman 1993).

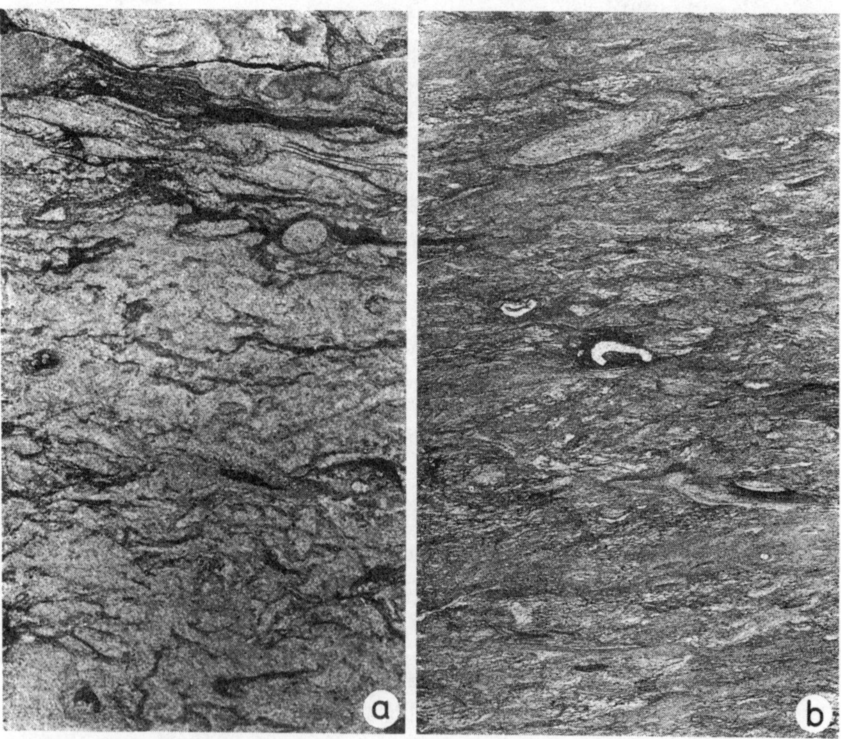

Figure 11.7 Two more ichnofabrics. (a) A nicely contrasted *Palaeophycus* fabric, dominated by this ichnogenus, from an inner-shelf environment. (b) From the outer shelf, a well-compacted, high-diversity assemblage containing *Rhizocorallium* isp. (top), many very squeezed *Teichichnus* spreiten, and tiny *Phycosiphon incertum* as the deepest tier. The white tubes at centre are *Terebellina* isp. Natural size.

Figure 11.8 From an inner-shelf setting, this ichnofabric shows ghost structures of a shallow-tier deposit-feeding activity cut by swarms of tiny black fills, possibly *Phycosiphon incertum*. Larger bodies may be the much re-used remains of *Thalassinoides* isp., and smaller ones *Planolites* isp. At the bottom are a few pale linings referable to *Terebellina* isp.

However, the name is commonly used for white-walled, brittle structures in core, even if their true origin is uncertain. Trace fossils referred to *Palaeophycus heberti* normally have thicker lining and less colour contrast with the matrix than *Terebellina* (Fig. 11.16).

Macaronichnus segregatis superficially resembles *Palaeophycus*. However, the outer sediment is a mantle, not a wall, and the structure is a backfill of a deposit feeder (section 8.3.2d, 9.4.4), not a domichnion like *Palaeophycus*. The zoned backfill has been created by particle segregation during the process of feeding, presumably on the basis of size, weight or shape. Thus, in a mixed sediment, dark particles tend to be concentrated in the mantle (not ingested?) and pale in the core (ingested?). If, however, the sediment has few dark particles, the contrast may be negligible and *M. segregatis* can be very difficult to see; it tends to be sediment-pervading, and produces an ichnofabric rather than individual trace fossils (Fig. 11.9).

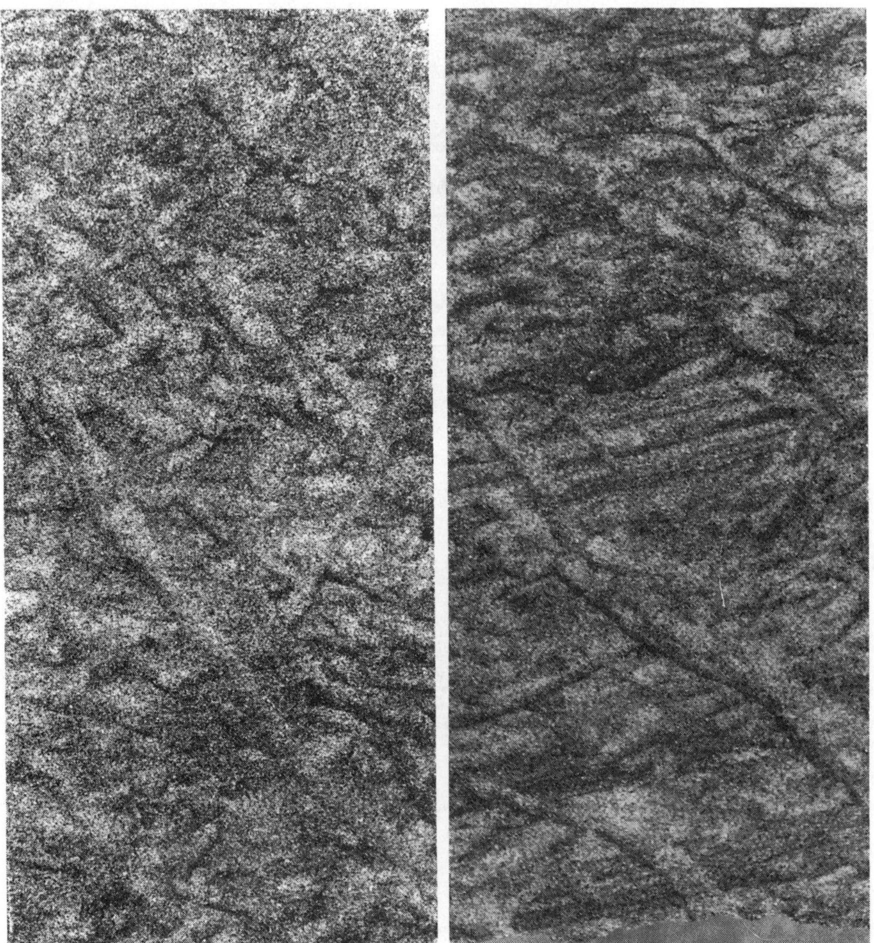

Figure 11.9 *Macaronichnus segregatis* ichnofabrics in Permian foreshore sand, offshore Norway. (a) Total bioturbation. (b) Incomplete bioturbation, leaving patches of primary lamination. The oblique burrows may represent up and down movements of animals in response to tidal shifting of the water table. Width of core 5.4 cm.

11.3.2 Phycosiphon incertum

This deposit-feeder structure has a dramatically different appearance in bedding-plane expression versus vertical section (Figs 11.10 and 11.11). The trace fossil is encountered abundantly in core and the two-dimensional view has been referred to *Helminthoida* by Chamberlain (1978), *Helminthopsis* in the first edition of this book, and given the new name *Anconichnus horizontalis* by Kern (1978). Examination of type material, however, has shown *Anconichnus* to be a junior synonym of *Phycosiphon* (Wetzel and Bromley 1994).

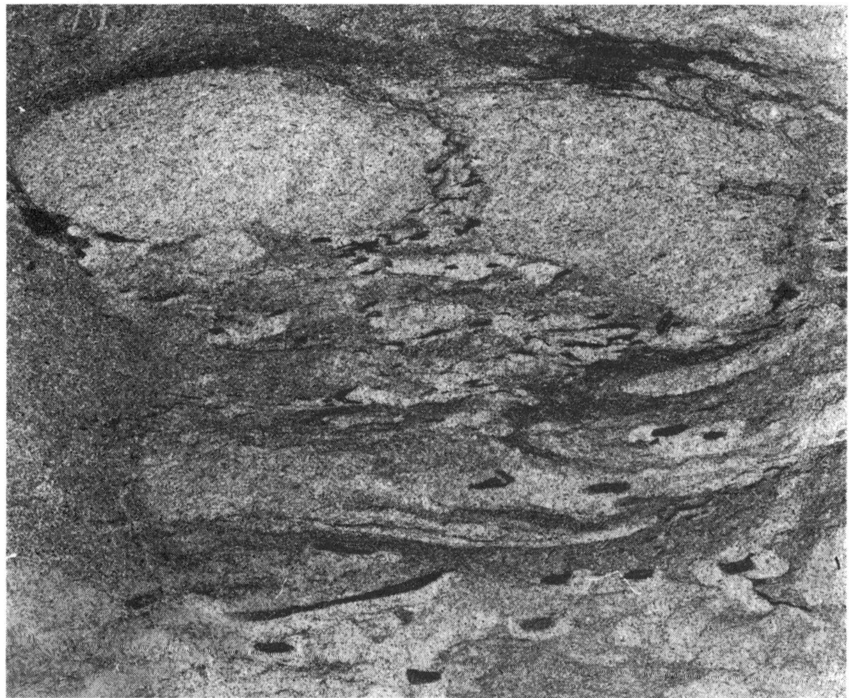

Figure 11.10 High diversity from the Carboniferous of the Barents Sea, offshore north Norway (x2). Two conspicuous rounded fills are referrable to *Thalassinoides* isp. These cut a mottled background fabric best seen at lower right, suggesting *Planolites* isp. The black-cored, pale-mantled *Phycosiphon incertum* avoids the *Thalassinoides* fills and represents a deep tier. At the upper right is a *Zoophycos* spreite, the deepest tier of the assemblage.

P. incertum is seen on bedding planes to be a complexly-lobed spreite fossil having a small-scale structure but covering large areas, following bedding-plane lamination and thus visible *in extenso* on cleavage surfaces. In more homogeneous or bioturbated sediment, however, the trace fossil does not spread horizontally but is tangled in three dimensions and the small spreite lobes commonly lie in a vertical plane (Kern 1978; Wetzel and Bromley 1994).

The whole structure is surrounded by a thin mantle of pale sediment. The core is of backfilled dark material, whereas the spreite is composed of the same pale material as the mantle. In section the dark fill is easily visible, but in a pale matrix the mantle and spreite are commonly difficult if not impossible to detect. In this case, the paired sections of fill where lobes are cut transversely, or 'fish-hooks' where cut longitudinally, are enough to reveal the presence of *P. incertum* (Fig. 10.21). The pale mantle is easily confused with a diagenetic halo (e.g. Tyszka 1994).

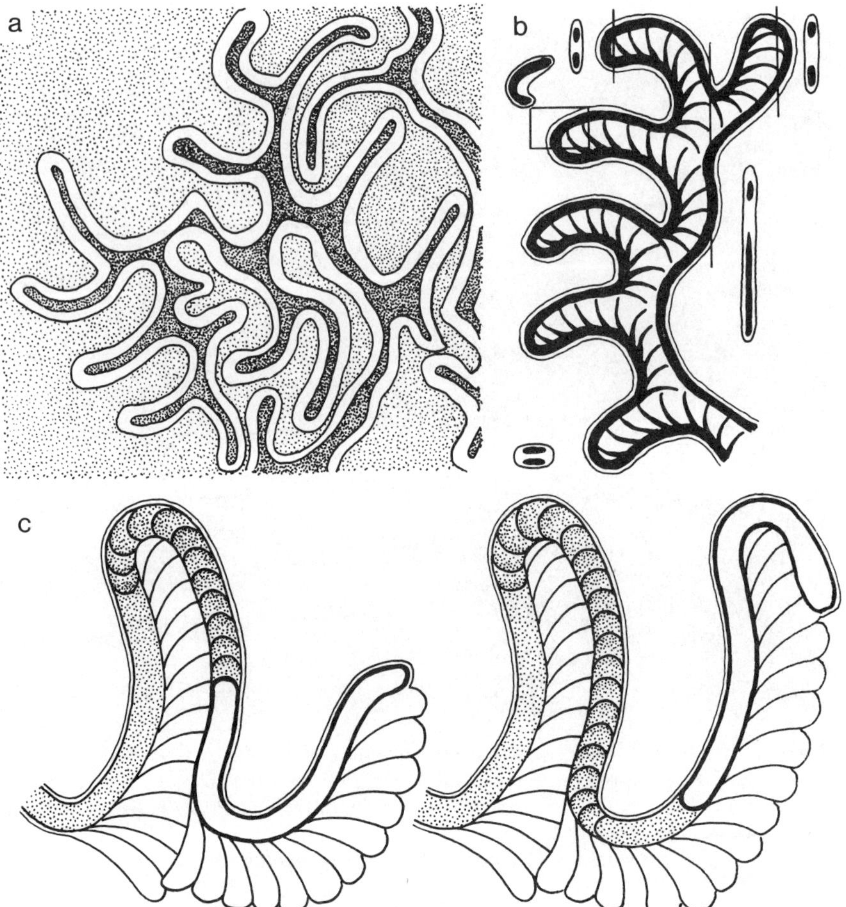

Figure 11.11 *Phycosiphon incertum*, lobes about 1 cm long. (a) Bedding-plane view of a horizontally orientated example. Spreite dark, marginal tube white (opposite of the usual colour-play). Note the rigorous thigmophobic behaviour: lobes never intersect. From a photograph (Häntzschel 1962). (b) Standard segment showing spreite structure, black core and white mantle. Some cross-sectional views are suggested: that at bottom left is of a compacted, vertically orientated lobe. Such cross-sections are visible in Fig. 10.21. (c) The core is alternately meniscate and homogeneous (Wetzel and Bromley 1994). The two sketches attempt to correlate structural details in a constructional 'probe-and-run' model: 19 probes and 19 meniscate packages. The worm has the length of a lobe.

11.3.3 Thalassinoides *and* Ophiomorpha

Where the diameter of cross-sectioned *Planolites* exceeds 1 cm, it becomes difficult to distinguish it from *Thalassinoides*. Because the diagnostic branching is so rarely displayed in core, *Thalassinoides* is an unsatisfactory ichnogenus to work with. Large, unlined fills may be referred to it,

Figure 11.12 A nice *Thalassinoides* ichnofabric cross-cut by tiny black fills probably representing *Phycosiphon incertum*, and by the feather-like spreite structure at upper right that is possibly *Lophoctenium* isp. (Fig. 11.17). Natural size.

although details of their true morphology are normally lacking. Despite the fact that this ichnogenus dominates shelf and nearby sediments throughout the Mesozoic and Cainozoic, it remains an unspectacular structure in core (Figs 11.5a, 11.10 and 11.12), and nearly always requires a question mark.

The opposite may be said of its cousin *Ophiomorpha*. Because its size does not range as large as *Thalassinoides*, and its wall structure is so characteristic, *O. nodosa* is immediately recognizable in core (Fig. 11.5b).

11.3.4 Teichichnus, Zoophycos *and* Rhizocorallium

These three ichnogenera, although quite distinctive in full view, are easily confused in section (Figs 11.4b and 11.13). Details of the spreite are commonly clearly visible; *Zoophycos* in particular can be subdivided into several types on this basis (Ekdale 1977), although formal subdivision of the ichnogenus into ichnospecies seriously needs attention.

Figure 11.13 A deposit-feeder community in an inner-shelf setting produced this ichnofabric, dominated by short *Teichichnus* spreiten. These cut across fills that may uncertainly be named *Thalassinoides* isp. or large *Planolites* isp. Some of the *Teichichnus* have been reworked in turn by minute *Phycosiphon incertum*. x2.

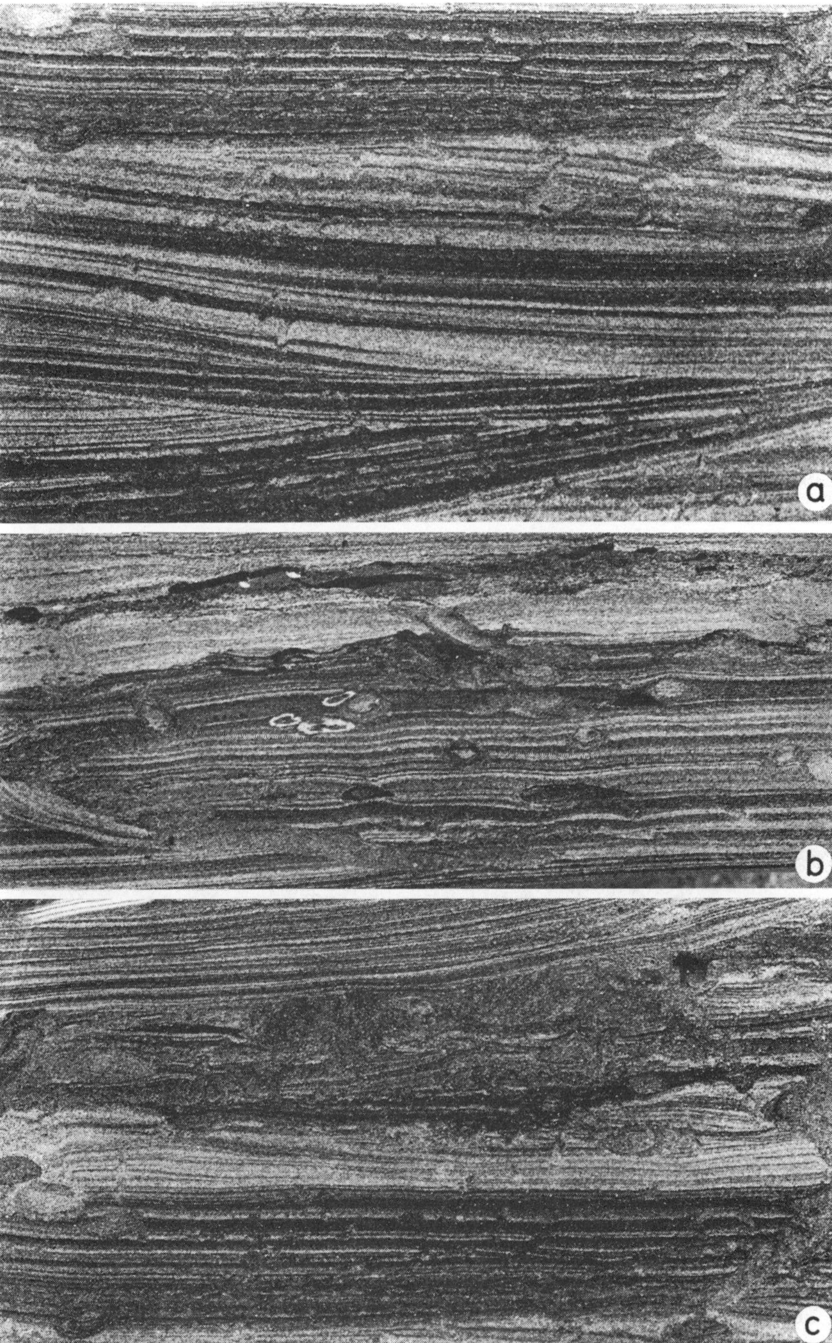

Figure 11.14 Three laminated units showing minimal biogenic disturbance. (a) If this sample were to be split along laminae, the trace fossils might be identifiable. In vertical section they are not. (b) Here, in particular, the larger structures appear to have multilaminar linings. Note the cluster of white-tubed *Terebellina* isp. (c) An erosional remnant of a more highly bioturbated top to a laminated unit (cf. Fig. 10.15b). Natural size.

11.3.5 Lost ichnogenera

While identification of the previously discussed ichnotaxa and many others can be problematic, at least their structure can be seen and interpreted. There are numerous ichnotaxa, however, that are important elements of assemblages as seen in outcrop, but which stand a negligible chance of being visible, let alone identified, in cross-section. Among these are such celebrities as the cubichnia *Rusophycus* and *Asteriacites*; the repichnia *Cruziana* and *Diplichnites*; and agrichnia such as *Paleodictyon* and *Cosmorhaphe*. These forms are largely dependent on semi-relief sandstone sole preservation which, in vertical section, appears at best as undiagnostic depressions in bed junctions (Fig. 11.14).

Figure 11.15 An ichnofabric largely composed of presumed deposit-feeder structures. A *Rhizocorallium* spreite is visible left of centre, and the other structures may be *Planolites* isp. and *Thalassinoides* isp. The two pairs of vertical tubes may be oblique sections of *Diplocraterion habichi*. Complete bioturbation, fairly high diversity: slow deposition of a sediment originally rich in organic material below the fair-weather wavebase is indicated. Natural size.

11.4 Ichnofabric and ichnodiversity

Yet another twist to the taphonomic filter has significance in the evaluation of ichnoassemblages in cores. The predominance of vertical planes of section clearly exaggerates the abundance of horizontal structures relative to vertical ones. Slender *Skolithos*, running parallel to the core, are unli-

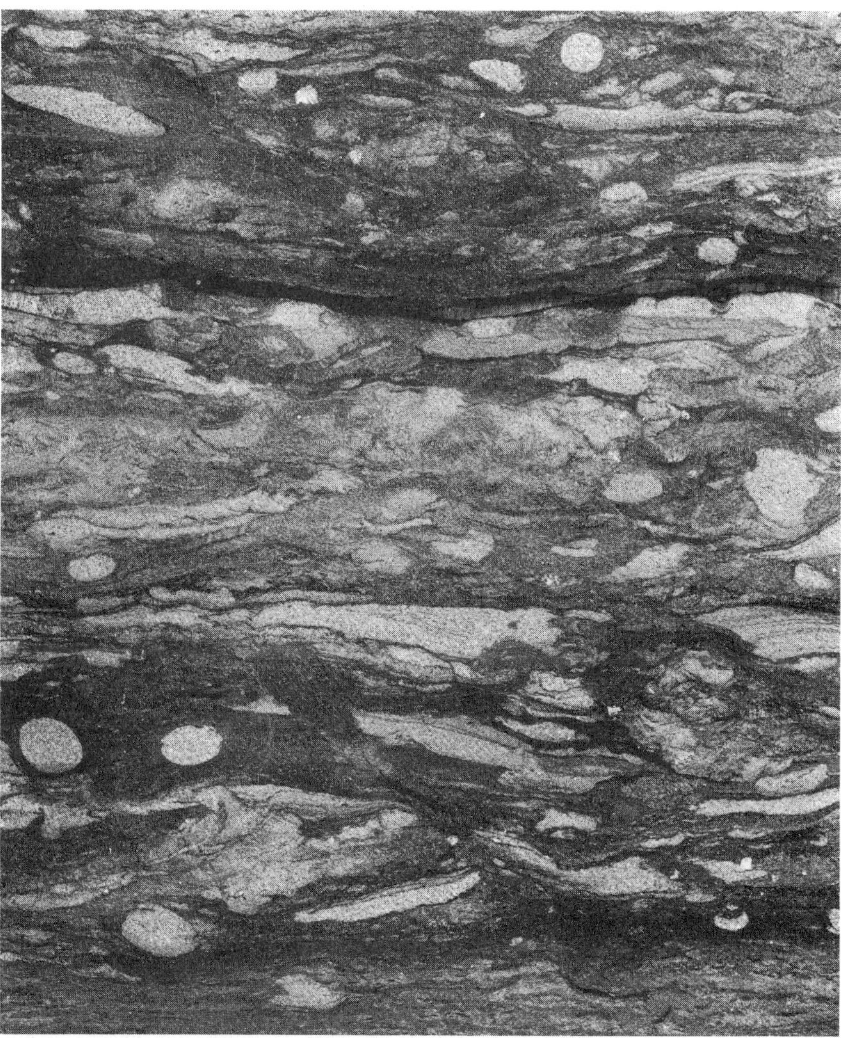

Figure 11.16 An ichnofabric dominated by large, thick-walled *Palaeophycus heberti*, suggesting domichnia of suspension feeders. These probably intersect the rather ill-defined fabric in the background. Bioturbation hardly exceeds 60 per cent. Although the individual trace fossils are mostly anonymous, a high diversity is present. Note the tiny *Phycosiphon incertum* in the bottom.

kely to be caught within the plane of section, whereas a horizontal section will catch them all. In vertical section, horizontal *Planolites* are seen wherever they occur. This discrepancy produces a bias in favour of deposit-feeder over suspension-feeder structures and also underplays ichnodiversity.

Despite this bias toward horizontal and smaller structures, and the problem of giving precise names to the trace fossils, there is rarely any diffi-

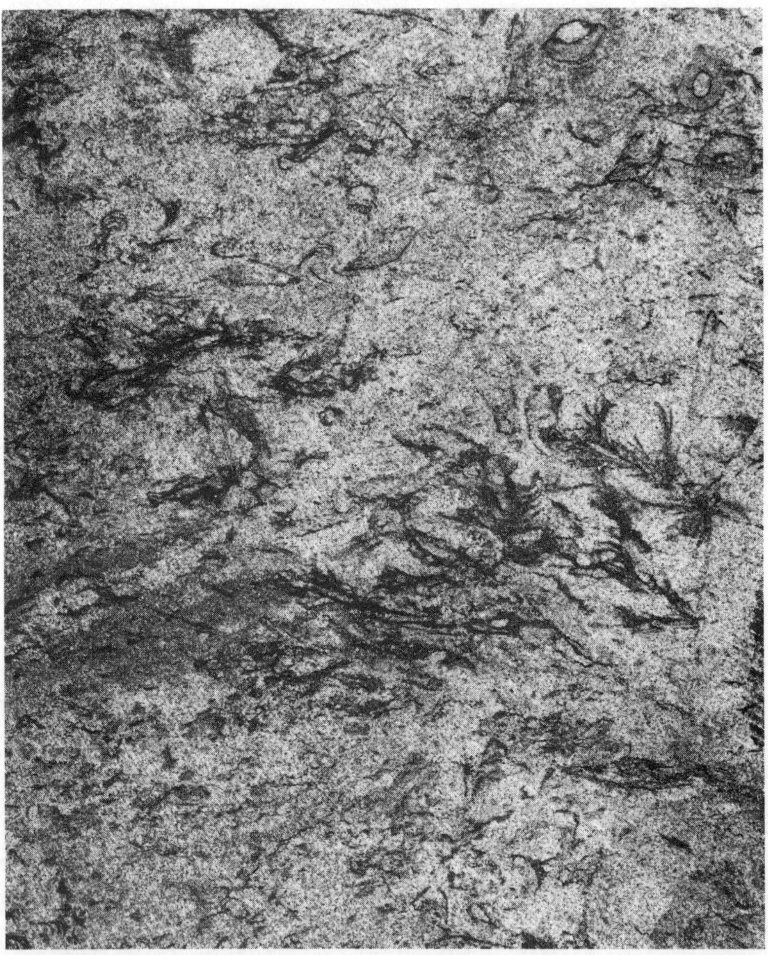

Figure 11.17 Totally bioturbated sand. Across a mottled backcloth suggesting centimetre-sized *Planolites* activity, well-preserved deeper-tier structures are visible. At top right, small *Palaeophycus heberti* form a group. Around the centre are the loose feathery spreiten of *Lophoctenium* isp. Tiny dark fills, possibly *Phycosiphon incertum*, speckle the fabric. A climax deposit-feeder community is indicated, suggesting reasonably slow deposition and stable, fully-marine conditions. Natural size.

culty encountered in obtaining detailed ichnological information. Accurate estimates of ichnodiversity of assemblages and density of bioturbation, bed by bed, give clear insight into the succession of facies and environments represented. This is why we work with ichnofabric (section 12.5).

Detail is the operative word. Compare the three examples of ichnofabrics in Figs 11.15–11.17, all from upper shoreface environments, with that in Fig. 11.18, from a foreshore setting.

Owing to the emphasis on well-preserved small detail, cut in awkward and undiagnostic section, studies of core material tend to underline ichnofabric analysis rather than ichnotaxonomy. This is especially the case in fully bioturbated sediments. A given assemblage might comprise eight distinct trace fossil types of which, for example, only one is identifiable to ichnospecies and three to ichnogeneric level. However, the cross-cutting relationships of nameable and anonymous structures alike may allow the tiering structure of the palaeoichnocoenosis to be assessed. Comparison

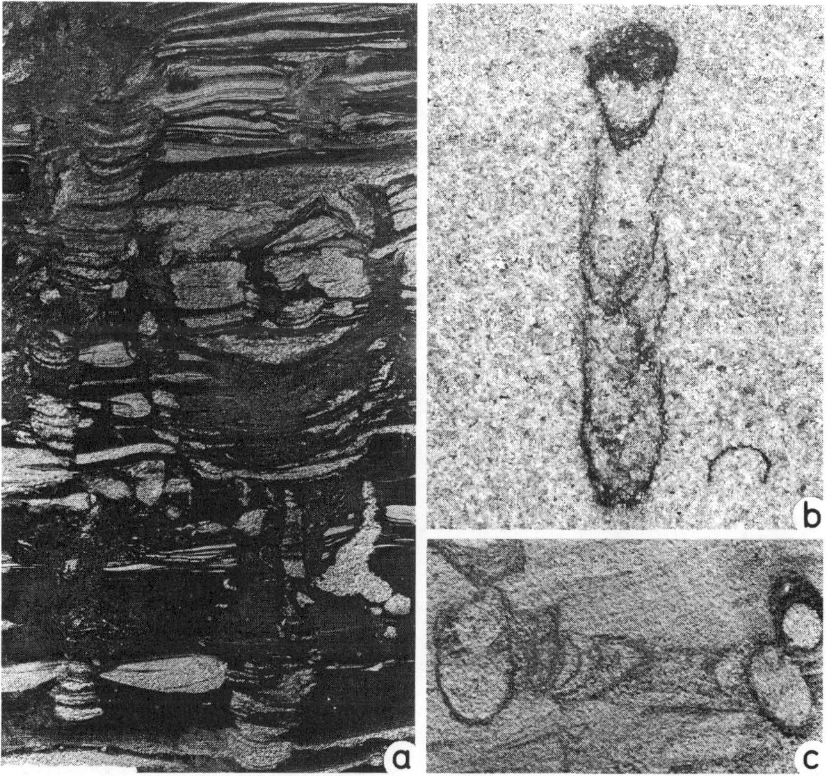

Figure 11.18 Three aspects of *Diplocraterion parallelum* in core: (a) a dense population in a heterolithite, x0.5; (b) vertical and (c) oblique sections, both natural size.

with neighbouring assemblages elsewhere in the core should emphasize differences in the bioturbation density of different tiers rather than the accidental discovery here and there of a particular ichnotaxon.

Succession of palaeoichnocoenoses, each represented by a change in ichnofabric, may clearly be documented, owing to the preservation of small detail. Thus, core is ideal for revealing the presence of several suites within an assemblage, identifying their chronological succession and relating them to possible environmental changes. These applications are discussed in the next chapter.

12 Solving problems with trace fossils

In the early days of ichnology, some very hopeful predictions were made of the new tool for sedimentology. Trace fossils seemed to provide an immediate key to palaeoenvironmental interpretation. Seilacher's (1967) ichnofacies were a fail-safe guide to palaeobathymetry and *Ophiomorpha nodosa* was a sure indicator of the littoral and shallow neritic environments (Weimer and Hoyt 1964).

There is much truth in these pronouncements, but the system has shown itself to be far more complicated and variable than was then envisaged and such statements are at best half-truths and over-generalizations. I would have liked to have ended this book with an environment-by-environment analysis on the basis of trace fossils, but because of the broad variation of each environment I shall follow another path. The reader is referred to Ekdale *et al.* (1984a) and Frey and Pemberton (1985) for environment-by-environment treatment.

Trace fossils effectively provide an *in situ* record of environment and environmental change based on factors that influence the individual and the community. I shall therefore discuss these factors in terms of ecological limits and disturbance events as they can be detected in the rock record. In a given stratigraphic succession these ecological indications can be pieced together to provide the local history of environmental change and basin evolution.

12.1 Identifying stress factors

Benthic communities are controlled in their development by a number of ecological limiting factors. The degree of ecological stress imposed on a community by these factors is recorded by a reduction of: (1) the diversity of species; (2) the size of individuals; and (3) the amount of endobenthic activity. All these changes are registered in the ichnofabric and trace fossils.

In aquatic benthic environments, the most important limiting factors are: (1) oxygen availability and (2) salinity. In addition, aspects of the depositional environment exercise an important control, including: (3) substrate consistency; (4) turbulence and bottom-water energy; (5) rate and style of deposition; (6) supply and type of organic matter; and (7) desiccation (e.g. Ekdale 1985, 1988; Oschmann 1988).

Furthermore, the benthic community is subject to several kinds of disturbance events: (1) turbidity currents; (2) storms; and (3) ash-falls. An understanding of the trace fossil assemblages and ichnofabrics associated with these limiting factors and disturbance events allows us to evaluate the more major environmental changes and in this way to identify the nature of the bounding surfaces that are of fundamental importance in sequence stratigraphy.

12.1.1 Oxygen

No sea-floor oxygen means no metazoan life and very slow breakdown of deposited organic matter. Thus, anoxia is of importance in hydrocarbon source-rock research, and there is a large literature. Recent important ichnologically orientated papers are Ekdale (1988), Oschmann (1988, 1991a, b, 1993), Savrda *et al.* (1991), Sageman *et al.* (1991), Wignall (1994), Erba and Primoli Silva (1994) and Savrda (in press).

A vital key for the identification of oxygen events lies in details of tiering (section 10.3.2). The fully aerobic biofacies is characterized by a fair degree of water turbulence and deep bioturbation. The endofauna is diverse and carbonate skeletons are well developed. With the reduction of dissolved oxygen below about 2.0–1.0 ml/l (opinions vary somewhat – Tyson and Pearson 1991) we pass into the upper dysoxic zone (Fig. 12.1). Owing to low turbulence, a nepheloid layer may develop, which will eliminate large endobenthos and suspension feeders. Bioturbation by deposit feeders continues but decreases in diversity, animal size and tier depth.

Below about 1.0–0.5 ml/l we enter the lower dysoxic zone. *Chondrites* and other probers penetrate below the mixed layer, and chemosymbiosis begins. There is little destruction of the microlamination, the motile bioturbators being chiefly restricted to nematodes. With further oxygen depletion there is a cessation of bioturbation, allowing the mud to be sealed by biomats. These provide a sharp separation between anoxic pore water containing H_2S, and the dysoxic bottom water. The bottom may be colonized by a chemosymbiotic shelly epifauna, representing the exaerobic biofacies of Savrda and Bottjer (1987a).

Sageman *et al.* (1991) preferred to see these large shelly colonists as opportunists during short oxic pulses, rather than chemosymbiotic K-strategists. But as oxygen concentration drops below 0.1 ml/l, effective anoxia is reached, and only microbes having anaerobic metabolism survive to produce biomats, and the primary microlamination of the sediment is preserved intact.

Wignall (1993) warned us that substrate softening from softground to soupground can produce similar reduction in tier depth, endobenthic biodiversity and animal size. However, this may be distinguished from an oxic–anoxic trend in that a zero point is not reached in the soupground, and species encrusting the shelly benthos remain unaffected.

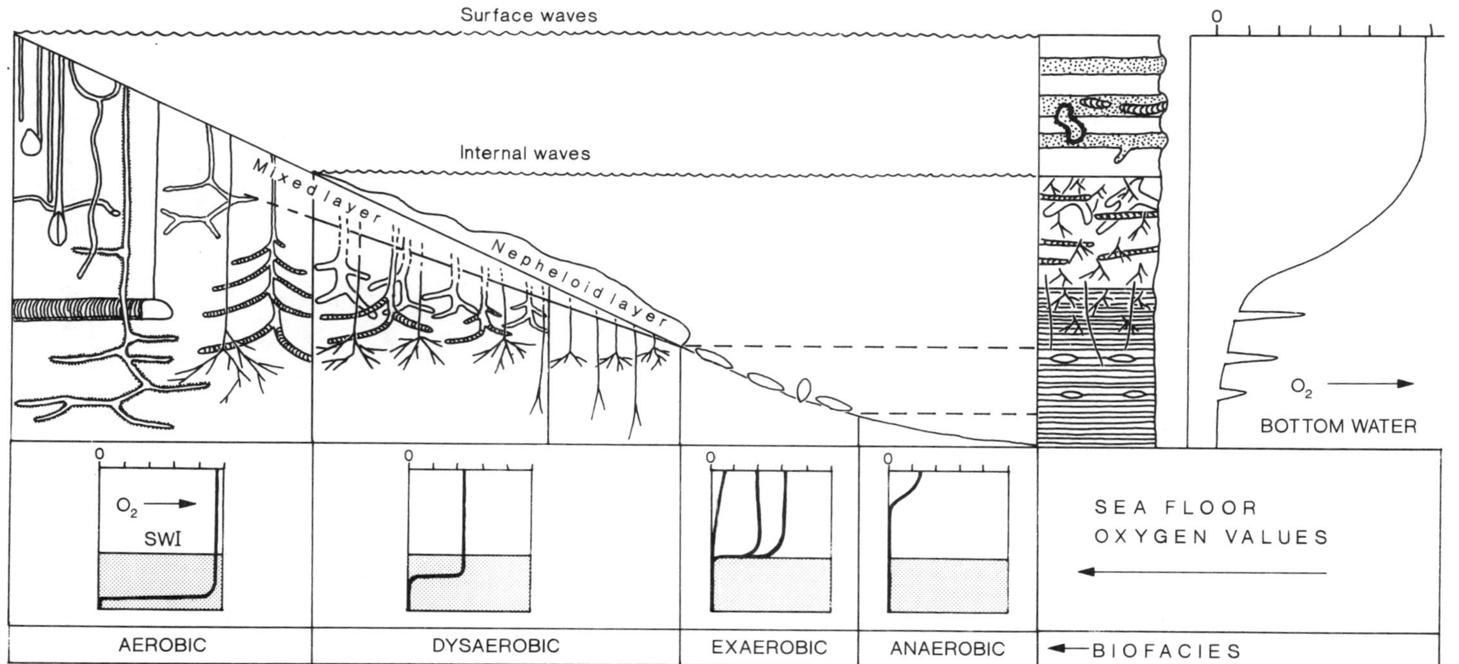

Figure 12.1 Oxygen-related biofacies scheme and related ichnofabrics. Dynamic vertical oxygen gradients across the sediment/water interface (SWI) are shown. Ichnogenera suggested are from left: *Skolithos*, bivalve burrows, callianassid burrow, *Planolites, Scolicia, Ophiomorpha, Thalassinoides, Chondrites, Zoophycos* and *Trichichnus*. Modified after Savrda *et al.* (1991).

Oschmann (1991a, b) pointed out that this model may be adequate for slope and deep-sea environments, where conditions show little short-term variation, but oxygen-controlled environments are also known in the open marine shelf and epeiric seas throughout earth history, and here the O_2–H_2S interface fluctuates in and out of the substrate. This may occur in a long-term cyclicity or as annual seasons.

In such settings a biofacies may be identified in which oxic and anoxic conditions alternate, allowing short-term colonization followed by suffocation. Oschmann (1991a) called this the poikiloaerobic biofacies. As anoxic conditions return and the O_2–H_2S interface rises to the sea floor, the burrowing opportunist bivalves move out of the substrate and die on the sea floor, and are not found in life position. Polychaete worms (Tyson and Pearson 1991) and spatangoid echinoids do likewise (Bromley *et al.* 1995). This opportunistic strategy demands that the anoxic periods are shorter than the life cycle of the bivalves and that they have planktonic larvae that live in the surface waters while the bottom is anoxic (Oschmann 1993).

Savrda and Bottjer (1986) and Savrda (1992) developed a model for reconstruction of oxygen-related ichnocoenoses based on ichnofabrics of

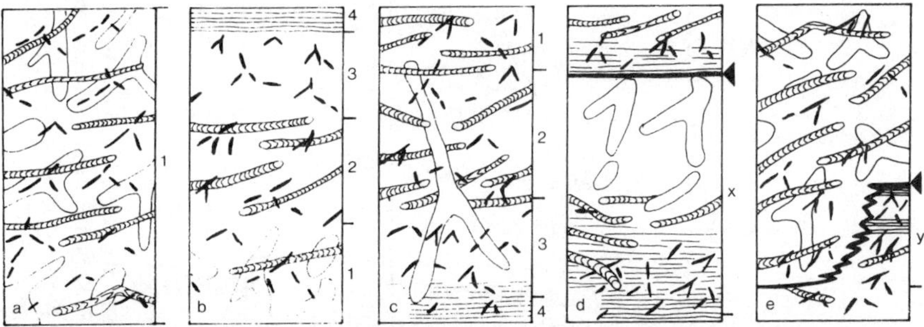

Figure 12.2 Diagrams illustrating tiering (cross-cutting) relationships expected in: (a) an initially stable, well-oxygenated sediment column; (b) conditions of gradually decreasing oxygenation; and (c) gradually increasing oxygenation. Numbered intervals delineate oxygen-related ichnocoenoses (ORI units), boundaries of which represent threshold levels at which certain ichnotaxa appear or disappear. Note that in (c), cross-cutting relationships appear to be the reverse of normal; this is due to the fact that as oxygenation increases, burrowing depth increases through time. In (d), abrupt decrease in oxygenation occurs at the arrow, allowing a frozen tiering profile to be preserved (x). In (e), an abrupt increase in oxygenation occurs at the arrow. Downward piping by the post-event community may partially or completely obliterate the low-oxygen ichnocoenosis immediately beneath the event surface, as is diagrammatically shown here. Trace fossils indicated are *Thalassinoides* (white), *Zoophycos* (spreite-structured) and *Chondrites* (black). Background is shown as homogenized by the mixed-layer tiers (white) or unbioturbated (laminated). Modified after Savrda and Bottjer (1986).

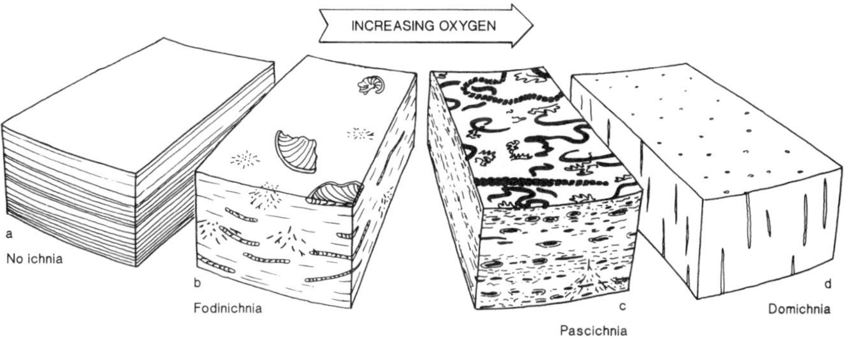

Figure 12.3 Ekdale–Mason model of oxygen-dependent trace fossil associations. (a) Primary lamination, abiotic situation. (b) So-called fodinichnia: *Zoophycos* and *Chondrites*. (c) Pascichnia dominated by *Scalarituba* (large, segmented), *Spirophycus* and *Phycosiphon* (minute), plus sparse *Chondrites*. (d) Full oxygenation of pore water and bottom water, dominated by *Skolithos*. Modified after Ekdale and Mason (1988).

Cretaceous chalks (of USA). Examples of cross-cutting relationships associated with different oxygen-related events are sketched in Fig. 12.2.

Ekdale and Mason (1988) presented a quite different model on the basis of bedding planes in a Carboniferous basin. Four trace fossil associations were considered to represent a gradient of increasing benthic oxygenation: (1) 'no ichnia'; (2) an association of fodinichnia, or traces of non-vagile deposit feeders; (3) an association of pascichnia, or traces of vagile deposit feeders; and (4) domichnia dominated by *Skolithos* (Fig. 12.3). At first sight the fodinichnia might seem to be out of place, but on closer examination these turn out to be the *Chondrites–Zoophycos* ichnoguild, and may in fact have more to do with chemosymbiosis than with deposit feeding. A somewhat similar trend is reported for Cretaceous turbidites (Buatois and Mángano 1992).

Both these models are from deep-water settings and, as Wheatcroft (1989) pointed out, do not compare closely with modern ocean community-recovery successions. The latter, at least in shallow water, normally begin with opportunistic domichnia producers (section 10.5.1).

More disparate results are being reported. Wignall (1991) described communities along an oxygen gradient from the Jurassic Kimmeridge Clay in which *Chondrites* was far from the last event, and a rigorous size reduction was not evident. In another Jurassic setting, the deepest tier of the most oxygen-restricted of six biofacies was occupied by the chemosymbiont *Solemya* sp. (Wignall and Pickering 1993).

However, Leszczinski (1991) and Leszczinski and Uchman (1993) are finding much in common between oxygen-stressed turbidite communities and chalk communities. We need more data before a general model can be constructed.

12.1.2 Salinity

Salinity is a critical factor for nearly all types of benthic life. Virtually no species are represented in both marine and freshwater environments. Brackish and hypersaline environments are colonized by organisms that are able to cope with changing salinity values, which means they can adjust to different osmotic pressures. Such organisms are called euryhaline, as opposed to stenohaline organisms, which include most marine and freshwater species. Stenohaline organisms can tolerate no or only minor deviations in salinity (see Fürsich 1994). Thus, some higher taxa are entirely, or almost entirely, restricted to these realms: brachiopods, echinoderms, polychaetes, sipunculans, echiurans, pogonophores, and prosobranch and opisthobranch snails are marine groups, as were the trilobites. Likewise, the insects, pulmonate snails and most oligochaetes characterize the freshwater realm.

Not so with behaviour. The salinity of the ambient water has little effect on the problems to be solved and the skills to be acquired for deposit feeding, prey catching, etc. Thus, there are many trace fossils that occur in both realms although their tracemakers were but distant relations.

Brackish and hypersaline waters are another matter. In most cases, both are subject to steep salinity gradients and fluctuations and can only be inhabited by relatively few euryhaline species. The brackish realm acts as a bottleneck between the marine and freshwater realms. Communities in brackish conditions show low diversity and small size of individuals. Environmental fluctuations, be they tidal or seasonal, maintain the communities at an early successional stage, dominated by opportunists, which may locally or temporarily reach high densities.

Freshwater bodies tend to be small, isolated and changeable. However, stability of salinity may allow larger diversity in more permanent lacustrine and fluviatile situations, and more mature and sophisticated communities may develop than in brackish conditions. Nevertheless, diversity and tiering do not approach those of climax marine communities.

Trace fossils show the same pattern. Ekdale *et al.* (1984a) expressed the great difficulty of summarizing the ichnology of environments of divergent salinity. Although much has now been published, the environments are so diverse, and temporarily and laterally variable, that there is little general pattern.

12.1.3 Brackish water

Pemberton and Wightman (1987) listed several generalities that would help identify brackish trace fossil assemblages. (1) Brackish waters are generally reduced in species numbers with respect to both freshwater and fully marine water. The species minimum occurs at about $5^{o}/_{oo}$, other major faunal breaks occurring at 18 and $30^{o}/_{oo}$. (2) Whereas the number of freshwater species decreases rapidly, even at slight increases in salinity, the reduction of marine species is more gradual. Therefore, the brackish water fauna can be considered more an impoverished marine assemblage than a true mixture of freshwater and marine elements. (3) Endobenthic animals are more abundant in low salinity waters than are epibenthic. The deep endobenthic habitat serves to buffer the animal against rapid and extreme salinity variations in the bottom waters. Thus, marine burrowing animals extend further up an estuary than marine epibenthic species. (4) With decreasing salinity, the reduction of species having calcareous skeleton is greater than of species lacking such a skeleton. (5) Many marine species display a size reduction of individuals with decreasing salinity. Conversely, however, freshwater organisms which are able to adapt to increasing salinities, show no changes in body size.

In their classic study of Georgia estuaries, Dörjes and Howard (1975) identified five animal communities on a traverse from freshwater to shallow shelf, and described their incipient trace fossils. Studies of ancient examples of estuarine environments have not recognized these communities. This is largely on account of such taphonomic factors as time-averaging of a highly variable depositional and ecological regime. Trace fossil assemblages therefore tend to be more diverse than expected and to include mixtures of Cruziana and Skolithos ichnofacies forms (Wightman *et al.* 1987; Greb and Chesnut 1994).

Pemberton and Wightman (1992) made some general conclusions on the basis of studies of several brackish settings of Cretaceous age (Beynon *et al.* 1988; Beynon and Pemberton 1992; Pattison 1992). They concluded that the general trace fossil assemblage reflects inherently fluctuating environmental parameters and is characterized by: (1) low diversity; (2) forms typically found in marine environments (*Skolithos, Arenicolites, Planolites, Teichichnus, Chondrites, Thalassinoides* and *Gyrolithes*); (3) simple structures constructed by trophic generalists; (4) suites that are commonly dominated by a single ichnotaxon; (5) mixtures of trace fossils characteristic of both Skolithos and Cruziana ichnofacies; and (6) some forms found in prolific numbers.

If we pass further into the geological past we meet similar patterns. In the Devonian, for instance, Miller (1991) found the spiral spreite fossil *Spirophyton* in estuarine settings to show a distribution indicating typical opportunistic strategy: the trace fossils occurred patchily in great abundance and almost, or completely, alone.

12.1.4 Freshwater

Freshwater trace fossil assemblages are produced in even more diverse environments than brackish assemblages, and the ecological transition from non-marine to marine is extremely varied (Fig. 12.4). It is therefore difficult to generalize.

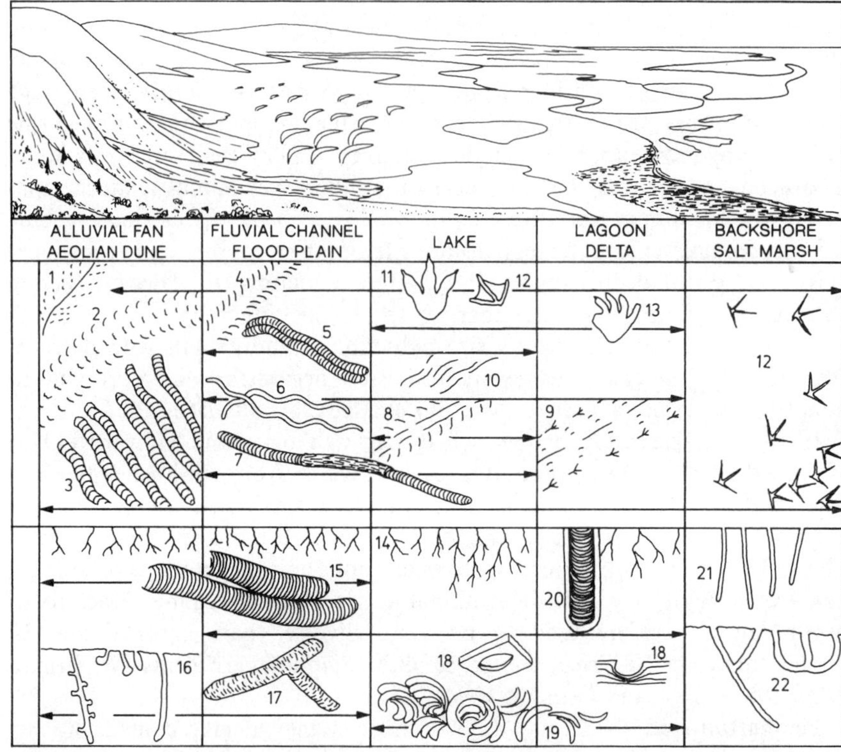

Figure 12.4 Distribution of some trace fossils in non-marine environments (modified after Pollard in Goldring 1991). 1, *Paleohelcura* (scorpion trackway); 2, *Mesichnium* (insect trackway); 3 *Entradichnus* (exogenic insect traces); 4, *Acripes* (crustacean trackway); 5, *Cruziana problematica* (branchiopod crustacean burrow); 6, *Cochlichnus*; 7, *Scoyenia gracilis*; 8, *Siskemia* (arthropod trackway); 9, *Kouphichnium* (xiphosuran trackway); 10, *Undichnus* (fish swimming traces); 11, reptile track; 12, bird tracks; 13, amphibian track; 14, roots; 15, *Beaconites*; 16, insect burrows; 17, *Spongeliomorpha carlsbergi* (insect burrow); 18, *Lockeia siliquaria*; 19, *Fuersichnus communis*; 20, *Diplocraterion parallelum*; 21, *Skolithos*; 22, *Psilonichnus* and other crab burrows.

Lacustrine sediments can be completely bioturbated by deposit feeders (Fig. 10.17b; McCall and Tevesz 1982) and, at the ichnogenus level, structures similar to marine trace fossils can be produced. Insect larvae construct *Arenicolites* and oligochaete worms produce both these and miniscule *Thalassinoides* networks, insects create *Spongeliomorpha* (Fig. 10.5) and notostracan crustaceans form *Cruziana* (Fig. 8.13). But, at the ichnospecies rank, these are different from marine forms, even though formal taxonomy of these structures is still very unsatisfactory (Bromley and Asgaard 1979; Tevesz and McCall 1982; Miller 1984; Ekdale and Picard 1985).

As with brackish assemblages, however, diversity tends to be very low, with heavy dominance by single ichnospecies, and tiering is insignificant.

Bipedal and quadrupedal vertebrate trackways are commonly associated with fluviatile and marginal lacustrine assemblages (Lockley 1991), and invertebrate trackways are normally abundant (Walter 1983; Walker 1985).

12.2 Interplay with depositional processes

Three fundamental sedimentological processes interact to produce sedimentary deposits: sediment accumulation, bioturbation and erosion. Each process operates at an independent rate, which may be constant or varying.

In very general terms, the character of any deposit depends on the balance between the rates of these processes. Where rate of sedimentation far exceeds that of bioturbation, primary stratification features and physical sedimentary structures will predominate. Where the reverse is the case, primary stratification will be obliterated and biogenic sedimentary structures will predominate (e.g. Howard and Reineck 1981).

Two general styles of deposition may be identified: (1) the slow aggradation that characterizes low-energy predictable environments; and (2) event-bed sedimentation. Assuming that the oxygen and salinity stress systems do not play a part, the following generalizations can be made.

12.2.1 Slow and predictable

Normally, the rate of bioturbation exceeds that of deposition and total bioturbation is achieved. Endobenthic communities become mature, tiering is well developed (in marine communities) and the structures of the deepest tiers dominate the ichnofabric as elite trace fossils (Fig. 10.19).

Nevertheless, deposition in low-energy environments may be rapid, and in itself represents an ecological limiting factor. Such a stressed community normally will show loss of the deeper tiers of specialists, and bioturbation will be incomplete.

12.2.2 Event deposition

The dynamics of event deposition produce an almost limitless variety of ichnological features. The events themselves range from the daily increments of tidal bundles to the once-a-millenium turbidity current, and have in common the sudden deposition of sediment (obrusion), sometimes associated with erosion. The thickness of the obrusion blanket, and any contrast in grain size compared with the background sediment, play important roles in the ichnological response.

In the case of minor depositional events, some of the endobenthos will be able to move up to the new depositional surface. Equilibrichnia or fugichnia will be produced according to the depth of burial (Fig. 10.25).

With deeper burial, however, the benthic community will be smothered and the new sea floor will be colonized afresh. Several paths of recruitment are possible. Pioneers may settle out as metamorphosing larvae from the plankton; or member species of the background community may be relocated by lateral migration of adults; or again, living endobenthic individuals may be introduced together with the event sediment (Föllmi and Grimm 1990; Frey and Goldring 1992).

Tempestites can be associated with both sudden erosion and accretion, and repeated events can produce complicated results (Fig. 10.15). In such cases it is important to attempt to identify the colonization surfaces from which trace fossils originate. In its simplest form, the colonization surface is the top of a laminated-to-scrambled unit (Figs 10.14 and 12.5; Orr 1994). However, it can be difficult to define this surface where: (1) deeper burrows of one community overlap upper tiers of another; (2) where the colonization surface has been removed by erosion; or (3) where storm-wane (decline) or post-storm sedimentation has caused upward relocation of individuals (Figs 10.23, 10.25 and 12.6). Nevertheless, the search for colonization surfaces helps to define these problems; it helps to find the top of distal mud-tempestites, to unravel amalgamated beds, and to evaluate the relative amounts of erosion (Aigner and Reineck 1982; Frey and Goldring 1992; Orr 1994). In fact, when recording these trace fossils during the logging of sedimentary successions, attempt should always be

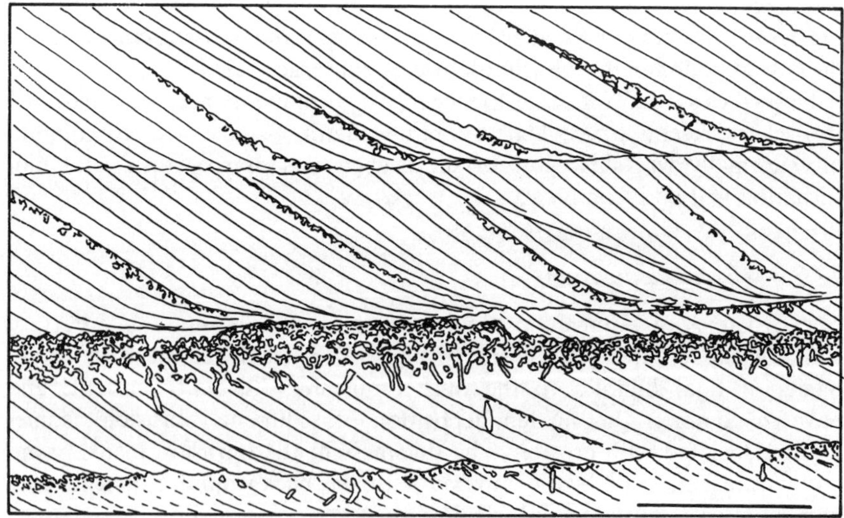

Figure 12.5 The colonization window. Long-term bioturbation in bottom set, and at periods of slack water in tidal cross-stratification. Based on Goldring (1991).

made to relate each trace fossil to an appropriate colonization surface. This practice increases the rigor of the documentation process.

12.2.3 The colonization window

Pollard *et al.* (1993) took this a step further. In higher energy environments (Skolithos ichnofacies), colonization of the shifting sands is impossible during energetic periods. The endobenthos probably gains access to the substrate during relatively brief periods of tranquility. *Ophiomorpha nodosa* is usually truncated upwards by erosion, but where apertures are preserved these often coincide with a mud horizon or lamina, indicating the colonization surface (Fig. 12.6). The importance of recognizing the

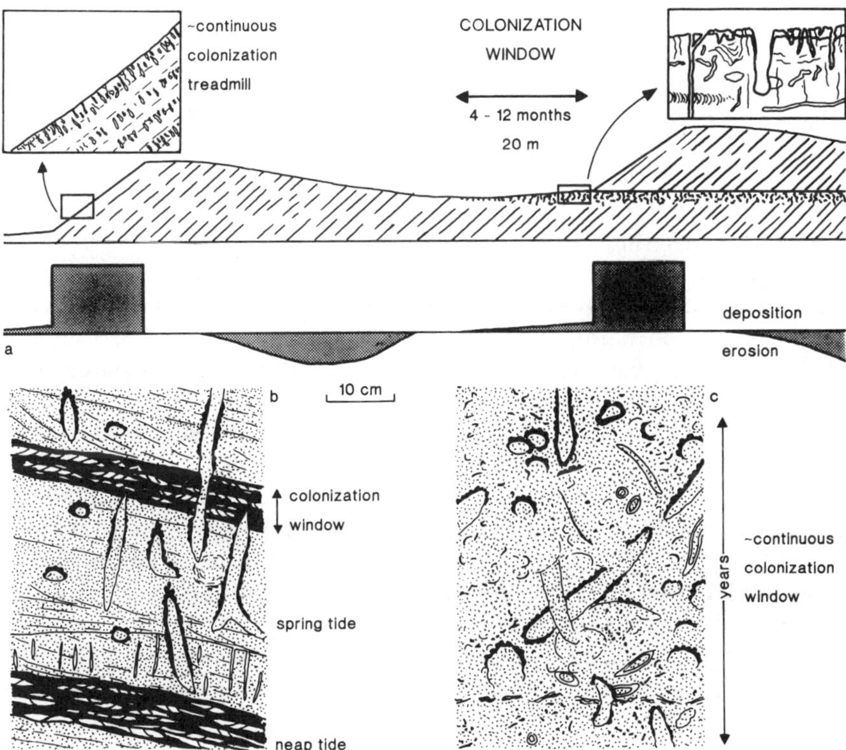

Figure 12.6 The colonization window. (a) Sand wave migration and opportunity for colonization of the trough and foresets. (b) Ichnofabric of an estuarine point bar where colonization by the producer of *Ophiomorpha nodosa* occurred only during periods of mud deposition. Note the narrow apertures in mud horizons, suggesting a colonization surface; but the *Ophiomorpha* producer has moved upwards in pace with increments of sedimentation. Low diversity and abundance. (c) Shoreface ichnofabric where physical restraints on colonization probably were limited to infrequent storms. High diversity, total bioturbation. Modified after Pollard *et al.* (1993).

colonization window lies in the fact that colonization events are linked to changes in the hydrodynamic regime.

The time available for colonization by a tracemaker represents the colonization window. The window may be more or less continuously open (as with slow accretion) or very briefly and seldom open (in high energy shifting sands or low energy poikiloaerobic facies).

Different tracemaking species having divergent colonization requirements will have different colonization windows. Thinking in these terms increases precision in observation, and aids in distinguishing narrow ichnocoenoses from compound assemblages. Do the *Skolithos* and *Ophiomorpha* of the *Skolithos–Ophiomorpha* ichnoguild (section 10.6.3) belong to the same ichnocoenosis or to two different suites representing subtly different subenvironments and colonization events?

12.2.4 Tubular tempestites

Tempestites commonly have a basal shell-bed conglomerate. Distal tempestites may be so thin that the recolonizing post-event endobenthos can reach through it and disrupt the shell bed (e.g. Fürsich *et al.* 1994). In proximal tempestites, by contrast, the pre-event burrows also may have a disrupting effect.

Callianassid shrimps may actively fill deep galleries of their burrow systems with gravel (Fig. 4.27b; Tudhope and Scoffin 1984; Scoffin 1992), but this process can also occur passively (gravitationally). Open *Thalassinoides* systems in hardgrounds have long been known to be refuges for skeletal material that has become trapped there during the period of non-deposition represented by the hardground (e.g. Voigt 1959).

Wanless *et al.* (1988) and Tedesco and Wanless (1991) described a situation where such a process of sediment trapping occurs in loosegrounds, and may therefore be repetitive, ultimately entirely to alter the fabric and lithofacies of the original substrate (Fig. 12.7). These authors founded their model on studies in Holocene carbonates of British West Indies and Florida in connection with hurricane deposition and involving burrows of callianassid shrimps. However, it is clear that, to varying degrees, 'tubular tempestites' must have been an important feature of sediments of many environments throughout the Phanerozoic (Tedesco and Wanless 1991).

Erosion during the early phase of the storm removes the restricted apertures and opens up the burrow shafts, allowing unhindered filling with sediment during the waning phase of the storm. Such is the capacity of the burrow systems that the entire storm lag may be swallowed by them and no basal bed developed at the sea floor. The repetitive coupling of excavation of deep burrow networks and their subsequent filling with sediment is a gradual subsurface process that transforms the original deposit into one of new compositional, textural and faunal attributes.

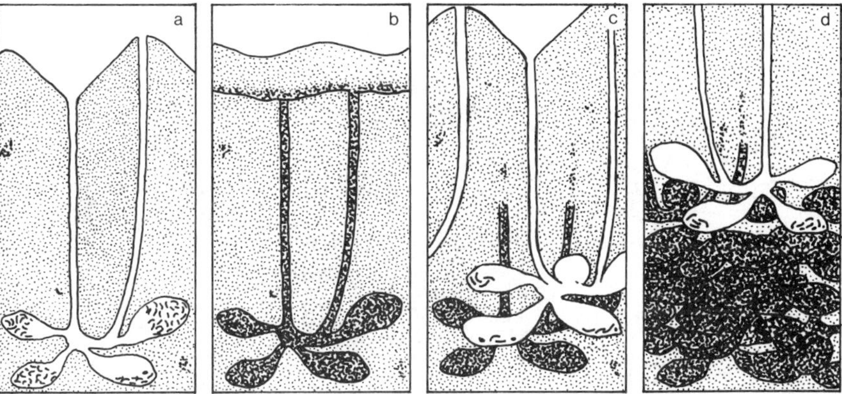

Figure 12.7 Diagrammatic sequence of stages of tubular tempestite modification of substrate. (a) Pre-storm, an open callianassid burrow in tidal flat sediment. Coarse grains are packed into the chambers by the crustacean burrower. (b) Mounded surface flattened by storm and deposition of storm layer. (c) Re-excavation of burrows, preferentially advecting fines up, coarse grains packed in chambers. (d) Repetition of these processes produces marked increase in grainsize of the substrate. Modified after Tedesco and Wanless (1991).

The biogenically generated lithology is an internal sediment, created subsurface, and is only indirectly related to the sea floor depositional environment.

12.3 Ichnology for sequence stratigraphy

Sequence stratigraphy recognizes a hierarchy of stratal units composed of genetically related facies and bounded by surfaces that are chronostratigraphically significant. It is based on physical relationships of the strata, the lateral continuity and geometry of the surfaces bounding the units and the nature of the facies on either side of the bounding surfaces.

As sensitive indicators of environmental change, trace fossils are therefore of value in detailed sequence analysis. Understanding the behavioural significance of individual trace fossils and the development of ichnofabrics allows key stratal surfaces to be identified, surfaces which may otherwise easily be overlooked. In some settings (e.g. lower coastal plain, offshore shelf), trace fossils are often the only indicators of stratal surfaces such as marine flooding surfaces. Also, in highly or completely bioturbated shallow marine sandstones, the study of ichnofabrics allows detailed palaeoenvironmental and sequence analysis (Bockelie 1991; Savrda 1991, in press; Taylor and Gawthorpe 1993).

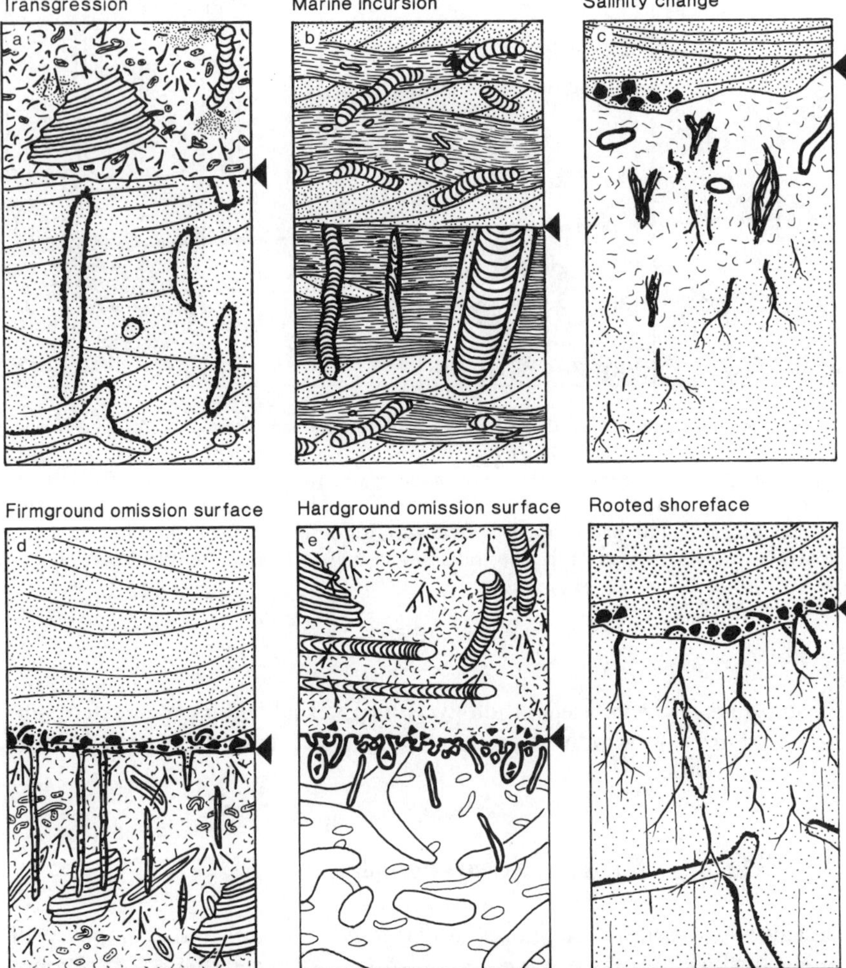

Figure 12.8 Cartoons showing some ways that trace fossils indicate key stratal surfaces. Note cross-cutting relationships of structures. (a) Change in facies: marine flooding surface. Completely bioturbated offshore mud containing *Chondrites, Teichichnus, Phycosiphon, Planolites* and ghosts of *Thalassinoides* overlies shoreface sands containing low-abundance *Ophiomorpha.* (b) Short-term salinity change: a brief colonization cross-cuts a non-marine *Taenidium* fabric in overbank heterolithics. (c) A marine incursion allowed colonization with *Rosselia, Planolites* and *Palaeophycus*, which cross-cut a plant-root fabric. Overlain again by fluvial channel sand. (d) Change in sediment consistency during a hiatus allowed colonization with firmground *Skolithos*, cross-cutting an offshore softground ichnocoenosis. The omission surface is overlain by shallow marine sand. (e) Cementation of an omission surface allows boring animals to colonize; their borings transect the work of a softground community (*Thalassinoides* and large *Planolites*). This is overlain by completely bioturbated offshore mud (*Chondrites, Zoophycos, Teichichnus, Planolites* and ghosts of (Thalassinoides). (f) An immature soil profile containing plant roots developed in an upper shoreface sand containing *Ophiomorpha* and *Skolithos.* It is overlain by fluvial sand. Based on Taylor and Gawthorpe (1993).

12.3.1 Bounding surfaces

Because trace fossils reveal sediment consistency, community structure, colonization patterns and bioturbation texture, they are particularly useful in defining environmental change at stratal surfaces (Fig. 12.8; Bromley 1975; Pemberton *et al.* 1992; MacEachern *et al.* 1992; Savrda in press). Some individual ichnotaxa are helpful in indicating the nature of bounding surfaces.

For example, *Diplocraterion parallelum* is a trustworthy indicator of marine flooding surfaces, and is found emplaced in all sorts of pre-flooding substrates (peat, mud, sand). *D. parallelum* commonly occurs alone in such situations as if tolerant of initial salinity fluctuations (Fig. 11.18a; e.g. Dam 1990b; Taylor and Gawthorpe 1993).

In Jurassic and Cretaceous sandstones, *D. habichi* commonly occurs at extensive omission surfaces (Heinberg and Birkelund 1984). These surfaces have been interpreted as transgression surfaces following minor lowering of sea-level (Dam 1990b; Surlyk 1991; Surlyk and Noe-Nygaard 1991).

This example also illustrates the value of trace fossils in identification of marker horizons. Bromley and Gale (1982) used the toponomy of trace fossils in correlating individual hardgrounds over large distances; the lithologies of the hardgrounds were similar, but each had a characteristic trace fossil assemblage.

The development of plant rooting structures at bounding surfaces, of course, is a valuable indicator of subaerial exposure (Bockelie 1994). Other exposure features, such as palaeosol development or karsting, are often eradicated by the following marine flooding, but plant roots may penetrate deeply enough to escape destruction. If marine burrows then cross-cut the root structures the case is closed.

When searching for surfaces it must be borne in mind that bioturbation can extend several metres beneath the colonization surface. In order to assess the environmental changes that occurred at the surface it will be necessary to distinguish the trace fossils that are connected with the omission surface from those of the pre-omission suite. There is commonly a lithofacies change at bounding surfaces, and the fill of at least the open burrows among the omission and post-omission trace fossils may reveal their nature by piping down the post-omission sediment (Fig. 10.20; Bromley 1975).

12.3.2 Sea-level changes

Although trace fossils provide no absolute bathymetric measuring-stick, they may give clear indications of water-depth changes and trends. Sensitivity of trace fossils to environmental changes related to water depth provide these indications. Changes in factors such as turbulence, organic matter deposition and oxygen supply are indicated by proximality trends, bioturbation patterns and burrow-size data (e.g. Easthouse and Driese 1988; Szulc 1990; Rioult *et al.* 1991; Bergman 1994; Ricketts 1994).

Individual ichnotaxa or trace fossil groups can be directly indicative of certain environments. Insect burrows identify terrestrial soil or freshwater deposits (Smith and Hein 1971; Retallak 1984; Hasiotis *et al.* 1993; Thackray 1994). Pollard *et al.* (1993) documented three ichnofabrics containing *Ophiomorpha nodosa* and *Macaronichnus segregatis* and related these to nearshore, shoreline and estuarine environments.

The problem with this approach, as with body fossil indication of palaeoenvironment, is the evolution of the biosphere. Insect burrows, *Ophiomorpha* and *Macaronichnus*, are hardly known from the Palaeozoic, just as *Cruziana* is hardly known after it. There are many examples of bathymetrically deepening trends of ichnotaxa through the Phanerozoic (e.g. Crimes and Fedonkin 1994). Several authors have hailed *Zoophycos* as an indicator of maximum flooding surfaces (Rioult *et al.* 1991; Savrda 1991) but this will not hold for the pre-Triassic! Savrda (in press) emphasizes that very different ichnofabrics have been associated with maximum flooding surfaces and highstand system tracts, and that much work needs to be done in this area of investigation.

12.4 Trace fossil analysis using assemblages

Trace fossil analysis can be approached from several directions, depending on the nature of the material, the type of project and the background of the ichnologist. The two main styles of analysis in use are: (1) that of ichnofacies, assemblages and ichnotaxa and (2) that of ichnofabric. The former is well suited to field exposures and bedding-plane material, the latter to core. Neither style, however, is exclusively restricted to these realms. In fact, they are really 'end-members' of a continuous spectrum of methods of analysis. For example, S. G. Pemberton and colleagues have great success using the assemblage method on core (Pemberton 1992), whereas Pollard *et al.* (1993) have successfully used ichnofabric in the field.

The use of assemblages in trace fossil analysis is very widespread. As an example I briefly outline the study by Dam (1990b) of the Lower Jurassic of East Greenland, not least because that is where I taught myself siliciclastic ichnology!

The shallow marine deposits of the Neill Klinter Formation contain a diverse assemblage of trace fossils. Dam (1990b) grouped the 34 ichnotaxa in a series of 11 'ichnocoenoses' (Fig. 12.9). As usual with this type of study, the so-called 'ichnocoenoses' are in fact clearly time-averaged amalgamations of the work of several related communities, as is expressed in Dam's block diagrams, and perhaps 'association' would be more accurate than ichnocoenosis. But this is a perennial problem.

Dam (1990b) interpreted the ichnocoenoses in the light of their trophic and ethological properties and found a strong correlation with the sedimentary environments. The distribution reflects changes in factors controlled by water depth, bottom-water oxygenation and environmental stability. The study emphasized the importance of collecting ichnological data in relation to sedimentological and palaeontological observations so as to produce an integrated analysis leading to reliable interpretations.

It would be desirable to relate Dam's ichnocoenoses to the ichnofabrics that are usually worked with in the offshore Jurassic of the North Sea (Bockelie 1991). However, the usual problems arise. The ichnocoenoses are largely based on ichnotaxa that do not provide distinctive images in vertical section. The key ichnogenera *Curvolithus, Phoebichnus, Gyrophyllites, Gyrochorte* and *Asteriacites* are notoriously difficult to identify in core.

12.5 Trace fossil analysis using ichnofabric

In order to promote the rigorous study of a core or outcrop succession, data may be collected in a variety of ways to provide as complete a picture as possible. The following pro forma schemes encourage the collection and manipulation of several forms of ichnological data simultaneously.

Figure 12.9 (Pages 292 and 293.) Some trace fossil associations from the marine Lower Jurassic of East Greenland recognized by Dam (1990b). (a) *Arenicolites* isp.2 association: aerated environment, continuous high-energy, large net sedimentation; proximal subaqueous fan delta. (b) *Arenicolites* isp.1 association: short aerated periods, no net sedimentation, high-viscosity gravity flows deposited between fair-weather and storm wavebases. (c) *Diplocraterion habichi* association: high-energy aerated conditions, omission suites at transgressive surfaces. (d) *Diplocraterion parallelum* association: high-energy, aerated shallow subtidal or intertidal environments. Above, in foreshore ridges; below, *D. parallelum* as omission suite in rooted coal at transgressive surface. (e) Same association in tidal sandwave fields. (f) *Curvolithus* association: aerated, low- to medium-energy, slow sedimentation, shelf environment; distal fan delta. (g) *Planolites* and *Taenidium* association: low-energy, oxygen-limited environment; storm-dominated inner shelf between fair-weather and storm wave bases. (h) *Cochlichnus* association: aerated, medium- to low-energy, shallow marine environment; rippled, tidally-influenced shelf, subaqueous Gilbert-type fan delta. A1, A3, *Arenicolites* ispp. 1 and 3; Aa, *Anchorichnus anchorichnus*; Al, *Asteriacites lumbricalis*; B, *Bergaueria* isp.; C, *Cochlichnus* isp.; Cm, *Curvolithus multiplex*; Dp, *Diplocraterion parallelum*; Gc, *Gyrochorte comosa*; Gk, *Gyrophyllites kwassizensis*; Gm, *Gordia marina*; Jh, *Jamesonichnites heinbergi*; Ls, *Lockeia siliquaria*; On, *Ophiomorpha nodosa*; P, *Palaeophycus* isp.; Pa, *Palaeophycus alternatus*; Pb, *Planolites beverleyensis*; Phb, *Phycodes bromleyi*; Pi, *Phycosiphon incertum*; Pt, *Phoebichnus trochoides*; Ri, *Rhizocorallium irregulare*; St, *Skolithos tentaculatum*; T, trackway; Th, *Thalassinoides* isp.; Ts, *Taenidium serpentinum*.

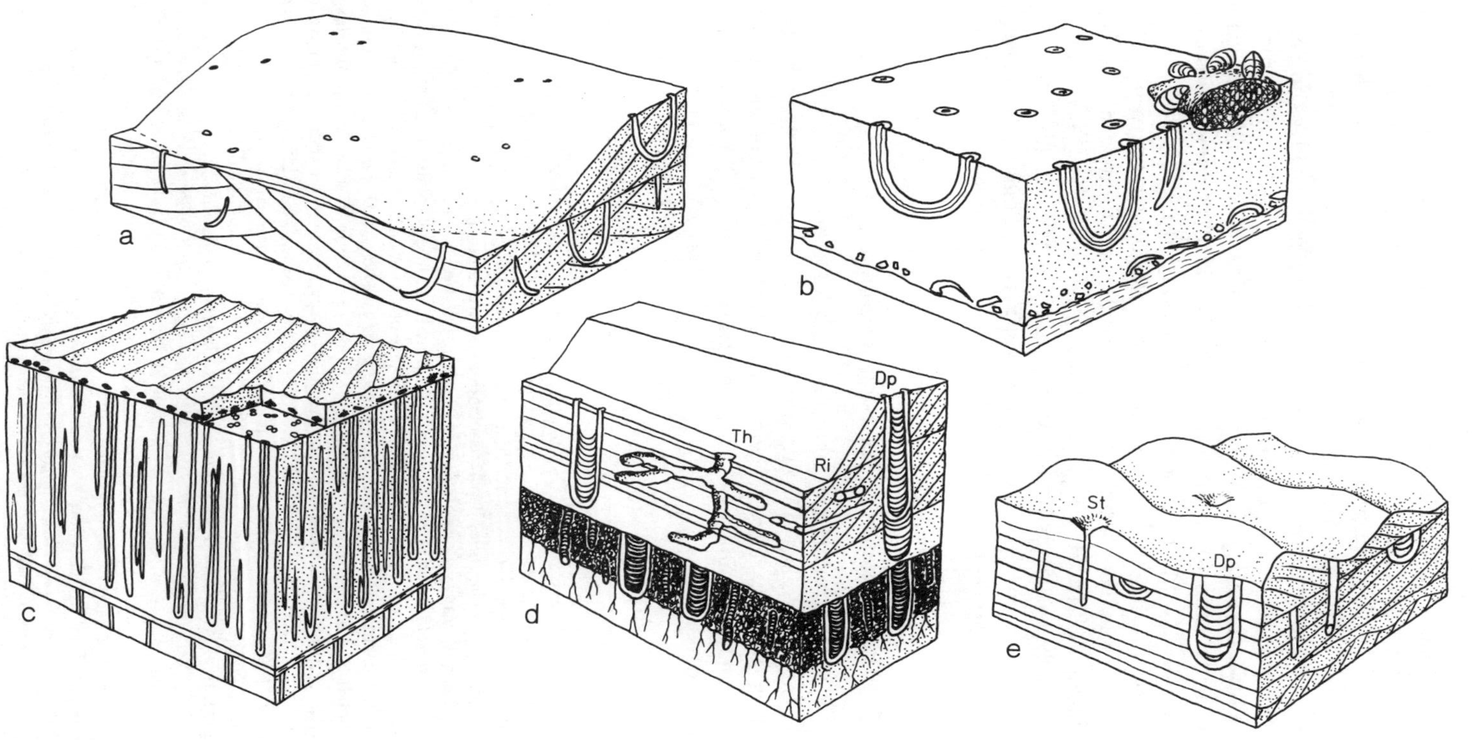
a
b
c
Th
Ri
Dp
d
St
Dp
e

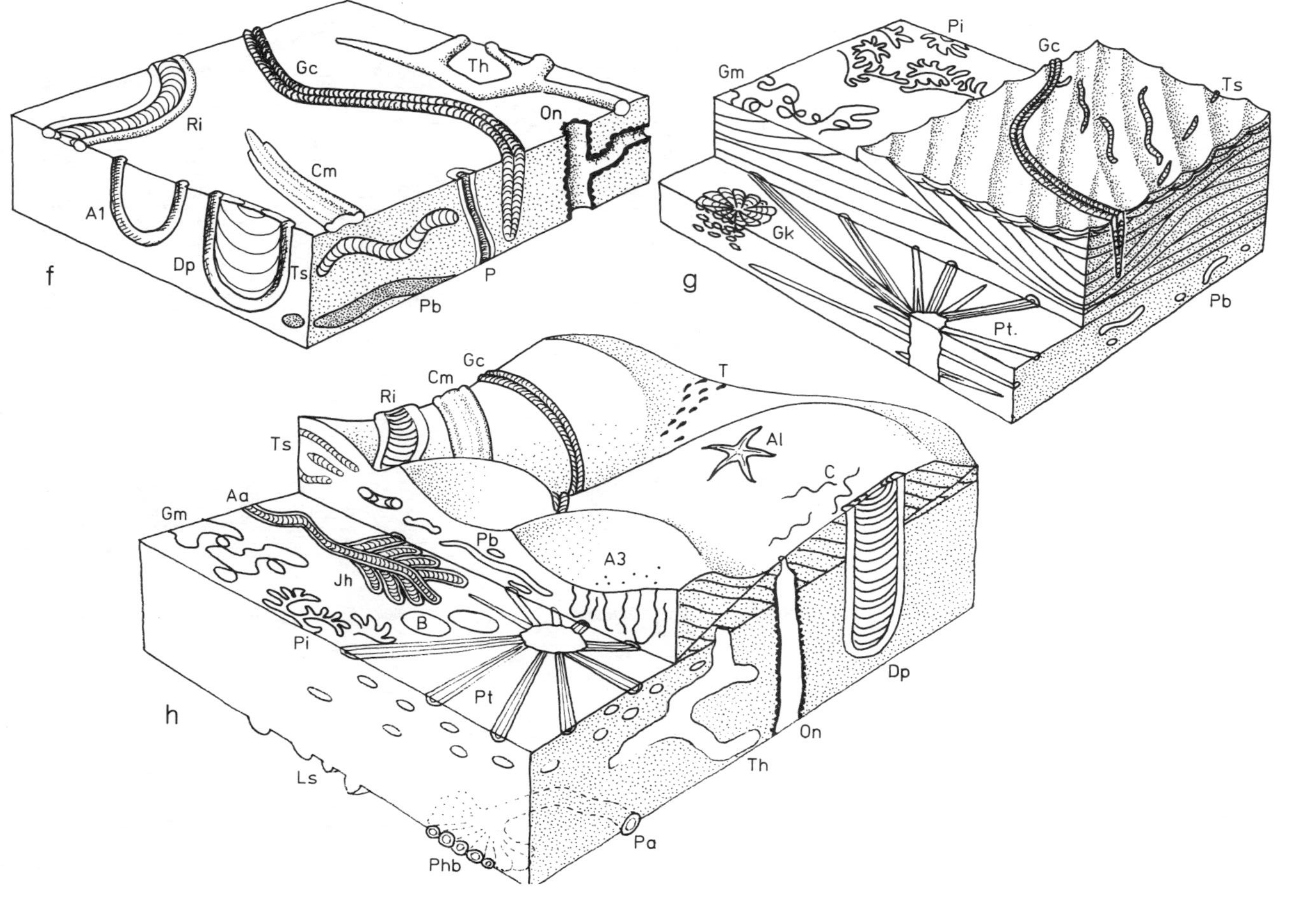
f
Ri
Gc
Th
On
Cm
A1
Dp
Ts
P
Pb
g
Gm
Pi
Gc
Ts
Gk
Pt.
Pb
h
Gc
Cm
Ri
T
Ts
Al
C
Aa
Gm
Pb
A3
Jh
B
Pi
Pt
Dp
On
Th
Ls
Pa
Phb

Ichnofabrics are most conveniently communicated in visual terms, and I find it useful to represent them with a cartoon or icon that symbolically sums up the visual expression of each ichnofabric. Thus, the sketch illustrates the texture and cross-cutting relationships, as well as the dominance, style of distribution or habit of the trace fossils present. The cross-cutting (or avoidance) relationships permit the construction of a tiering diagram. It should now be possible to reconstruct the abundance of trace fossils tier by tier in a bar diagram; this can rarely be done with great accuracy, but an approximation is nevertheless useful. Where key

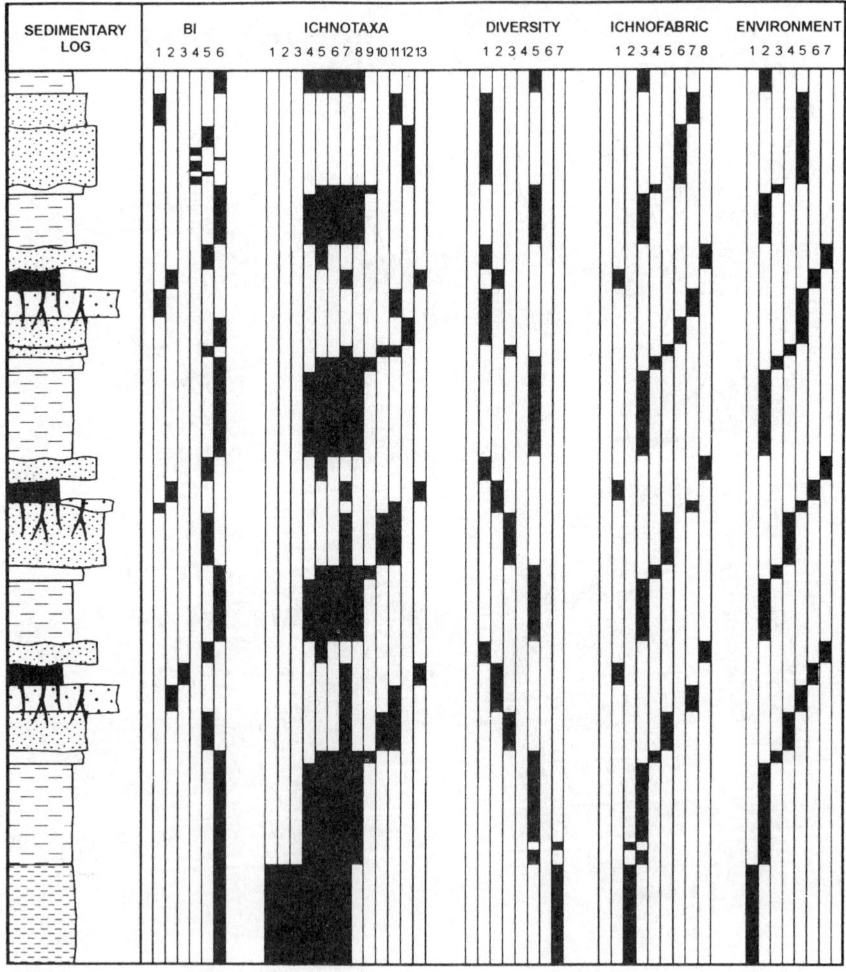

Figure 12.10 Compilation of ichnological data from a sedimentary log, comprising bioturbation index, the ichnotaxa recognized, trace fossil diversity, ichnofabrics defined (see Fig. 12.11) and suggested palaeoenvironments.

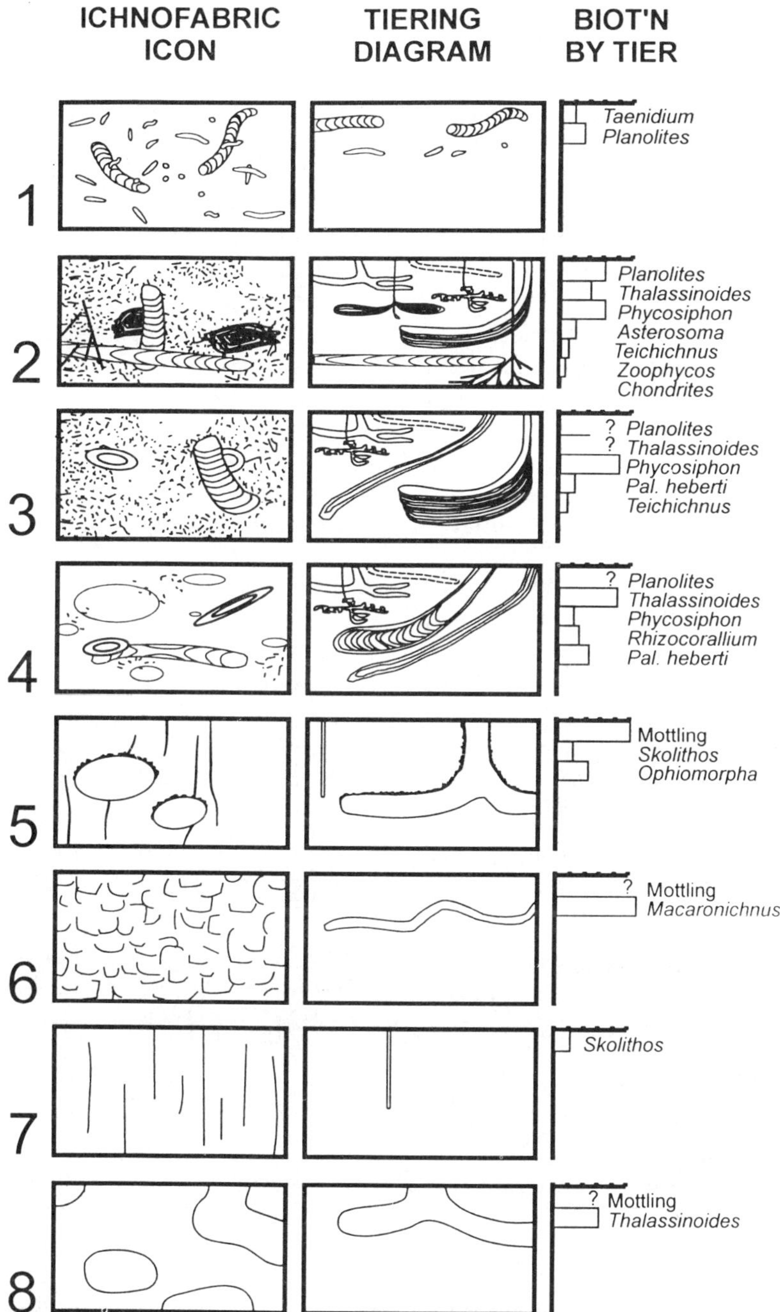

Figure 12.11 Definition of the ichnofabrics used in Fig. 12.10. Each ichnofabric is represented by a sketch indicating textural properties, a tiering diagram, and an attempt at representing the amount of bioturbation in each tier.

bioturbators have been active in deeper tiers, the amount of activity in shallower tiers will be difficult to judge.

To this compilation can be added the distribution of any ichnotaxa that can be identified, as well as the Bioturbation Index or Ichnofabric Index, according to the possibilities of the material (Figs 12.10 and 12.11).

Another type of visual representation of ichnofabric has been offered by Taylor and Goldring (1993): the ichnofabric constituent diagram (ICD; Fig. 12.12). The advantages of the ICD are that: (1) in incompletely bioturbated substrates, non-biogenic structures may be included; (2) the size of the structures is shown by the thickness of the bar; (3) tier succession is considered as time (ordering), so that in composite fabrics, later overprinting by recolonization events following erosion, to form palimpsest fabrics, may be included; (4) the bar length, indicating quantity, is given a logarithmic scale. This scale emphasizes the presence of minor constituents. Note that this quantity represents the surviving area of structure on a section surface, not the interpreted original quantity as in Fig. 12.10.

A disadvantage of the ICD is that each tier may contain only a single trace fossil. Tiers commonly contain several trace fossils of different sizes. Also, the 'size' of a structure may be ambiguous. In horizontal trace fossils diameter is chosen, and preferably also for vertical ones (or 5 m

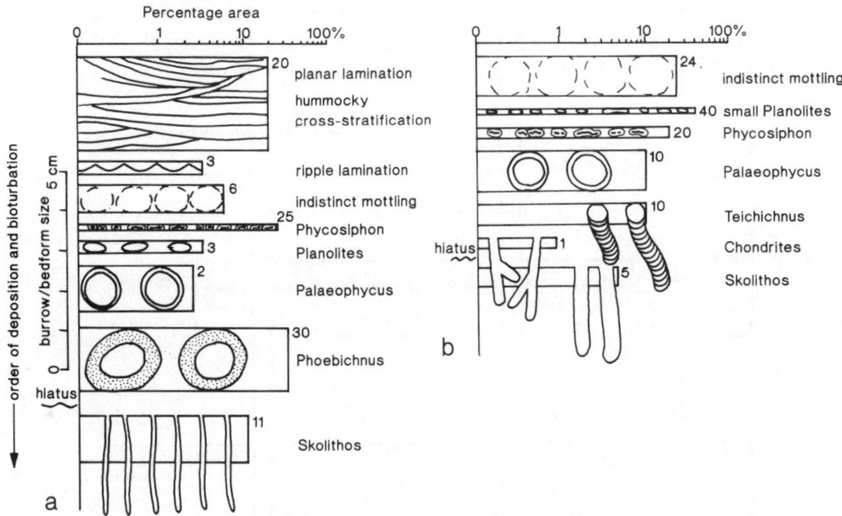

Figure 12.12 Ichnofabric constituent diagrams. (a) ICD of a fabric similar to Fig. 11.6. First event is taken to be the physical lamination; last event is a suite of *Skolithos* following a hiatus (not present in Fig. 11.6). The *Skolithos* are not drawn to scale. *Phoebichnus* occupies the deepest tier, and represents 30 per cent of the fabric, and may be considered an elite trace fossil. Modified after Taylor and Goldring (1993). (b) ICD of the fabric in Fig. 12.8d bottom.

Ophiomorpha will pose a problem!), whereas for physical structure height is used. But the individual worker must solve these problems as they arise.

Worksheet and questionnaire approaches such as these are invaluable for sharpening the observation process and maintaining rigor; the individual worker can modify and combine methods according to the suitability of the material.

13 Conclusion

Study of living burrowing animals not only shows how they gain a living from an endobenthic mode of life. It also reveals some remarkably sophisticated animal–sediment relationships. Perhaps disturbingly, it shows also how easily many individuals can turn from one sophisticated trophic strategy to a completely different one as the local environment changes. This introduces some uncertainty into trophic interpretation of trace fossils but it helps to explain how two or more ichnotaxa commonly occur together as parts of a single structure.

Study of burrowing animals also underlines how biased trace fossil assemblages are by emphasizing the extreme differences in preservation potential the different biogenic structures possess. Upward conveyor activity is of dominating importance in many communities today, being practised by numerous species and representing considerable biomass. Reverse conveyors are comparatively rare: why pass sediment down from the sea floor, anyway? Yet upward conveyors produce little more than a biodeformational fabric and no incipient trace fossil with high potential for preservation. The rare reverse conveyors, on the other hand, produce ready-made trace fossils, actively packed, having good diagenetic potential, and their structures are abundant in the fossil record.

Spreite structures show the same bias. They are important trace fossils, dominating many assemblages. Observations of spreiten being produced today are negligible. The fossilization barrier is a serious problem.

Nevertheless, had we no understanding of living endobenthos and their sedimentary strategies, we would still be calling trace fossils fucoid algae. *Chondrites* would still be misinterpreted as a deposit-feeder trace fossil had we not discovered chemosymbiosis in living animal–bacterial relationships. Some knowledge of neoichnology is essential for the interpretation of trace fossils.

The study of trace fossil assemblages has for years produced increasing understanding of individual trace fossil autecology and taphonomy, and improved their value in palaeoenvironmental interpretation. Originally, these trace fossils were examined on bedding planes, where they can be exposed *in extenso*, and they stole the limelight from the majority of trace fossils, which are hidden in their matrix.

The hidden trace fossils were released from neglect as the ichnofabric concept evolved, an essential spin-off from core logging. Study of trace fossils in two dimensions boosted the concept of tiering, which in turn has led to the recognition of ichnoguilds and has revolutionized the analysis of ichnofabrics.

The ichnological discipline now encompasses a wealth of facets, avenues of study and areas of application. Many individual ichnotaxa urgently require revision; this will greatly refine their use as tools of sedimentological interpretation. Neoichnological studies give insight into trophic interpretations of trace fossils. Ichnocoenoses are analysed as representatives of endobenthic palaeocommunities and their trophic structure examined. Ichnofabric applications continue to grow and to change.

New methods of analysing ichnofabrics are making them easier to use, and new techniques for their study promise rapid development in the near future. The use of computerized tomography to produce images and the further enhancement of these using image analysis techniques are research fields that have only just opened.

But all this richness of opportunity does not detract from the most exiting moment of all. Sitting in the mud on a Greek beach for four days with your dedicated colleagues, gently, painstakingly coaxing a more or less complete *Zoophycos*, 2 m in diameter, three whorls high, out of the bathyal Pliocene clay.

References

Abel, O. (1922) *Lebensbilder aus der Tierwelt der Vorzeit*, Fischer, Jena.

Abel, O. (1935) *Vorzeitliche Lebensspuren*, Fischer, Jena.

Aceñolaza, F. G. and Buatois, L. A. (1991) Trazas fosiles del Paleozoico superior continental Argentino. *Ameghiniana*, **28**, 89–108.

Aceñolaza, F. G. and Buatois, L. A. (1993) Nonmarine perigondwanic trace fossils from the late Paleozoic of Argentina. *Ichnos*, **2**, 183–201.

Ager, D. V. and Wallace, P. (1970) The distribution and significance of trace fossils in the uppermost Jurassic rocks of the Boulonnais, northern France, in *Trace Fossils* (eds T. P. Crimes and J. C. Harper), *Geological Journal Special Issues*, **3**, 1–18.

Aigner, T. and Reineck, H.-E. (1982) Proximality trends in modern storm sands from the Helgoland Bight (North Sea) and their implications for basin analysis. *Senckenbergiana Maritima*, **14**, 183–215.

Alexander, R. R., Stanton, R. J. and Dodd, J. R. (1993) Influence of sediment grain size on the burrowing of bivalves: correlation with distribution and stratigraphic persistence of selected Neogene clams. *Palaios*, **8**, 289–303.

Allen, J. A. (1958) On the basic form and adaptations to habitat in the Lucinacea (Eulamellibranchia). *Philosophical Transactions of the Royal Society of London*, **B 241**, 421–84.

Allen, J. A. (1983) The ecology of deep-sea molluscs, in *The Mollusca, 6: Ecology* (ed. W. D. Russel-Hunter), Academic Press, Orlando, pp. 38–75.

Aller, R. C. (1978) Experimental studies in changes produced by deposit feeders on pore water, sediment, and overlying water chemistry. *American Journal of Science*, **278**, 1185–234.

Aller, R. C. (1980) Relationships of tube-dwelling benthos with sediment and overlying water chemistry, in *Marine Benthic Dynamics* (eds K. R. Tenore and B. C. Coull), University of South Carolina, Chapel Hill, pp. 285–301.

Aller, R. C. (1982) The effects of macrobenthos on chemical properties of marine sediment and overlying water, in *Animal–Sediment Relationships* (eds P. L. McCall and M. J. S. Tevesz), Plenum, New York, pp. 52–102.

Aller, R. C. (1983) The importance of the diffusive permeability of animal burrow linings in determining marine sediment chemistry. *Journal of Marine Research*, **41**, 299–322.

Aller, R. C. and Cochran, J. K. (1976) $^{234}Th/^{238}U$ disequilibrium in nearshore sediment; particle reworking and diagenetic time scales. *Earth and Planetary Science and Letters*, **29**, 37–50.

Aller, R. C. and Dodge, R. E. (1974) Animal–sediment relations in a tropical lagoon, Discovery Bay, Jamaica. *Journal of Marine Research*, **32**, 209–32.

Aller, R. C. and Yingst, J. Y. (1978) Biogeochemistry of tube-dwellings: a study of the sedentary polychaete *Amphitrite ornata* (Leidy). *Journal of Marine Research*, **36**, 201–54.

Alpert, S. P. (1977) Trace fossils and the basal Cambrian boundary, in *Trace Fossils 2* (eds T. P. Crimes and J. C. Harper). *Geological Journal Special Issues*, **9**, 1–8.

Andersen, F. Ø. and Kristensen, E. (1991) Effects of burrowing macrofauna on organic matter decomposition in coastal marine sediments, in *The Environmental*

Impact of Burrowing Animals and Animal Burrows (eds P. S. Meadows and A. Meadows). *Symposium of the Zoological Society of London*, **63**, 69–88.

Anderson, J. G. and Meadows, P. S. (1978) Microenvironments in marine sediments. *Proceedings of the Royal Society of Edinburgh*, **76B**, 1–16.

Ansell, A. D. and Trueman, E. R. (1967) Observations on burrowing in *Glycymeris glycymeris* (L.) (Bivalvia, Arcacea). *Journal of Experimental Marine Biology and Ecology*, **1**, 65–75.

Ansell, A. D. and Trueman, E. R. (1968) The mechanism of burrowing in the anemone *Peachia hastata* Gosse. *Journal of Experimental Marine Biology and Ecology*, **2**, 124–34.

Archer, A. and Maples, C. G. (1984) Trace-fossil distribution across a marine-to-nonmarine gradient in the Pennsylvanian of southwestern Indiana. *Journal of Paleontology*, **58**, 448–66.

Armentrout, J. M. (1980) *Ophiomorpha* from upper bathyal Eocene subsea fan facies, northwestern Washington. *Bulletin of the American Association of Petroleum Geologists*, **64**, 670–1.

Armstrong, L. R. (1965) Burrowing limitations in Pelecypoda. *Veliger*, **7**, 195–200.

Asgaard, U. and Bromley, R. G. (1983) Palaeolimnichnology: state of the art. *1st. International Congress on Paleoecology, Lyon, Abstracts*, 6.

Ashworth, R. B. and Cormier, M. J. (1967) Isolation of 2,6-dibromophenol from the marine hemichordate *Balanoglossus biminiensis*. *Science*, **155**, 1558–9.

Atkinson, R. J. A. (1974a) Spatial distribution of *Nephrops* burrows. *Estuarine and Coastal Marine Science*, **2**, 171–6.

Atkinson, R. J. A. (1974b) Behavioural ecology of the mud-burrowing crab *Goneplax rhomboides*. *Marine Biology*, **25**, 239–52.

Atkinson, R. J. A. (1986) Mud-dwelling megafauna of the Clyde Sea area. *Proceedings of the Royal Society of Edinburgh*, **90B**, 351–61.

Atkinson, R. J. A., Moore, P. G. and Morgan, P. J. (1982) The burrows and burrowing behaviour of *Maera loveni* (Crustacea: Amphipoda). *Journal of Zoology, London*, **198**, 399–416.

Atkinson, R. J. A. and Nash, R. D. M. (1990) Some preliminary observations on the burrows of *Callianassa subterranea* (Montagu) (Decapoda: Thalassinidea) from the west coast of Scotland. *Journal of Natural History*, **24**, 403–13.

Atkinson, R. J. A. and Taylor, A. C. (1991) Burrows and burrowing behaviour of fish, in *The Environmental Impact of Burrowing Animals and Animal Burrows* (eds P. S. Meadows and A. Meadows), *Symposium of the Zoological Society, London*, **63**, 133–55.

Ausich, W. I. (1979) *Hondichnus monroensis* n. gen. n. sp. a new Early Mississippian trace fossil. *Journal of Paleontology*, **53**, 1155–9.

Baird, G. C. and Brett, C. E. (1986) Erosion on an anaerobic seafloor: significance of reworked pyrite deposits from the Devonian of New York State. *Palaeogeography, Palaeoclimatology, Palaeoecology*, **57**, 157–93.

Bambach, R. K. (1983) Ecospace utilization and guilds in marine communities through the Phanerozoic, in *Biotic Interactions in Recent and Fossil Benthic Communities* (eds M. J. S. Tevesz and P. L. McCall), Plenum, New York, pp. 719–46.

Barnes, R. D. (1964) Tube building and feeding in the chaetopterid polychaete, *Spiochaetopterus oculatus*. *Biological Bulletin of the Marine Laboratory, Woods Hole*, **127**, 397–412.

Barnes, R. D. (1965) Tube building and feeding in chaetopterid polychaetes. *Biological Bulletin of the Marine Laboratory, Woods Hole*, **129**, 217–33.

Barnes, R. D. (1980) *Invertebrate Zoology*, 4th edition, Saunders College, Philadelphia.

Bassan, P. B. (1979) Trace fossil nomenclature: the developing picture. *Palaeogeography, Palaeoclimatology, Palaeoecology*, **28**, 143–6.

Baumfalk, Y. A. (1979) Heterogeneous grain size distribution in tidal flat sediment caused by bioturbation activity of *Arenicola marina* (Polychaeta). *Netherlands Journal of Sea Research*, **13**, 428–40.

Bayer, U., Altheimer, E. and Deutschle, W. (1985) Environmental evolution in shallow epicontinental seas: sedimentary cycles and bed formation. *Lecture Notes in Earth Science*, **1**, 347–81.

Belt, E. S., Frey, R. W. and Welch, J. S. (1983) Pleistocene coastal marine and estuarine sequences, Lee creek Mine. *Smithsonian Contributions in Paleobiology*, **53**, 229–63.

Bender, K. and Davis, W. R. (1984) The effect of feeding by *Yoldia limatula* on bioturbation. *Ophelia*, **23**, 91–100.

Berger, W. H., Ekdale, A. A. and Bryant, P. P. (1979) Selective preservation of burrows in deep-sea carbonates. *Marine Geology*, **32**, 205–30.

Berger, W. H. and Heath, G. R. (1968) Vertical mixing in pelagic sediments. *Journal of Marine Research*, **26**, 134–43.

Berger, W. H., Johnson, R. F. and Killingley, J. S. (1977a) 'Unmixing' of the deep-sea record and the deglacial meltwater spike. *Nature*, **269**, 661–3.

Berger, W. H., Johnson, T. C. and Hamilton, E. L. (1977b) Sedimentation on Ontong Java Plateau: observations on a classical 'carbonate monitor', in *The Fate of Fossil Fuel CO_2 in the Oceans* (eds N. R. Andersen and A. Malahoff), Plenum Press, New York, pp. 543–67.

Bergman, K. M. (1994) Shannon Sandstone in Hartzog Draw – Heldt Draw fields (Cretaceous, Wyoming, USA) reinterpreted as lowstand shoreface deposits. *Journal of Sedimentary Research*, **B64**, 184–201.

Bergström, J. (1973) Organization, life, and systematics of trilobites. *Fossils and Strata*, **2**.

Bett, B. J. and Rice, A. L. (1993) The feeding behaviour of an abyssal echiuran revealed by *in situ* time-lapse photography. *Deep-Sea Research*, **40**, 1767–99.

Beynon, B. M. and Pemberton, S. G. (1992) Ichnological signature of a brackish water deposit: an example from the Lower Cretaceous Grand Rapids Formation, Cold Lake Oil Sands area, Alberta, in *Applications of Ichnology to Petroleum Exploration* (ed. S. G. Pemberton), *SEPM Core Workshop*, **17**, 199–221.

Beynon, B. M., Pemberton, S. G., Bell, D. D. and Logan, C. A. (1988) Environmental implications of ichnofossils from the Lower Cretaceous Grand Rapids Formation, Cold Lake Oil Sands Deposit. *Canadian Society of Petroleum Geologists, Memoirs*, **15**, 275–90.

Billett, D. S. M., Lampitt, R. S., Rice, A. L. and Mantoura, R. F. C. (1983) Seasonal sedimentation of phytoplankton to the deep-sea benthos. *Nature*, **302**, 520–2.

Bockelie, J. F. (1991) Ichnofabric mapping and interpretation of Jurassic reservoir rocks of the Norwegian North Sea. *Palaios*, **6**, 206–15.

Bockelie, J. F. (1994) Plant roots in core, in *The Palaeobiology of Trace Fossils* (ed. S. K. Donovan), Wiley, Chichester, pp. 177–99.

Bottjer, D. J., Arthur, M. A., Dean, W. E., Hattin, D. E. and Savrda, C. E. (1986) Rhythmic bedding produced in Cretaceous pelagic carbonate environments: sensitive recorders of climatic cycles. *Paleoceanography*, **1**, 467–81.

Bottjer, D. J. and Ausich, W. I. (1982) Tiering and sampling requirements in paleocommunity reconstruction. *Proceedings of the Third North American Paleontologic Convention*, **1**, 57–9.

Bottjer, D. J. and Droser, M. L. (1994) The history of Phanerozoic bioturbation, in *The Palaeobiology of Trace Fossils* (ed. S. K. Donovan), Wiley, Chichester, pp. 155–76.

Bottjer, D. J., Droser, M. L. and Jablonski, D. (1987) Bathymetric trends in the history of trace fossils, in *New Concepts in the Use of Biogenic Sedimentary Structures for Paleoenvironmental Interpretation* (ed. D. J. Bottjer), Pacific Section SEPM, Los Angeles, 57–65.

Bottjer, D. J., Droser, M. L. and Jablonski, D. (1988) Palaeoenvironmental trends in the history of trace fossils. *Nature*, **333**, 252–5.

Boucot, A. J. (1990) *Evolutionary Paleobiology of Behavior and Coevolution*, Elsevier, Amsterdam.

Boudreau, B. P. (1986a) Mathematics of tracer mixing in sediments: I. Spatially-dependent, diffuse mixing. *American Journal of Science*, **286**, 161–98.

Boudreau, B. P. (1986b) Mathematics of tracer mixing in sediments: II. Nonlocal mixing and biological conveyor-belt phenomena. *American Journal of Science*, **286**, 199–238.

Bown, T. R. and Ratcliffe, B. C. (1988) The origin of *Chubutolithes* Ihering, ichnofossils from the Eocene and Oligocene of Chubut Province, Argentina. *Journal of Paleontology*, **62**, 163–7.

Bradley, J. (1973) *Zoophycos* and *Umbellula* (Pennatulacea): their synthesis and identity. *Palaeogeography, Palaeoclimatology, Palaeoecology*, **13**, 103–28.

Bradley, J. (1980) *Scolicia* and *Phycodes*, trace fossils of *Renilla* (Pennatulacea). *Pacific Geology*, **14**, 73–86.

Bradley, J. (1981) *Radionereites, Chondrites* and *Phycodes*; trace fossils of anthoptiloid sea pens. *Pacific Geology*, **15**, 1–16.

Braithwaite, C. J. R. and Talbot, M. R. (1972) Crustacean burrows in the Seychelles, Indian Ocean. *Palaeogeography, Palaeoclimatology, Palaeoecology*, **11**, 265–85.

Brambell, F. W. R. and Cole, H. A. (1939) *Saccoglossus cambrensis*, sp. n., an enteropneust occurring in Wales. *Proceedings of the Zoological Society, London*, **109B**, 211–36.

Brambell, F. W. R. and Goodhart, C. B. (1941) *Saccoglossus horsti* sp. n., an enteropneust occurring in the Solent. *Journal of the Marine Biological Association, UK*, **24**, 283–301.

Bramlette, M. N. and Bradley, W. H. (1942) Geology and biology of North Atlantic deep-sea cores between Newfoundland and Ireland. *US Geological Survey Professional Papers*, **196**.

Branch, G. M. and Pringle, A. (1987) The impact of the sand prawn *Callianassa kraussi* Stebbing on sediment turnover and on bacteria, meiofauna, and benthic microflora. *Journal of Experimental Marine Biology and Ecology*, **107**, 219–35.

Brenchley, G. A. (1981) Disturbance and community structure: an experimental study of bioturbation in marine soft-bottom environments. *Journal of Marine Research*, **39**, 767–90.

Brenchley, G. A. (1982) Mechanisms of spatial competition in marine soft-bottom communities. *Journal of Experimental Marine Biology and Ecology*, **60**, 17–33.

Brett, C. E., Miller, K. B. and Baird, G. C. (1990) A temporal hierarchy of paleoecologic processes within a Middle Devonian epeiric sea, in *Paleocommunity Temporal Dynamics: the Long-Term Development of Multispecies Assemblies* (ed. W. Miller III), *Paleontological Society Special Publications*, **5**, 178–209.

Bright, D. B. and Hogue, C. L. (1972) A synopsis of the burrowing land crabs of the world and list of their arthropod symbionts and burrow associates. *Contributions in Science from the Natural History Museum, Los Angeles County*, **220**.

Brodie, I. and Kemp, A. E. S. (1995) Pelletal structures in Peruvian upwelling sediments. *Journal of the Geological Society, London*, **152**, 141–50.

Bromley, R. G. (1967) Some observations on burrows of thalassinidean Crustacea in chalk hardgrounds. *Quarterly Journal of the Geological Society, London*, **123**, 157–82.

Bromley, R. G. (1975) Trace fossils at omission surfaces, in *The Study of Trace Fossils* (ed. R. W. Frey), Springer-Verlag, New York, pp. 399–428.

Bromley, R. G. (1981a) Enhancement of visibility of structures in marly chalk: modification of the Bushinsky oil technique. *Bulletin of the Geological Society of Denmark*, **29**, 111–18.

Bromley, R. G. (1981b) Concepts in ichnotaxonomy illustrated by small round holes in shells. *Acta Geològica Hispànica*, **16**, 55–64.

Bromley, R. G. (1991) *Zoophycos*: strip mine, refuse dump, cache or sewage farm? *Lethaia*, **24**, 460–2.

Bromley, R. G. (1993) Predation habits of octopus past and present and a new ichnospecies, *Oichnus ovalis*. *Bulletin of the Geological Society of Denmark*, **40**, 167–73.

Bromley, R. G. (1994) The palaeoecology of bioerosion, in *The Palaeobiology of Trace Fossils* (ed. S. K. Donovan), Wiley, Chichester, pp. 134–54.

Bromley, R. G. and Allouc, J. (1992) Trace fossils in bathyal hardgrounds, Mediterranean Sea. *Ichnos*, **2**, 43–54.

Bromley, R. G. and Asgaard, U. (1972a) Freshwater *Cruziana* from the Upper Triassic of Jameson Land, East Greenland. *Grønlands Geologiske Undersøgelse Rapport*, **49**, 7–13.

Bromley, R. G. and Asgaard, U. (1972b) The burrows and microcoprolites of *Glyphea rosenkrantzi*, a Lower Jurassic palinuran crustacean from Jameson Land, East Greenland. *Grønlands Geologiske Undersøgelse Rapport*, **49**, 15–21.

Bromley, R. G. and Asgaard, U. (1972c) A large radiating burrow-system in Jurassic micaceous sandstones of Jameson Land, East Greenland. *Grønlands Geologiske Undersøgelse Rapport*, **49**, 23–30.

Bromley, R. G. and Asgaard, U. (1975) Sediment structures produced by a spatangoid echinoid: a problem of preservation. *Bulletin of the Geological Society of Denmark*, **24**, 261–81.

Bromley, R. G. and Asgaard, U. (1979) Triassic freshwater ichnocoenoses from Carlsberg Fjord, East Greenland. *Palaeogeography, Palaeoclimatology, Palaeoecology*, **28**, 39–80.

Bromley, R. G. and Asgaard, U. (1990) *Solecurtus strigilatus*: a jet-propelled burrowing bivalve, in *The Bivalvia – Proceedings of a Memorial Symposium in Honour of Sir Charles Maurice Yonge, Edinburgh* (ed. B. Morton), Hong Kong University Press, Hong Kong, pp. 313–20.

Bromley, R. G. and Asgaard, U. (1991) Ichnofacies: a mixture of taphofacies and biofacies. *Lethaia*, **24**, 153–63.

Bromley, R. G. and Asgaard, U. (1993a) Endolithic community replacement on a Pliocene rocky coast. *Ichnos*, **2**, 93–116.

Bromley, R. G. and Asgaard, U. (1993b) Two bioerosion ichnofacies produced by early and late burial associated with sea-level change. *Geologische Rundschau*, **82**, 276–80.

Bromley, R. G., Curran, H. A., Frey, R. W., Gutschick, R. C. and Suttner, L. J. (1975a) Problems in interpreting unusually large burrows, in *The Study of Trace Fossils* (ed. R. W. Frey), Springer-Verlag, New York, pp. 351–76.

Bromley, R. G. and Ekdale, A. A. (1984a) *Chondrites*: a trace fossil indicator of anoxia in sediments. *Science*, **224**, 872–4.

Bromley, R. G. and Ekdale, A. A. (1984b) Trace fossil preservation in flint in the European chalk. *Journal of Paleontology*, **58**, 298–311.

Bromley, R. G. and Ekdale, A. A. (1986) Composite ichnofabrics and tiering of burrows. *Geological Magazine*, **123**, 59–65.

Bromley, R. G. and Ekdale, A. A. (in press) Ichnofabrics in condensed carbonates, Ordovician of Sweden, in *Atlas of Ichnofabrics* (eds J. E. Pollard, H. A. Curran and R. G. Bromley). *SPEM*..

Bromley, R. G. and Frey, R. W. (1974) Redescription of the trace fossil *Gyrolithes* and taxonomic evaluation of *Thalassinoides*, *Ophiomorpha* and *Spongeliomorpha*. *Bulletin of the Geological Society of Denmark*, **23**, 311–35.

Bromley, R. G. and Fürsich, F. T. (1980) Comments on the proposed amendments to the international Code of zoological nomenclature regarding ichnotaxa. Z.N.(S.) 1973. *Bulletin of Zoological Nomenclature*, **37**, 6–10.

Bromley, R. G. and Gale, A. S. (1982) The lithostratigraphy of the English Chalk Rock. *Cretaceous Research*, **3**, 273–306.

Bromley, R. G. and Goldring, R. (1992) The palaeoburrows at the Cretaceous to Palaeocene firmground unconformity in southern England. *Tertiary Research*, **13**, 95–102.

Bromley, R. G. and Hanken N.-H. (1991) The growth vector in trace fossils: examples from the Lower Cambrian of Norway. *Ichnos*, **1**, 261–75.

Bromley, R. G., Jensen, M. and Asgaard, U. (1995) Spatangoid echinoids: deep-tier trace fossils and chemosymbiosis. *Neues Jahrbuch für Geologie und Paläontologie, Abhandlungen*, **195**, 25–35.

Bromley, R. G., Pemberton, S. G. and Rahmani, R. A. (1984) A Cretaceous woodground: the *Teredolites* ichnofacies. *Journal of Paleontology*, **58**, 488–98.

Bromley, R. G., Schulz, M.-G. and Peake, N. B. (1975b) Paramoudras: giant flints, long burrows and the early diagenesis of chalks. *Kongelige Danske Videnskabernes Selskab, Biologiske Skrifter*, **20** (10).

Brunch and Pringle (1987) to follow.

Buatois, L. A. and Mángano, M. G. (1990) Una asociacion de trazas fosiles del Carbonico lacustre del area de Los Jumes, Caramarca, Argentina: su comparacion con la icnofacies de *Scoyenia*. *5, Congreso Argentino de Paleontologia y Bioestratigrafia, Actas*, **1**, 77–81.

Buatois, L. A. and Mángano, M. G. (1992) La oxigenacion como factor de control en la distribution de asociaciones de trazas fosiles, formation Kotick Point, Cretacico de Antartida. *Ameghiniana*, **29**, 69–84.

Buatois, L. A. and Mángano, M. G. (1993) Trace fossils from a Carboniferous turbiditic lake: implications for the recognition of additional nonmarine ichnofacies. *Ichnos*, **2**, 237–58.

Buch, J. (1980) *Muldvarpen*, Rhodos, Copenhagen.

Buchanan, J. B. (1963) The biology of *Calocaris macandreae* (Crustacea: Thalassinidea). *Journal of the Marine Biological Association, UK*, **43**, 729–47.

Buchanan, J. B. (1966) The biology of *Echinocardium cordatum* (Echinodermata: Spatangoidea) from different habitats. *Journal of the Marine Biological Association, UK*, **46**, 97–114.

Buchholz, H. (1986) Die Höhle eines Spechtvogels aus dem Eozän von Arizona, U.S.A. (Aves, Piciformes). *Verh. naturwiss. Ver. Hamburg (N. F.)*, **28**, 5–25.

Buckman, J. O. (1992) Palaeoenvironment of a Lower Carboniferous sandstone succession northwest Ireland: ichnological and sedimentological studies, in *Basins on the Atlantic Seaboard: Petroleum Sedimentology and Basin Evolution* (ed. J. Parnell), *Geological Society of London Special Publications*, **62**, 217–41.

Burdon-Jones, C. (1951) Observations on the spawning behaviour of *Saccoglossus horsti* Brambell and Goodhart, and of other Enteropneusta. *Journal of the Marine Biological Association, UK*, **29**, 625–38.

Burdon-Jones, C. (1956) Observations on the enteropneust, *Protoglossus kochleri* (Caullery and Mesnil). *Proceedings of the Zoological Society, London*, **127**, 35–58.

Byers, C. W. (1977) Biofacies patterns in euxenic basins: a general model. *SEPM Special Publications*, **25**, 5–17.

Byers, C. W. (1982) Geological significance of marine biogenic sedimentary structures, in *Animal–Sediment Relations* (eds P. L. McCall and M. J. S.Tevesz), Plenum Press, New York, pp. 221–56.

Cadée, G. C. (1976) Sediment reworking by *Arenicola marina* on tidal flats in the Dutch Wadden Sea. *Netherlands Journal of Sea Research*, **10**, 440–60.

Cadée, G. C. (1979) Sediment reworking by the polychaete *Heteromastus filiformis* on a tidal flat in the Dutch Wadden Sea. *Netherlands Journal of Sea Research*, **13**, 441–56.

Cadée, G. C. (1984) 'Opportunistic feeding', a serious pitfall in trophic structure analysis of (paleo)faunas. *Lethaia*, **17**, 289–92.

Caldwell, R. L. and Dingle, H. (1976) Stomatopods. *Scientific American*, **234**, 80–9.

Calzada, R. S. (1981) Revisión del icno *Spongeliomorpha iberica* Saporta, 1887 (Mioceno de Alcoy, España). *Boletín de la Real Sociedad Española de Historia Natural (Geologia)*, **79**, 189–95.

Campbell, K. A. (1992) Recognition of a Mio-Pliocene cold seep setting from the northwest Pacific convergent margin, Washington, U.S.A. *Palaios*, **7**, 422–33.

Carney, R. S. (1981) Bioturbation and biodeposition, in *Principles of Benthic Marine Paleoecology* (ed. A. J. Boucot), Academic Press, New York, pp. 357–99.

Carney, R. S. (1989) Examining relationships between organic carbon flux and deep-sea deposit feeding, in *Ecology of Marine Deposit Feeders* (eds G. Lopez, G. Taghon and J. Levinton), *Lecture Notes on Coastal and Estuarine Studies*, **31**, 24–59.

Caster, K. E. (1938) A restudy of the tracks of *Paramphibius. Journal of Paleontology*, **12**, 3–60.

Caster, K. E. (1939) Were *Micrichnus scotti* Abel and *Artiodactylus sinclairi* Abel of the Newark Series (Triassic) made by vertebrates or limuloids? *American Journal of Science*, **237**, 786–97.

Caster, K. E. (1941) Die sogenannten 'Wirbeltierspuren' und die *Limulus*-Fährten der Solnhofener Plattenkalke. *Paläontologische Zeitschrift*, **22**, 12–29.

Caster, K. E. (1944) Limuloid trails from the Upper Triassic (Chinle) of the Petrified Forest National Monument, Arizona. *American Journal of Science*, **242**, 74–84.

Chamberlain, C. K. (1971) Morphology and ethology of trace fossils from the Ouachita Mountains, southeast Oklahoma. *Journal of Paleontology*, **45**, 212–46.

Chamberlain, C. K. (1975) Trace fossils in DSDP cores of the Pacific. *Journal of Paleontology*, **49**, 1074–96.

Chamberlain, C. K. (1978) Recognition of trace fossils in cores. *SEPM Short Course*, **5**, 119–66.

Chapman, C. J. and Rice, A. L. (1971) Some direct observations on the ecology and behaviour of the Norway lobster *Nephrops norvegicus. Marine Biology*, **10**, 321–9.

Chapman, G. (1949) The thixotropy and dilatancy of a marine soil. *Journal of the Marine Biological Association, UK*, **28**, 123–40.

Chapman, G. and Newell, G. E. (1947) The role of the body fluid in relation to movement in soft-bodied invertebrates. I: The burrowing of *Arenicola. Proceedings of the Royal Society of London*, **B 134**, 431–55.

Chesher, R. H. (1963) The morphology and function of the frontal ambulacrum of *Moira atropos* (Echinoidea: Spatangoida). *Bulletin of Marine Science of the Gulf and Caribbean*, **13**, 549–73.

Chesher, R. H. (1968) The systematics of sympatric species in West Indian spatangoids: a revision of the genera *Brissopsis, Plethotaenia, Paleopneustes* and *Saviniaster. Studies in Tropical Oceanography*, **7**.

Chesher, R. H. (1969) Contributions to the biology of *Meoma ventricosa* (Echinoidea: Spatangoida). *Bulletin of Marine Science*, **19**, 72–110.

Christensen, Aa. M. (1970) Feeding biology of the sea-star *Astropecten irregularis* Pennant. *Ophelia*, **8**, 1–134.

Christensen, O. (1971) Notes on the biology of Foraminifera. *Vie et Milieu, Supplement*, **22**, 465–77.

Chuang, S. H. (1962) Feeding mechanism of the echiuroid, *Ochetostoma erythrogrammon* Leuckart and Rueppell, 1828. *Biological Bulletin*, **123**, 80–5.

Clark, G. R. and Ratcliffe, B. C. (1989) Observations on the tunnel morphology of *Heterocerus brunneus* Melsheimer (Coleoptera: Heteroceridae) and its paleoecological significance. *Journal of Paleontology*, **63**, 228–32.

Clark, R. M. (1964) *Dynamics in Metazoan Evolution. The origin of the coelom and segments*, Clarendon Press, Oxford.

Clayton, C. J. (1986) The chemical environment of flint formation in Upper Cretaceous chalks, in *The Scientific Study of Flint and Chert* (eds G. de G. Sieveking and M. B. Hart), Cambridge University Press, Cambridge, pp. 43–54.

Clifton, H. E. and Thompson, J. K. (1978) *Macaronichnus segregatis*: a feeding structure of shallow marine polychaetes. *Journal of Sedimentary Petrology*, **48**, 1293–301.

Codez, J. and de Saint-Seine, R. (1958) Révision des cirripèdes acrothoraciques fossiles. *Bulletin de la Société Géologiques de France*, (6) **7**, 699–719.

Colella, A. and D'Alessandro, A. (1988) Sand waves, *Echinocardium* traces and their bathyal depositional setting (Monte Torre Palaeostrait, Plio-Pleistocene, southern Italy). *Sedimentology*, **35**, 219–37.

Colin, P. L., Suchanek, T. H. and McMurtry, G. (1986) Water pumping and particulate resuspension by callianassids (Crustacea: Thalassinidea) at Eniwetak and Bikini Atolls, Marshall Islands. *Bulletin of Marine Science*, **38**, 19–24.

Collins, D. (1987) Life in the Cambrian seas. *Nature*, **326**, 127.

Conway Morris, S. (1977) Fossil priapulid worms. *Special Papers in Palaeontology*, **20**.

Conway Morris, S. (1979) The Burgess Shale (Middle Cambrian) fauna. *Annual Review of Ecology and Systematics*, **10**, 327–49.

Conway Morris, S. (1985) Cambrian Lagerstätten: their distribution and significance, in *Extraordinary Fossil Biotas: their Ecological and Evolutionary Significance* (eds H. B. Whittington and S. Conway Morris), *Philosophical Transactions of the Royal Society of London*, **B311**, 49–65.

Conway Morris, S., Peel, J. S., Higgins, A. K., Soper, N. J. and Davis, N. C. (1987) A Burgess Shale-like fauna from the Lower Cambrian of North Greenland. *Nature*, **326**, 181–3.

Cory, R. L. and Pierce, E. L. (1967) Distribution and ecology of lancelets (order Amphioxi) over the continental shelf of the southeastern United States. *Limnology and Oceanography*, **12**, 650–6.

Crimes, T. P. (1977) Trace fossils of an Eocene deep-sea sand fan, northern Spain, in *Trace Fossils 2* (eds T. P. Crimes and J. C. Harper), *Geological Journal, Special Issues*, **9**, 71–90.

Crimes, T. P. (1994) The period of early evolution failure and the dawn of evolutionary success: the record of biotic changes across the Precambrian–Cambrian boundary, in *The Palaeobiology of Trace Fossils* (ed. S. K. Donovan), Wiley, Chichester, pp. 105–33.

Crimes, T. P. and Fedonkin, M. A. (1994) Evolution and dispersal of deepsea traces. *Palaios*, **9**, 74–83.

Crimes, T. P., Hildago, J. F. G. and Poire, D. G. (1992) Trace fossils from Arenig flysch sediments of Eire and their bearing on the early colonization of the deep seas. *Ichnos*, **2**, 61–77.

Crimes, T. P., Legg, I., Marcos, A. and Arboleya, M. (1977) ?Late Precambrian–low Lower Cambrian trace fossils from Spain. *Geological Journal Special Issues*, **9**, 91–138.

Crimes, T. P. and McCall, G. J. H. (1995) A diverse ichnofauna from Eocene-Miocene rocks of the Makran Range (S.E. Iran). *Ichnos*, **3**, 231–58.

Crimes, T. P. and Uchman, A. (1993) A concentration of exceptionally well-preserved large tubular foraminifera in the Eocene Zumaya flysch, northern Spain. *Geological Magazine*, **130**, 851–3.

Cullen, D. J. (1973) Bioturbation of superficial marine sediments by interstitial meiobenthos. *Nature*, **242**, 323–4.

Cuomo, M. C. and Rhoads, D. C. (1987) Biogenic sedimentary fabrics associated with pioneering polychaete assemblages: modern and ancient. *Journal of Sedimentary Petrology*, **57**, 537–43.

Curran, H. A. (1985) The trace fossil assemblage of a Cretaceous nearshore environment: Englishtown formation of Delaware, USA. *SEPM Special Publications*, **35**, 261–76.

Curran, H. A. (1994) The palaeobiology of ichnocoenoses in Quaternary, Bahamian-style carbonate environments: the modern to fossil transition, in *The Palaeobiology of Trace Fossils* (ed. S. K. Donovan), Wiley, Chichester, pp. 83–104.

Curran, H. A. and Frey, R. W. (1977) Pleistocene trace fossils from North Carolina (USA), and their Holocene analogues. *Geological Journal Special Issues*, **9**, 139–62.

D'Alessandro, A. and Bromley, R. G. (1987) Meniscate trace fossils and the *Muensteria–Taenidium* problem. *Palaeontology*, **30**, 743–63.

D'Alessandro, A., Bromley, R. G. and Stemmerik, L. (1987) *Rutichnus*: a new ichnogenus for branched, walled, meniscate trace fossils. *Journal of Paleontology*, **61**, 1112–19.

D'Alessandro, A., Ekdale, A. A. and Sonnino, M. (1986) Sedimentologic significance of turbidite ichnofacies in the Saraceno formation (Eocene), southern Italy. *Journal of Sedimentary Petrology*, **56**, 294–306.

Dam, G. (1990a) Taxonomy of trace fossils from the shallow marine Lower Jurassic Neill Klinter Formation, East Greenland. *Bulletin of the Geological Society of Denmark*, **38**, 119–44.

Dam, G. (1990b) Palaeoenvironmental significance of trace fossils from the shallow marine Lower Jurassic Neill Klinter Formation, East Greenland. *Palaeogeography, Palaeoclimatology, Palaeoecology*, **79** 221–48.

Dando, P. R. and Southward, A. J. (1986) Chemoautotrophy in bivalve molluscs of the genus *Thyasira*. *Journal of the Marine Biological Association, UK*, **66**, 915–29.

Dando, P. R., Southward, A. J. and Southward, E. C. (1986) Chemoautotrophic symbionts in the gills of the bivalve mollusc *Lucinoma borealis*, and the sediment chemistry of its habitat. *Proceedings of the Royal Society of London*, **B227**, 227–47.

Dapples, E. C. (1942) The effect of macro-organisms upon near-shore sediments. *Journal of Sedimentary Petrology*, **12**, 118–26.

Darwin, C. (1881) *The Formation of Vegetable Mould through the Action of Worms*, Murray, London.

Darwin, H. (1901) On the small vertical movements of a stone laid on the surface of the ground. *Proceedings of the Royal Society of London*, **68**, 253–61.

Davidson, C. (1891) On the amount of sand brought up by lobworms to the surface. *Geological Magazine*, **8** (3), 489–93.

Dennell, R. (1933) The habits and feeding mechanism of the amphipod *Haustorius arenarius* Slabber. *Journal of the Linnaean Society, London*, **38**, 363–88.

De Ridder, C. and Jangoux, M. (1993) The digestive tract of the spatangoid echinoid *Echinocardium cordatum* (Echinodermata): morphofunctional study. *Acta Zoologica*, **74**, 337–51.

De Ridder, C., Jangoux, M. and v. Impe, E. (1985) Food selection and absorption efficiency in the spatangoid echinoid *Echinocardium cordatum* (Echinodermata). *Proceedings of the 5th International Echinoderm Conference, Galway*, 245–51.

de Vaugelas, J. (1989) Deep-sea lebensspuren: remarks on some echiuran traces in the Porcupine Seabight, northeast Atlantic. *Deep-Sea Research*, **36**, 975–82.

Devine, C. E. (1966) Ecology of *Callianassa filholi* Milne-Edwards 1878 (Crustacea, Thalassinidea). *Transactions of the Royal Society of New Zealand, Zoology*, **8**, 93–110.

de Wilde, P. A. W. J. (1976) The benthic boundary layer from the point of view of a biologist, in *The Benthic Boundary Layer* (ed. I. N. McCave), Plenum Press, New York, pp. 81–94.

de Wilde, P. A. W. J. (1991) Interactions in burrowing communities and their effects on the structure of marine benthic ecosystems, in *The Environmental Impact of Burrowing Animals and Animal Burrows* (eds P. S. Meadows and A. Meadows), *Symposium of the Zoological Society of London*, **63**, 107–17.

Dinamani, P. (1964) Burrowing behavior of *Dentalium*. *Biological Bulletin*, **126**, 28–32.

Dodd, J. R. and Stanton, J. S. (1990) *Paleoecology: Concepts and Applications*, 2nd edn, Wiley, New York.

Doering, P. H. (1981) Observations on the behavior of *Asterias forbesi* feeding on *Mercenaria mercenaria*. *Ophelia*, **20**, 169–77.

Dörjes, J. and Hertweck, G. (1975) Recent biocoenoses and ichnocoenoses in shallow water marine environments, in *The Study of Trace Fossils* (ed. R. W. Frey), Springer-Verlag, New York, pp. 459–91.

Dörjes, J. and Howard, J. D. (1975) Estuaries of the Georgia coast, U.S.A.: sedimentology and biology. IV. Fluvial–marine transition indicators in an estuarine environment, Ogeechee River–Ossabaw Sound. *Senckenbergiana Maritima*, **7**, 137–79.

Donovan, S. K. (1994) Insects and other arthropods as tracemakers in nonmarine environments and palaeoenvironments, in *The Palaeobiology of Trace Fossils* (ed. S. K. Donovan), Wiley, Chichester. pp. 200–20.

Donselaar, M. E. (1989) The Cliff House Sandstone, San Juan Basin, New Mexico: model for the stacking of 'transgressive' barrier complexes. *Journal of Sedimentary Petrology*, **59**, 13–27.

Droser, M. L. (1991) Ichnofabric of the Paleozoic *Skolithos* ichnofacies and the nature and distribution of *Skolithos* piperock. *Palaios*, **6**, 316–25.

Droser, M. L. and Bottjer, D. J. (1986) A semiquantitative field classification of ichnofabric. *Journal of Sedimentary Petrology*, **56**, 558–9.

Droser, M. L. and Bottjer, D. J. (1987) Development of ichnofabric indices for strata deposited in high-energy nearshore terrigenous clastic environments, in *New Concepts in the Use of Biogenic Sedimentary Structures for Paleoenvironmental Interpretation* (ed. D. J. Bottjer), SEPM Pacific Section, Los Angeles, pp. 29–33.

Droser, M. L. and Bottjer, D. J. (1989) Ichnofabric of sandstones deposited in high energy nearshore environments: measurement and utilization. *Palaios*, **4**, 598–604.

Droser, M. L., Hughes, N. C. and Jell, P. A. (1994) Infaunal communities and tiering in lower Palaeozoic nearshore clastic environments: trace fossil evidence from the Cambro–Ordovician of New South Wales. *Lethaia*, **27**, 273–83.

Dubiel, R. F., Blodgett, R. H. and Bown, T. M. (1987) Lungfish burrows in the Upper Triassic Chinle and Dolores Formations, Colorado Plateau. *Journal of Sedimentary Petrology*, **57**, 512–21.

Dubiel, R. F., Blodgett, R. H. and Bown, T. M. (1988) Lungfish burrows in the Upper Triassic Chinle and Dolores Formations, Colorado Plateau – a reply. *Journal of Sedimentary Petrology*, **58**, 367–9.

Dubiel, R. F., Blodgett, R. H. and Bown, T. M. (1989) Lungfish burrows in the Upper Triassic and Dolores Formations, Colorado Plateau – a reply. *Journal of Sedimentary Petrology*, **59**, 876–8.

Duncan, P. B. (1987) Burrow structure and burrowing activity of the funnel-feeding enteropneust *Balanoglossus aurantiacus* in Bogue Sound, North Carolina, U.S.A. *P.S.Z.N.I: Marine Ecology*, **8**, 75–95.

Dworschak, P. C. (1983) The biology of *Upogebia pusilla* (Petagna) (Decapoda, Thalassinidea). 1. Burrows. *Marine Ecology*, **4**, 19–43.

Dworschak, P. C. (1987a) Feeding behaviour of *Upogebia pusilla* and *Callianassa tyrrhena* (Crustacea, Decapoda, Thalassinidea). *Investigación Pesquera, Barcelona*, **51** (1), 421–9.

Dworschak, P. C. (1987b) Burrows of *Solecurtus strigilatus* (Linné) and *S. multistriatus* (Scacchi). *Senckenbergiana maritima*, **19**, 131–47.

Dworschak, P. C. and Ott, J. A. (1993) Decapod burrows in mangrove-channel and back-reef environments at the Atlantic Barrier Reef, Belize. *Ichnos*, **2**, 277–90.

Dworschak, P. C. and Pervesler, P. (1988) Burrows of *Callianassa bouvieri* Nobili 1904 from Safaga (Egypt, Red Sea) with some remarks on the biology of the species. *Senckenbergiana maritima*, **20**, 1–17.

Dybern, B. I. (1973) Lobster burrows in Swedish waters. *Helgoländer wissenschaftliche Meeresuntersuchungen*, **24**, 401–14.

Dybern, B. I. and Høisæter, T. (1965) The burrows of *Nephrops norvegicus* (L.). *Sarsia*, **21**, 49–55.

Easthouse, K. A. and Driese, S. G. (1988) Paleobathymetry of a Silurian shelf system: applications of proximality trends and trace fossil distributions. *Palaios*, **3**, 473–86.

Eckman, J. E. and Nowell, A. R. M. (1984) Boundary skin friction and sediment transport about an animal-tube mimic. *Sedimentology*, **31**, 851–62.

Ekdale, A. A. (1977) Abyssal trace fossils in worldwide Deep Sea Drilling Project cores. *Geological Journal Special Issues*, **9**, 163–82.

Ekdale, A. A. (1978) Trace fossils in Leg 42A cores. *Initial Reports of the Deep Sea Drilling Project*, **42**, 821–7.

Ekdale, A. A. (1980) Graphoglyptid burrows in modern deep-sea sediment. *Science*, **207**, 304–6.

Ekdale, A. A. (1985) Paleoecology of the marine endobenthos. *Palaeogeography, Paleoclimatology, Palaeoecology*, **50**, 63–81.

Ekdale, A. A. (1988) Pitfalls of paleobathymetric interpretations based on trace fossil assemblages. *Palaios*, **3**, 464–72.

Ekdale, A. A. and Berger, W. H. (1978) Deep-sea ichnofacies: modern organism traces on and in pelagic carbonates of the western equatorial Pacific. *Palaeogeography, Palaeoclimatology, Palaeoecology*, **23**, 263–78.

Ekdale, A. A. and Bromley, R. G. (1983) Trace fossils and ichnofabric in the Kjølby Gaard Marl, uppermost Cretaceous, Denmark. *Bulletin of the Geological Society of Denmark*, **31**, 107–19.

Ekdale, A. A. and Bromley, R. G. (1991) Analysis of composite ichnofabrics: an example in uppermost Cretaceous chalk of Denmark. *Palaios*, **6**, 232–49.

Ekdale, A. A., Bromley, R. G. and Pemberton, S. G. (1984a) Ichnology – the use of trace fossils in sedimentology and stratigraphy. *SEPM Short Course*, **15**.

Ekdale, A. A. and Lewis, D. W. (1993) Sabellariid reefs in Ruby Bay, New Zealand: a modern analogue of *Skolithos* 'piperock' that is *not* produced by burrowing activity. *Palaios*, **8**, 614–20.

Ekdale, A. A. and Mason, T. R. (1988) Characteristic trace fossil associations in oxygen-poor sedimentary environments. *Geology*, **16**, 720–3.

Ekdale, A. A., Muller, L. N. and Novak, M. T. (1984b) Quantitative ichnology of modern pelagic deposits in the abyssal Atlantic. *Palaeogeography, Palaeoclimatology, Palaeoecology*, **45**, 189–223.

Ekdale, A. A. and Picard, M. D. (1985) Trace fossils in a Jurassic eolianite, Entrada Sandstone, Utah, U.S.A. *SEPM Special Publications*, **35**, 3–12.

Ekdale, A. A. and Pollard, J. E. (1991) Ichno-this and ichno-that. *Palaios*, **6**, 197–8.

Elder, H. Y. (1973) Direct peristaltic progression and the functional significance of the dermal connective tissues during burrowing in the polychaete *Polyphysia crassa* (Oersted). *Journal of Experimental Biology*, **58**, 637–55.

Elder, H. Y. and Hunter, R. D. (1980) Burrowing of *Priapulus caudatus* (Vermes) and the significance of the direct peristaltic wave. *Journal of Zoology, London*, **191**, 333–51.

Elders, C. A. (1975) Experimental approaches in neoichnology, in *The Study of Trace Fossils* (ed. R. W. Frey), Springer-Verlag, New York, pp. 513–36.

Enders, H. E. (1908) Observations on the formation and enlargement of the tubes of the marine annelid, (*Chaetopterus variopedatus*). *Proceedings of the Indiana Academy of Science* (for 1907), 128–35.

Enos, P. (1977) Tamabra limestone of the Poza Rica trend, Cretaceous, Mexico. *SEPM Special Publications*, **25**, 273–314.

Enos, P. (1983) Shelf environment. *American Association of Petroleum Geologists, Memoirs*, **33**, 267–95.

Erba, E. and Primoli Silva, I. (1994) Orbitally driven cycles in trace-fossil distribution from the Piobbico core (late Albian, central Italy). *International Association of Sedimentologists, Special Publications*, **19**, 211–25.

Ericson, D. B., Ewing, M. and Wollin, G. (1963) Pliocene–Pleistocene boundary in deep-sea sediments. *Science*, **139**, 727–37.

Ewing, M. and Davis, R. A. (1967) Lebensspuren photographed on the ocean floor, in *Deep-Sea Photography* (ed. J. B. Hersey), Johns Hopkins Press, Baltimore, pp. 259–94.

Fager, E. W. (1964) Marine sediments: effects of a tube-building polychaete. *Science*, **143**, 356–9.

Farrow, G. E. (1971) Back-reef and lagoonal environments of Aldabra Atoll distinguished by their crustacean burrows. *Symposia of the Zoological Society of London*, **28**, 455–500.

Farrow, G. E. (1975) Techniques for the study of fossil and recent traces, in *The Study of Trace Fossils* (ed. R. W. Frey), Springer-Verlag, New York, pp. 537–54.

Fauchald, K. (1974) Polychaete phylogeny: a problem in protostome evolution. *Systematic Zoology*, **23**, 493–506.

Fauchald, K. and Jumars, P. A. (1979) The diet of worms: a study of polychaete feeding guilds. *Oceanography and Marine Biology Annual Review*, **17**, 193–284.

Featherstone, R. P. and Risk, M. J. (1977) Effect of tube-building polychaetes on intertidal sediments of the Minas Basin, Bay of Fundy. *Journal of Sedimentary Petrology*, **47**, 446–50.

Felbeck, H. (1983) Sulfide oxidation and carbon fixation by the gutless clam *Solemya reidi*: an animal–bacteria symbiosis. *Journal of Comparative Physiology*, **B152**, 3–11.

Felbeck, H., Childress, J. J. and Somero, G. N. (1984) Calvin–Benson cycle and sulphide oxidation enzymes in animals from sulphide-rich habitats. *Nature*, **293**, 291–3.

Fenchel, T. (1969) The ecology of the marine microbenthos, 4. *Ophelia*, **6**, 1–182.

Fenchel, T. and Riedl, R. J. (1970) The sulfide system: a new biotic community underneath the oxidized layer of marine sand bottoms. *Marine Biology*, **7**, 255–68.

Figuier, L. (1866) *La Terre avant la Déluge*. 5th edition, Paris. (Fide Abel 1935.)

Fischer, W. K. (1946) Echiuroid worms of the north Pacific Ocean. *Proceedings of the United States National Museum*, **96**, 215–92.

Fischer, W. K. and MacGinitie, G. E. (1928) The natural history of an echiuroid worm. *Annals and Magazine of Natural History*, (10) **1**, 204–13.

Fisher, J. B., Lick, W. J., McCall, P. L. and Robbins, J. A. (1980) Vertical mixing of lake sediments by tubificid oligochaetes. *Journal of Geophysical Research*, **85**, 3997–4006.

Fisher, M. R. and Hand, S. C. (1984) Chemoautotrophic symbionts in the bivalve *Lucina floridana* from eelgrass beds. *Biological Bulletin*, **167**, 445–59.

Föllmi, K. B. and Grimm, K. A. (1990) Doomed pioneers: gravity-flow deposition and bioturbation in marine oxygen-deficient environments. *Geology*, **18**, 1069–72.

Forbes, T. L. (1989) The importance of size-dependent processes in the ecology of deposit-feeding benthos, in *Ecology of Marine Deposit Feeders* (eds G. Lopez, G. Taghon and J. Levinton), *Lecture Notes on Coastal and Estuarine Studies*, **31**, 171–200.

Frankel, L. and Mead, D. J. (1973) Mucilaginous matrix of some estuarine sands in Connecticut. *Journal of Sedimentary Petrology*, **43**, 1090–5.

Frankenberg, D. and Smith, K. L. (1968) Coprophagy in marine animals. *Limnology and Oceanography*, **13**, 443–50.

Frey, R. W. (1968) The Lebensspuren of some common marine invertebrates near Beaufort, North Carolina. 1, Pelecypod burrows. *Journal of Paleontology*, **42**, 570–4.

Frey, R. W. (1970) The Lebensspuren of some common marine invertebrates near Beaufort, North Carolina. 2, Anemone burrows. *Journal of Paleontology*, **44**, 308–11.

Frey, R. W. (1973) Concepts in the study of biogenic sedimentary structures. *Journal of Sedimentary Petrology*, **43**, 6–19.

Frey, R. W. (1975) The realm of ichnology, its strengths and limitations, in *The Study of Trace Fossils* (ed. R. W. Frey), Springer-Verlag, New York, pp. 13–38.

Frey, R. W. (1990) Trace fossils and hummocky cross-stratification, Upper Cretaceous of Utah. *Palaios*, **5**, 203–18.

Frey, R. W. and Bromley, R. G. (1985) Ichnology of American chalks: the Selma Group (Upper Cretaceous), western Alabama. *Canadian Journal of Earth Sciences*, **22**, 801–28.

Frey, R. W. and Goldring, R. (1992) Marine event beds and recolonization surfaces as revealed by trace fossil analysis. *Geological Magazine*, **129**, 325–35.

Frey, R. W. and Howard, J. D. (1969) A profile of biogenic structures in a Holocene barrier island–salt marsh complex, Georgia. *Transactions of the Gulf Coast Association of Geological Societies*, **19**, 427–44.

Frey, R. W. and Howard, J. D. (1972) Radiographic study of sedimentary structures made by beach and offshore animals in aquaria. *Senckenbergiana Maritima*, **4**, 169–82.

Frey, R. W. and Howard, J. D. (1975) Endobenthic adaptations of juvenile thalassinidean shrimp. *Bulletin of the Geological Society of Denmark*, **24**, 283–97.

Frey, R. W. and Howard, J. D. (1981) *Conichnus* and *Schaubcylindrichnus*: redefined trace fossils from the Upper Cretaceous of the Western Interior. *Journal of Paleontology* **55**, 800–4.

Frey, R. W. and Howard, J. D. (1986) Mesotidal estuarine sequences: a perspective from the Georgia Bight. *Journal of Sedimentary Petrology*, **56**, 911–24.

Frey, R. W. and Howard, J. D. (1990) Trace fossils and depositional sequences in a clastic shelf setting, Upper Cretaceous of Utah. *Journal of Paleontology*, **64**, 803–20.

Frey, R. W. and Mayou, T. V. (1971) Decapod burrows in Holocene barrier island beaches and washover fans, Georgia. *Senckenbergiana Maritima*, **3**, 53–77.

Frey, R. W. and Pemberton, S. G. (1984) Trace fossil facies models, in *Facies Models*, 2nd edn (ed. R. G. Walker), Geoscience Canada Reprint Series, pp. 189–207.

Frey, R. W. and Pemberton, S. G. (1985) Biogenic structures in outcrops and cores. I. Approaches to ichnology. *Bulletin of Canadian Petroleum Geology*, **33**, 72–115.

Frey, R. W. and Pemberton, S. G. (1987) The *Psilonichnus* ichnocoenose, and its relationship to adjacent marine and nonmarine ichnocoenoses along the Georgia coast. *Bulletin of Canadian Petroleum Geology*, **35**, 333–57.

Frey, R. W., Pemberton, S. G. and Saunders, T. D. (1990) Ichnofacies and bathymetry: a passive relationship. *Journal of Paleontology*, **64**, 155–8.

Frey, R. W. and Seilacher, A. (1980) Uniformity in marine invertebrate ichnology. *Lethaia*, **13**, 183–207.

Frey, R. W., Howard, J. D. and Pryor, W. A. (1978) *Ophiomorpha*: its morphologic, taxonomic, and environmental significance. *Palaeogeography, Palaeoclimatology, Palaeoecology*, **23**, 199–229.

Fricke, H. W. (1973) Behaviour as part of ecological adaptation. In situ studies in the coral reef. *Helgoländer wissenschaftliche Meeresuntersuchungen*, **24**, 120–44.

Fricke, H. and Kacher, H. (1982) A mound-building deep water sand tilefish of the Red Sea: *Hoplolatilus geo* n. sp. (Perciformes: Branchiostegidae). Observations from a research submersible. *Senckenbergiana maritima*, **14**, 245–59.

Friedrich, H. and Langeloh, H.-P. (1936) Untersuchungen zur Physiologie der Bewegung und des Hauptmuskelschlauches bei *Halicryptus spindulosus* und *Priapulus caudatus*. *Biologische Zentralblatt*, **56**, 249–60.

Fu, S. (1991) Funktion, Verhalten und Einteilung fucoider und lophocteniider Lebensspuren. *Courier Forschungs-Institut Senckenberg*, **135**, 1–79.

Fu, S. and Werner, F. (1994) Distribution and composition of biogenic structures on the Iceland-Faeroe Ridge: relation to different environments. *Palaios*, **9**, 92–101.

Fu, S. and Werner, F. (1995) Is *Zoophycos* a feeding trace? *Neues Jahrbuch für Geologie und Paläontologie, Abhandlungen*, **195**, 37–47.

Fu, S., Werner, F. and Brossmann, J. (1994) Computed tomography: application in studying biogenic structures in sedimentary cores. *Palaios*, **9**, 116–19.

Fuchs, T. (1895) Studien über Fucoiden und Hieroglyphen. *Denkschrift der kaiserlichen Akademie der Wissenschaft in Wien, mathematisch-naturwissenschaftliche Classe*, **62**, 369–448.

Fuchs, T. (1909) Über einiger neuere Arbeiten zur Aufklärung der Natur der Alectoruriden. *Mitteilungen der Geologischen Gesellschaft in Wien*, **2**, 335–50.

Fuglewicz, R., Ptaszyński, T. and Rdzanek, K. (1990) Lower Triassic footprints from the Swietokrzyskie (Holy Cross) Mountains, Poland. *Acta Palaeontologica Polonica*, **35**, 109–64.

Fürsich, F. T. (1973) A revision of the trace fossils *Spongeliomorpha, Ophiomorpha* and *Thalassinoides*. *Neues Jahrbuch für Geologie und Paläontologie, Monatshefte*, 719–35.

Fürsich, F. T. (1974a) Ichnogenus *Rhizocorallium. Paläontologische Zeitschrift*, **48**, 16–28.

Fürsich, F. T. (1974b) On *Diplocraterion* Torell 1870 and the significance of morphological features in vertical, spreite-bearing, U-shaped trace fossils. *Journal of Paleontology*, **48**, 952–62.

Fürsich, F. T. (1978) The influence of faunal condensation and mixing on the preservation of fossil benthic communities. *Lethaia*, **11**, 243–50.

Fürsich, F. T. (1981) Invertebrate trace fossils from the Upper Jurassic of Portugal. *Communicacöes dos Servicos Geológicos de Portugal*, **67**, 153–68.

Fürsich, F. T. (1994) Palaeoecology and evolution of Mesozoic salinity-controlled benthic macroinvertebrate associations. *Lethaia*, **26**, 327–46.

Fürsich, F. T. and Bromley, R. G. (1985) Behavioural interpretation of a rosetted spreite trace fossil: *Dactyloidites ottoi* (Geinitz). *Lethaia*, **18**, 199–207.

Fürsich, F. T., Kennedy, W. J. and Palmer, T. J. (1981) Trace fossils at a regional discontinuity surface: the Austin/Taylor (Upper Cretaceous) contact in central Texas. *Journal of Paleontology*, **55**, 537–51.

Fürich, F. T. and Mayr, H. (1981) Non-marine *Rhizocorallium* (trace fossil) from the Upper Freshwater Molasse (Upper Miocene) of southern Germany. *Neues Jahrbuch für Geologie und Paläontologie, Monatshefte*, 321-333.

Fürsich, F. T., Pandey, D. K., Oschmann, W., Jaitly, A. K. and Singh, I. B. (1994) Ecology and adaptive strategies of corals in unfavourable environments: examples from the Middle Jurassic of the Kachchh Basin, western India. *Neues Jahrbuch für Geologie und Paläontologie, Abhandlungen*, **194**, 269–303.

Fürsich, F. T., Oschmann, W., Singh, I. B. and Jaitly, A. K. (1992) Hardgrounds, reworked concretion levels and condensed horizons in the Jurassic of western India: their significance for basin analysis. *Journal of the Geological Society, London*, **149**, 313–31.

Gage, J. D. and Tyler, P. A. (1991) *Deep-Sea Biology*, Cambridge University Press, Cambridge.

Gaillard, C. (1988) Bioturbation récente au large de la Nouvelle-Calédonie. Premiers résultats de la campagne Biocal. *Oceanologica Acta*, **11**, 389–99.

Gaillard, C. (1991) Recent organism traces and ichnofacies on the deep-sea floor off New Caledonia, southwestern Pacific. *Palaios*, **6**, 302–15.

Gans, C. (1960) Studies on amphisbaenids (Amphisbaenia, Reptilia). I. A taxonomic revision of the Trogonophinae and a functional interpretation of the amphisbaenid adaptive pattern. *American Museum of Natural History, Bulletin*, **119**, 129–207.

Gans, C. (1968) Relative success of divergent pathways in amphisbaenian specialization. *American Naturalist*, **102**, 345–62.

Gand, G. (1994) Ichnocoenoses à *Isopodichnus furcosus* nov. ichnosp. dans le Permien du Bassin de Lodève (Massif Central, France). *Geobios*, **27**, 73–86.

Genise, J. F. and Bown, T. M. (1994a) New Miocene scarabeid and hymenopterous nests and Early Miocene (Santacrucian) paleoenvironments, Patagonian Argentina. *Ichnos*, **3**, 107–17.

Geneise, J. F. and Bown, T. M. (1994b) New trace fossils of termites (Insecta: Isopoda) from the late Eocene–early Miocene of Egypt, and the reconstruction of ancient isopteran social behavior. *Ichnos*, **3**, 155–83.

Genise, J. F. and Cladera, G. (1995) Application of computerized tomography to study insect traces. *Ichnos*, **4**, 77–81.

Gierlowski-Kordesch, E. (1991) Ichnology of an ephemeral lacustrine/alluvial plain system: Jurassic East Berlin Formation, Hartford Basin, USA. *Ichnos*, **1**, 221–32.

Gislén, T. (1940) Investigations on the ecology of *Echiurus*. *Lunds Universitets Årsskrifter, Nye Fölge*, **36** (2), (10).

Glass, B. P. (1969) Reworking of deep-sea sediments as indicated by the vertical dispersion of the Australasian and Ivory Coast mikrotektite horizons. *Earth and Planatary Science Letters*, **6**, 409–15.

Goldring, R. (1964) Trace fossils and the sedimentary surface in shallow water marine sediments. *Developments in Sedimentology*, **1**, 136–43.

Goldring, R. (1965) Sediments into rock. *New Scientist*, June 24, 863–5.

Goldring, R. (1985) The formation of the trace fossil *Cruziana*. *Geological Magazine*, **122**, 65–72.

Goldring, R. (1991) *Fossils in the Field*, Longman, Harlow.

Goldring, R. (1993) Ichnofacies and facies interpretation. *Palaios*, **8**, 403–5.

Goldring, R. (1995a) Organisms and the substrate: response and effect, in *Marine Palaeoenvironmental Analysis from Fossils* (eds D. W. J. Bosence and P. A. Allison), *Geological Society, London, Special Publications*, **83**, 151–80.

Goldring, R. (1995b) Book review: *The Palaeobiology of Trace Fossils* edited by S. K. Donovan. *Historical Biology*, **9**, 335–7.

Goldring, R. and Kazmierczak, J. (1974) Ecological succession in intraformational hardground formation. *Palaeontology*, **17**, 949–62.

Goldring, R. and Pollard, J. E. (1993) Organisms and sediments: relationships and applications. *Journal of the Geological Society, London*, **150**, 137–9.

Goldring, R., Pollard, J. E. and Taylor, A. M. (1991) *Anconichnus horizontalis*: a pervasive ichnofabric-forming trace fossil in post-Paleozoic offshore siliciclastic facies. *Palaios*, **6**, 250–63.

Gooday, A. J. (1986) Meiofaunal foraminiferans from the bathyal Porcupine Seabight (northeast Atlantic): size structure, standing stock, taxonomic composition, species diversity and vertical distribution in the sediment. *Deep-Sea Research*, **33**, 1345–73.

Gooday, A. J., Levin, L. A., Thomas, C. L. and Hecker, B. (1992) The distribution and ecology of *Bathysiphon filiformis* Sars and *B. major* de Folin (Protista, Foraminifera) on the continental slope off North Carolina. *Journal of Foraminiferal Research*, **22**, 129–46.

Gordon, D. C., Jr (1966) The effects of the deposit feeding polychaete *Pectinaria gouldii* on the intertidal sediments of Barnstable harbor. *Limnology and Oceanography*, **11**, 327–32.

Gradziński, R. and Uchman, A. (1994) Trace fossils from interdune deposits – an example from the Lower Triassic aeolian Tumlin Sandstone, central Poland. *Palaeogeography, Palaeoclimatology, Palaeoecology*, **108**, 121–38.

Graf, G. (1989) Benthic–pelagic coupling in a deep-sea benthic community. *Nature*, **341**, 437–9.

Greb, S. F. and Chesnut, D. R. (1994) Paleoecology of an estuarine sequence in the Breathitt Formation (Pennsylvanian), central Appalachian Basin. *Palaios*, **9**, 388–402.

Gregory, M. R., Ballance, P. F., Gibson, G. W. and Ayling, A. M. (1979) On how some rays (Elasmobranchia) excavate feeding depressions by jetting water. *Journal of Sedimentary Petrology*, **49**, 1125–30.

Griffis, R. B. and Suchanek, T. H. (1991) A model of burrow architecture and trophic modes in thalassinidean shrimp (Decapoda: Thalassinidea). *Marine Ecology Progress Series*, **79**, 171–83.

Griggs, G. B., Carey, A. G. and Kulm, L. D. (1969) Deep-sea sedimentation and sediment–fauna interaction in Cascadia Channel and on Cascadia Abyssal Plain. *Deep-Sea Research*, **16**, 157–70.

Gripp, K. (1927) Über einen 'gefürte Mäander' erzeugenden Bewohner des Ostsee-Litorals. *Senckenbergiana*, **9**, 93–9.

Gruszczyński, M. (1979) Ecological succession in Upper Jurassic hardgrounds from central Poland. *Acta Palaeontologica Polonica*, **24**, 429–50.

Gruszczyński, M. (1986) Hardgrounds and ecological succession in the light of early diagenesis (Jurassic, Holy Cross Mts., Poland. *Acta Palaeontologica Polonica*, **31**, 163–212.

Guinasso, N. L. and Schink, D. R. (1975) Quantitative estimates of biological mixing rates in abyssal sediments. *Journal of Geophysical Research*, **80**, 3032–43.

Häntzschel, W. (1939) Die Lebens-Spuren von *Corophium volutator* (Pallas) und ihre paläontologische Bedeutung. *Senckenbergiana*, **21**, 215–27.

Häntzschel, W. (1962) Trace fossils and problematica, in *Treaise on Invertebrate Paleontology* (ed. R. C. Moore), **W**, 177–245, Geological Society of America and Kansas University Press, New York and Lawrence.

Häntzschel, W. (1965) Vestigia invertebratorum et problematica. *Fossilium Catalogus I: Animalia*, **108**, W. Junk, s'Gravenhage.

Häntzschel, W. (1970) Star-like trace fossils, in *Trace Fossils* (eds T. P. Crimes and J. C. Harper), *Geological Journal Special Volumes*, **3**, 201–14.

Häntzschel, W. (1975) Trace fossils and problematica, in *Treatise on Invertebrate Paleontology* (ed. C. Teichert), **W**, Geological Society of America and Kansas University Press, Boulder and Lawrence.

Häntzschel, W. and Kraus, O. (1972) Names based on trace fossils (ichnotaxa): request for a recommendation Z.N.(S.) 1973. *Bulletin of Zoological Nomenclature*, **29**, 137–41.

Hagmeier, A. and Hinrichs, J. (1931) Bemerkungen über die Ökologie von *Branchiostoma lanceolatum* (Pallas) und das Sediment seines Wohnortes. *Senckenbergiana* **13**, 255–67.

Hakes, W. G. (1976) Trace fossils and depositional environment of four clastic units, Upper Pennsylvanian megacyclothems, northeast Kansas. *University of Kansas Paleontological Contributions*, **63**.

Hallam, A. (1975) Preservation of trace fossils, in *The Study of Trace Fossils* (ed. R. W. Frey), Springer-Verlag, New York, pp. 55–63.

Hammond, R. D. (1970) The burrowing of *Priapulus caudatus*. *Journal of Zoology, London*, **162**, 469–80.

Hand, S. C. and Somero, G. N. (1983) Energy metabolism pathways of hydrothermal vent animals: adaptations to a food-rich and sulfide-rich deep-sea environment. *Biological Bulletin*, **165**, 167–81.

Hanor, J. S. and Marshall, N. F. (1971) Mixing of sediment by organisms. *School of Geoscience Miscellaneous Publications, Louisiana State University*, **71-1**, 127–35.

Hansen, J. M. (1977) Sedimentary history of the island of Læsø, Denmark. *Bulletin of the Geological Society of Denmark*, **26**, 217–36.

Hasiotis, S. T., Aslan, A. and Bown, T. M. (1993) Origin, architecture, and paleoecology of the early Eocene continental ichnofossil *Scaphichnium hamatum* – integration of ichnology and paleopedology. *Ichnos*, **3**, 1–9.

Hasiotis, S. T. and Bown, T. M. (1992) Invertebrate trace fossils: the backbone of continental ichnology, in *Trace Fossils* (eds C. G. Maples and R. R. West), *Paleontological Society Short Courses in Paleontology*, **5**, 64–104.

Hasiotis, S. T. and Mitchell, C. E. (1989) Lungfish burrows in the Upper Triassic Chinle and Dolores Formations, Colorado Plateau – discussion: new evidence suggests origin by a burrowing decapod crustacean. *Journal of Sedimentary Petrology*, **59**, 871–5.

Hasiotis, S. T., Mitchell, C. E. and Dubiel, R. F. (1993) Application of morphologic burrow interpretations to discern continental burrow architects: lungfish or crayfish? *Ichnos*, **2**, 315–33.

Hattin, D. E. (1981) Petrology of Smoky Hill member, Niobrara Chalk (Upper Cretaceous), in type area, western Kansas. *American Association of Petroleum Geologists, Bulletin*, **65**, 831–49.

Hauksson, E. (1979) Feeding biology of *Stichopus tremulus*, a deposit-feeding holothurian. *Sarsia*, **64**, 155–60.

Haven, D. S. and Morales-Alamo, R. (1966) Aspects of biodeposition by oysters and other invertebrate filter feeders. *Limnology and Oceanography*, **11**, 487–98.

Haven, D. S. and Morales-Alamo, R. (1968) Occurrence and transport of faecal pellets in suspension in a tidal estuary. *Sedimentary Geology*, **2**, 141–51.

Haven, D. S. and Morales-Alamo, R. (1972) Biodeposition as a factor in sedimentation of fine suspended solids in estuaries. *Geological Society of America, Memoirs*, **133**, 121–30.

Hayasaka, I. (1935) The burrowing activities of certain crabs and their geological significance. *American Midland Naturalist*, **16**, 99–103.

Hayward, B. W. (1976) Lower Miocene bathyal and submarine canyon ichnocoenoses from Northland, New Zealand. *Lethaia*, **9**, 149–62.

Heer, O. (1877) *Die vorweltliche Flora der Schweiz*. J. Wurster, Zürich.

Heezen, B. C. and Hollister, C. D. (1971) *The Face of the Deep*, Oxford University Press, Oxford.

Heinberg, C. (1970) Some Jurassic trace fossils from Jameson Land (East Greenland). *Geological Journal, Special Issues*, **3**, 227–34.

Heinberg, C. (1974) A dynamic model for a meniscus filled tunnel (*Ancorichnus* n. ichnogen.) from the Jurassic *Pecten* Sandstone of Milne Land, East Greenland. *Grønlands Geologiske Undersøgelse, Rapporter*, **62**.

Heinberg, C. and Birkelund, T. (1984) Trace fossil assemblages and basin evolution of the Vardekløft formation (Middle Jurassic, central East Greenland). *Journal of Paleontology*, **58**, 362–97.

Hertweck, G. (1970a) The animal community of a muddy environment and the development of biofacies as effected by the life cycle of the characteristic species. *Geological Journal, Special Issues*, **3**, 235–42.

Hertweck, G. (1970b) Die bewohner des Wattenmeeres in ihren Auswirkung auf das Sediment, in *Das Watt* (ed. H.-E. Reineck), Kramer, Frankfurt am Main, pp. 106–27.

Hertweck, G. (1972) Georgia coast region, Sapelo Island, USA: sedimentology and biology 5. Distribution and environmental significance of Lebensspuren and in-situ skeletal remains. *Senckenbergiana Maritima*, **4**, 125–67.

Hester, N. C. and Pryor, W. A. (1972) Blade-shaped crustacean burrows of Eocene age: a composite form of *Ophiomorpha*. *Geological Society of America, Bulletin*, **83**, 677–88.

Higgs, R. (1988) Fish trails in the Upper Carboniferous of south-east England. *Palaeontology*, **31**, 255–72.

Hill, G. W. and Hunter, R. E. (1973) Burrows of the ghost crab *Ocypode quadrata* (Fabricius) on the barrier islands, south-central Texas coast. *Journal of Sedimentary Petrology*, **43**, 24–30.

Hill, G. W. and Hunter, R. E. (1976) Interaction of biological and geological processes in the beach and nearshore environments, northern Padre Island, Texas. *SEPM, Special Publications*, **24**, 169–87.

Hogue, C. L. and Bright, D. B. (1971) Observations on the biology of land crabs and their burrow associates on the Kenya coast. *Contributions in Science, Los Angeles County Museum*, **210**.

Hollister, C. D., Heezen, B. C. and Nafe, K. E. (1975) Animal traces on the deep-sea floor, in *The Study of Trace Fossils*, (ed. R. W. Frey), Springer-Verlag, New York, pp. 493–510.

Hongguang, M., Zhiying, Y. and Cadée, G. C. (In press) Macrofauna distribution and bioturbation on tidal confluences of the Dutch Wadden Sea. Netherlands *Journal of Aquatic Ecology*.

Hovland, M. and Thomsen, E. (1989) Hydrocarbon-based communities in the North Sea? *Sarsia*, **74**, 29–42.

Howard, J. D. (1968) X-ray radiography for examination of burrowing in sediments by marine invertebrate organisms. *Sedimentology*, **11**, 249–58.

Howard, J. D. (1978) Sedimentology and trace fossils, in *Trace Fossil Concepts* (ed. P. B. Basan), *SEPM Short Course no.5*, 11–42.

Howard, J. D. and Elders, C. A. (1970) Burrowing patterns of haustoriid amphipods from Sapelo Island, Georgia. *Geological Journal, Special Issues*, **3**, 243–62.

Howard, J. D. and Frey, R. W. (1975) Regional animal–sediment characteristics of Georgia estuaries. *Senckenbergiana Maritima*, **7**, 33–103.

Howard, J. D. and Reineck, H.-E. (1972) Physical and biogenic sedimentary structures of the nearshore shelf. *Senckenbergiana Maritima*, **4**, 81–123.

Howard, J. D. and Reineck, H.-E. (1981) Depositional facies of high-energy beach-to-offshore sequence: comparison with low energy sequence. *American Association of Petroleum Geologists, Bulletin*, **65**, 807–30.

Hughes, D. J., Ansell, A. D., Atkinson, R. J. A. and Nickell, L. A. (1993) Underwater television observations of surface activity of the echiuran worm *Maxmuelleria lankesteri* (Echiura: Bonelliidae). *Journal of Natural History*, **27**, 219–48.

Hughes, R. N. (1969) A study of feeding in *Scrobicularia plana*. *Journal of the Marine Biological Association, UK*, **49**, 805–23.

Hughes, R. N. and Crisp, D. J. (1976) A further description of the echiuran *Prashadus pirotansis*. *Journal of Zoology, London*, **180**, 233–42.

Hunt, A. P., Chin, K. and Lockley, M. G. (1994) The palaeobiology of vertebrate coprolites, in *The Palaeobiology of Trace Fossils* (ed. S. K. Donovan), Wiley, Chichester, pp. 221–40.

Hunter, R. D. and Elder, H. Y. (1967) Analysis of burrowing mechanism in *Leptosynapta tenuis* and *Goldfingia gouldi*. *Biological Bulletin of the Marine Biological Laboratory, Woods Hole*, **133**, 470.

Hylleberg, J. (1975) Selective feeding by *Abarenicola pacifica* with notes on *Abarenicola vagabunda* and a concept of gardening in lugworms. *Ophelia*, **14**, 113–37.

Ingle, R. W. (1966) An account of the burrowing behaviour of the amphipod *Corophium arenarium* Crawford (Amphipoda: Corophiidae). *Annals and Magazine of Natural History*, **9** (13), 309–17.

Ivanov, A. V. (1960) Embranchement des Pogonophores, in *Traité de Zoologie, 5* (ed. P.-P. Grassé), Masson, Paris, pp. 1521–622.

Jaccarini, V. and Schembri, P. J. (1977) Feeding and particle selection in the echiuran worm *Bonellia viridis* Rolando (Echiura, Bonelliidae). *Journal of Experimental Marine Biology and Ecology*, **28**, 163–81.

Jacobsen, V. H. (1967) The feeding of the lugworm, *Arenicola marina* (L.). Quantitive studies. *Ophelia*, **4**, 91–109.

Jannasch, H. W. (1984) Chemosymbiosis: the nutritional basis for life at deep-sea vents. *Oceanus*, **27** (3), 73–8.

Jarvis, I. (1992) Sedimentology, geochemistry and origin of phosphatic chalks: the Upper Cretaceous deposits of NW Europe. *Sedimentology*, **39**, 55–97.

Jenkins, R. J. F. (1975) The fossil crab *Ommatocarcinus corioensis* (Cresswell) and a review of related Australasian species. *National Museum of Victoria, Memoirs*, **36**, 33–62.

Jensen, P. (1992) *Cerianthus vogti* Danielssen, 1890 (Anthozoa: Ceriantharia). A species inhabiting an extended tube system deeply buried in deep-sea sediments off Norway. *Sarsia*, **77**, 75–80.

Jensen, P. (1992) 'An enteropneust's nest': result of the burrowing traits by the deep-sea acorn worm *Stereobalanus canadensis* (Spengel). *Sarsia*, **77**, 125–9.

Jensen, P., Emrich, R. and Weber, K. (1992) Brominated metabolites and reduced numbers of meiofauna organisms in the burrow wall lining of the deep-sea enteropneust *Stereobalanus canadensis*. *Deep-Sea Research*, **39**, 1247–53.

Jensen, S. (1990) Predation by early Cambrian trilobites on infaunal worms – evidence from the Swedish Mickwitzia Sandstone. *Lethaia*, **23**, 29–42.

Jewell, P. A. (1958) Natural history and experiment in archaeology. *Advance of Science*, **15**, 165–72.

Johnson, E. W., Briggs, D. E. G., Suthren, R. J., Wright, J. L. and Tunnicliff, S. P. (1994) Non-marine arthropod traces from the subaerial Ordovician Borrodale Volcanic Group, English Lake District. *Geological Magazine*, **131**, 395–406.

Johnson, R. G. (1971) Animal–sediment relationships in shallow water benthic communities. *Marine Geology*, **11**, 93–104.

Johnson, R. G. (1972) Conceptual models of benthic marine communities, in *Models in Paleobiology* (ed. T. J. M. Schopf), Freeman, Cooper and Co., San Francisco.

Johnson, R. G. (1977) Vertical variation in particulate matter in the upper twenty centimeters of marine sediments. *Journal of Marine Research*, **35**, 273–82.

Jones, S. E. and Jago, C. F. (1987) Geophysical assessment of sediment bioturbation in some Welsh estuaries. *Proceedings of the Geologists' Association*, **98**, 409–12.

Jumars, P. A. (1978) Spatial autocorrelation with RUM (Remote Underwater Manipulator): vertical and horizontal structure of a bathyal benthic community. *Deep-Sea Research*, **25**, 589–604.

Jumars, P. A. and Fauchald, K. (1977) Between-community contrasts in successful polychaete feeding strategies, in *Ecology of Marine Benthos* (ed. B. C. Coull), University of Carolina Press, Columbia, pp. 1–20.

Jumars, P. A., Nowell, A. R. M. and Self, R. F. L. (1981) A simple model of flow–sediment–organism interaction. *Marine Geology*, **42**, 155–72.

Jumars, P. A., Self, R. F. L. and Nowell, A. R. M. (1982) Mechanics of particle selection by tentaculate deposit-feeders. *Journal of Experimental Marine Biology and Ecology*, **64**, 47–70.

Kanazawa, K. (1991) Burrowing mechanism and test profile in spatangoid echinoids, in *Biology of Echinodermata* (eds Yanagisawa, Yasumasu, Oguro, Suzuki and Motokawa), Balkema, Rotterdam, pp. 147–51.

Kanazawa, K. (1992) Adaptation of test shape for burrowing and locomotion in spatangoid echinoids. *Palaeontology*, **35**, 733–50.

Karplus, I., Szlep, R. and Tsurnamal, M. (1972) Associative behavior of the fish *Cryptocentrus cryptocentrus* (Gobiidae) and the pistol shrimp *Alpheus djiboutensis* (Alpheidae) in artificial burrows. *Marine Biology*, **15**, 95–104.

Karplus, I., Szlep, R. and Tsurnamal, M. (1974) The burrows of alpheid shrimp associated with gobiid fish in the northern Red Sea. *Marine Biology*, **24**, 259–68.

Keighley, D. G. and Pickerill, R. K. (1994) The ichnogenus *Beaconites* and its distinction from *Ancorichnus* and *Taenidium*. *Palaeontology*, **37**, 305–37.

Keighley, D. G. and Pickerill, R. K. (1995) The ichnotaxa *Palaeophycus* and*Planolites*: historical perspectives and recommendations. *Ichnos*, **3**, 301–9.

Keller, G. H., Richards, A. F. *et al.* (1976) Sea-floor deposition, erosion, and transportation, in *The Benthic Boundary Layer* (ed. I. N. Cave), Plenum Press, New York, pp. 247–60.

Kelly, S. R. A. (1990) Trace fossils, in *Palaeobiology, a Synthesis*, (eds D. E. G. Briggs and P. R. Crowther), Blackwell, Oxford, pp. 423–5.

Kennedy, W. J. (1967) Burrows and surface traces from the Lower Chalk of southern England. *Bulletin of the British Museum (Natural History), Geology*, **15**, 127–67.

Kennedy, W. J., Jakobsen, M. E. and Johnson, R. T. (1969) A *Favreina–Thalassinoides* association from the Great Oolite of Oxfordshire. *Palaeontology*, **12**, 549–54.

Kern, J. P. (1978) Paleoenvironment of new trace fossils from the Eocene Mission Valley formation, California. *Journal of Paleontology*, **52**, 186–94.

Kern, J. P. (1980) Origin of trace fossils in Polish Carpathian flusch. *Lethaia*, **13**, 347–62.

Kern, J. P. and Warme, J. E. (1974) Trace fossils and bathymetry of the Upper Cretaceous Point Loma formation, San Diego, California. *Geological Society of America, Bulletin*, **85**, 893–900.

Kershaw, P. J., Swift, D. J., Pentreath, R. J. and Lovett, M. B. (1983) Plutonium redistribution by biological activity in Irish Sea sediments. *Nature*, **306**, 774–5.

Kidwell, S. M. and Aigner, T. (1985) Sedimentary dynamics of complex shell beds: implications for ecologic and evolutionary patterns. *Lecture Notes in Earth Science*, **1**, 382–95.

Kidwell, S. M. and Bosence, D. W. (1991) Taphonomy and time averaging of marine shelly faunas, in *Taphonomy: Releasing the Data Locked in the Fossil Record* (eds P. A. Allison and D. E. G. Briggs), Plenum Press, New York, pp. 115–209.

Kieth, A. (1942) A postscript to Darwin's 'Formation of vegetable mould through the action of worms'. *Nature*, **149**, 716–20.

Kilpper, K. (1962) *Xenohelix* Mansfield 1927 aus der miozänen Niederrheinischen Braunkohlenformation. *Paläontologische Zeitschrift*, **36**, 55–8.

King, A. F. (1965) Xiphosurid trails from the Upper Carboniferous of Bude, north Cornwall. *Proceedings of the Geological Society, London*, **1626**, 162–5.

King, G. M. (1986) Inhibition of microbial activity in marine sediments by a bromophenol from a hemichordate. *Nature*, **323**, 257–9.

Klausewitz, W. (1962) Röhrenaale im Roten Meer. *Natur und Volk*, **92**, 95–8.

Knight-Jones, E. W. (1953) Feeding in *Saccoglossus* (Enteropneusta). *Proceedings of the Zoological Society, London*, **123**, 637–54.

Kotake, N. (1989) Paleoecology of the *Zoophycos* producers. *Lethaia*, **22**, 327–41.

Kotake, N. (1991) Packing process for the filling material in *Chondrites*. *Ichnos*, **1**, 277–85.

Kotake, N. (1992) Deep-sea echiurans: possible producers of *Zoophycos*. *Lethaia*, **25**, 311–16.

Kranz, P. M. (1974) The anastrophic burial of bivalves and its paleoecological significance. *Journal of Geology*, **82**, 237–65.

Krüger, F. (1959) Zur Ernährungsphysiologie von *Arenicola marina* L. *Zoologischer Anzeiger*, **22** (Supplement), 115–20.

Kudenov, J. D. (1978) The feeding ecology of *Axiothella rubrocincta* (Johnson) (Polychaeta: Maldanidae). *Journal of Experimental Marine Biology and Ecology*, **31**, 209–21.

Lampitt, R. S. (1985) Evidence for a seasonal deposition of detritus to the deep-sea floor and its subsequent resuspension. *Deep-Sea Research*, **32**, 885–97.

Landing, E. and Brett, C. E. (1987) Trace fossils and regional significance of a Middle Devonian (Givetian) disconformity in southwestern Ontario. *Journal of Paleontology*, **61**, 205–30.

Lemche, H. (1973) Comments on the application concerning trace fossils. *Bulletin of Zoological Nomenclature*, **30**, 70.

Lemche, H., Hansen, B., Madsen, F. J., Tendal, O. S. and Wolff, T. (1976) Hadal life as analysed from photographs. *Videnskabelige Meddelelser af den Danske Naturhistoriske Forening*, **139**, 263–336.

Leszczyński, S. (1991) Oxygen-related controls on predepositional ichnofacies in turbidites, Guipuzcoan Flysch (Albian–Lower Eocene), northern Spain. *Palaios*, **6**, 271–80.

Leszczyński, S. and Seilacher, A. (1991) Ichnocoenoses of a turbidite sole. *Ichnos*, **1**, 293–303.

Leszczyński, S. and Uchman, A. (1993) Biogenic structures of organics-poor siliciclastic sediments: examples from Paleogene variegated shales, Polish Carpathians. *Ichnos*, **2**, 267–75.

Levin, L. A. (1994) Paleoecology and ecology of xenophyophores. *Palaios*, **9**. 32–41.

Levinsen, G. (1884) Systematisk-geografisk Oversigt over de nordiske Annulata, Gephyrea, Chaetognathi og Balanoglossi. *Videnskabelige Meddelelser af den Historiske Forening, København*, **5** (4), 92–350.

Levinton, J. S. (1972) Stability and trophic structure in deposit-feeding and suspension-feeding communities. *American Naturalist*, **106**, 472–86.

Levinton, J. S. (1977) Ecology of shallow water deposit-feeding communities, Quisset Harbor, Massachusetts, in *Ecology of Marine Benthos* (ed. B. C. Coull), University of South Carolina Press, Columbia, pp. 191–227.

Levinton, J. S. (1979) Deposit-feeders, their resources, and the study of resource limitation. *Marine Science*, **10**, 117–41.

Levinton, J. S. (1989) Deposit feeding and coastal oceanography, in *Ecology of Marine Deposit Feeders* (eds G. Lopez, G. Taghon and J. Levinton), *Lecture Notes on Coastal and Estuarine Studies*, **31**, 1–23.

Levinton, J. S. and Bambach, R. K. (1975) A comparative study of Silurian and recent deposit-feeding bivalve communities. *Paleobiology*, **1**, 97–124.

Liljedahl, L. (1992) The Silurian *Ilionia prisca*, oldest known deep-burrowing suspension-feeding bivalve. *Journal of Paleontology*, **66**, 206–10.

Linke, O. (1939) Die Biota des Jadebusenwattes. *Helgoländer wissenschaftliche Meeresuntersuchungen*, **1**, 201–348.

Lipps, J. H. (1983) Biotic interactions in benthic foraminifera, in *Biotic Interactions in Recent and Fossil Benthic Communities* (eds M. J. S. Tevesz and P. L. McCall), Plenum Press, New York, pp. 331–76.

Lockley, M. [G.] (1991) *Tracking Dinosaurs*, Cambridge University Press, New York.

Lockley, M. G. (1993) Ichnotopia. The Paleontology Society short course on trace fossils. *Ichnos*, **2**, 337–42.

Lockley, M. G., Hunt, A. P. and Meyer, C. A. (1994a) Vertebrate tracks and the ichnofacies concept: implications for palaeoecology and palichnostratigraphy, in *The Palaeobiology of Trace Fossils* (ed. S. K. Donovan), Wiley, Chichester, pp. 241–68.

Lockley, M. G., Logue, T. J., Moratalla, J. J., Hunt, A. P., Schultz, R. J. and Robinson, J. W. (1995) The fossil trackway *Pteraichnus* is pterosaurian, not crocodilian: implications for the global distribution of pterosaur tracks. *Ichnos*, **4**, 7–20.

Lockley, M. G., Novikov, V., Dos Santos, V. F., Nessov, L. A. and Forney, G. (1994b) 'Pegados de Mula': an explanation for the occurrence of Mesozoic traces that resemble mule tracks. *Ichnos*, **3**, 125–33.

Lockley, M. G., Rindsberg, A. K. and Zeiler, R. M. (1987) The paleoenvironmental significance of the nearshore *Curvolithus* ichnofacies. *Palaios*, **2**, 255–62.

Lomnicki, A. M. (1886) Slodkowodny utwor trzeciorzedny na Podolu galicyjskiem. *Akademii Umiejetnosci w Krakowie, Sprawozdania Komisji Fizyograficznej*, **20** (2), 48–119.

Longbottom, M. R. (1970) The distribution of *Arenicola marina* (L.) with particular reference to the effects of particle size and organic matter of the sediment. *Journal of Experimental Marine Biology and Ecology*, **5**, 138–57.

Lund, E. J. (1957) Self-silting by the oyster and its significance for sedimentation geology. *Publications of the Institute of Marine Science, Port Arkansas, Texas*, **4**, 320–7.

Lutze, J. (1938) Über Systematik, Entwicklung und Ökologie von *Callianassa*. *Helgoländer wissenschaftliche Meeresuntersuchungen*, **1**, 162–99.

MacEachern, J. A., Raychaudhuri, I. and Pemberton, S. G. (1992) Stratigraphic applications of the *Glossifungites* ichnofacies: delineating discontinuities in the rock record, in *Applications of Ichnology to Petroleum Exploration* (ed. S. G. Pemberton), *SEPM Core Workshop*, **17**, 169–98.

MacGinitie, G. E. (1930) The natural history of the mud shrimp *Upogebia pugettensis* (Dana). *Annals and Magazine of Natural History*, **6** (10), 36–44.

MacGinitie, G. E. (1932) The role of bacteria as food for bottom animals. *Science*, **76**, 490.

MacGinitie, G. E. (1934) The natural history of *Callianassa californiensis* Dana. *American Midland Naturalist*, **15**, 166–77.

MacGinitie, G. E. (1939) The method of feeding of *Chaetopterus*. *Biological Bulletin of the Marine Biological Laboratory, Woods Hole*, **77**, 115–18.

MacGinitie, G. E. (1945) The size of the mesh openings in mucous feeding nets of marine animals. *Biological Bulletin of the Marine Biological Laboratory, Woods Hole*, **88**, 107–11.

MacGinitie, G. E. and MacGinitie, N. (1949) *Natural History of Marine Animals*, McGraw-Hill, New York.

MacNaughton, R. B. and Pickerill, R. K. (1995) Invertebrate ichnology of the nonmarine Lepreau Formation (Triassic), southern New Brunswick, eastern Canada. *Journal of Paleontology*, **69**, 160–71.

Magnus, D. B. E. (1967) Zur Ökologie sedimentbewohnende *Alpheus*-Garnelen (Decapoda, Nanantia) des Roten Meeres. *Helgoländer wissenschaftliche Meeresuntersuchungen*, **15**, 506–22.

Magwood, J. P. A. and Ekdale, A. A. (1994) Computer-aided analysis of visually complex ichnofabrics in deep-sea sediments. *Palaios*, **9**, 102–15.

Maillard, G. (1887) Considérations sur les fossiles décrits comme algues. *Société Paléontologique Suisse, Mémoires*, **14**.

Manfrin, G. and Piccinetti, C. (1970) Osservazioni etologiche su *Squilla mantis* L. *Note del Laboratorio di Biologia Marina de Pesca Fano*, **3**, 93–104.

Mángano, M. G. and Buatois, L. A. (1991) Discontinuity surfaces in the Lower Cretaceous of the High Andes (Mendoza, Argentina): trace fossils and environmental implications. *Journal of South American Earth Sciences*, **4**, 215–29.

Mangum, C. P. (1964) Studies on speciation in maldanid polychaetes of the North American Atlantic coast. II. Distribution and competitive interaction of five sympatric species. *Limnology and Oceanography*, **9**, 12–26.

Mangum, D. C. (1970) Burrowing behavior of the sea anemone *Phyllactis*. *Biological Bulletin of the Marine Biological Laboratory, Woods Hole*, **138**, 316–25.

Manning, R. B. and Felder, D. L. (1986) The status of the callianassid genus *Callichirus* Stimpson, 1866 (Crustacea: Decapoda: Thalassinidea). *Proceedings of the Biological Society of Washington*, **99**, 437–43.

Maples, C. G. and West, R. R. (eds) (1992) Trace fossils. *Short Courses in Paleontology*, Paleontological Society, **5**.

Marintsch, E. J. and Finks, R. M. (1982) Lower Devonian ichnofacies at Highland Mills, New York and their gradual replacement across environmental gradients. *Journal of Paleontology*, **56**, 1050–78.

Mariscal, R. N., Conklin, E. J. and Bigger, C. H. (1977) The ptychocyst, a major new category of cnida used in tube construction by the cerianthid anemone. *Biological Bulletin*, **152**, 392–405.

Martinsson, A. (1965) Aspects of a Middle Cambrian thanatotope on Öland. *Geologiska Förening in Stockholm Förhandlingar*, **87**, 181–230.

Martinsson, A. (1970) Toponomy of trace fossils, in *Trace Fossils* (eds T. P. Crimes and J. C. Harper), *Geological Journal Special Issues*, **3**, 323–30.

Mauviel, A., Juniper, S. K. and Sibuet, M. (1987) Discovery of an enteropneust associated with a mound-burrows trace in the deep sea: ecological and geochemical implications. *Deep-Sea Research*, **34**, 329–35.

Mauzey, K. P., Birkland, C. and Dayton, P. K. (1968) Feeding behavior of asteroids and escape responses of their prey in the Puget Sound region. *Ecology*, **49**, 603–19.

McAllister, J. A. (1988) Lungfish burrows in the Upper Triassic Chinle and Dolores Formations, Colorado Plateau – comments on the recognition criteria of fossil lungfish burrows. *Journal of Sedimentary Petrology*, **58**, 365–7.

McBride, E. F. and Picard, M. D. (1991) Facies implications of *Trichichnus* and *Chondrites* in turbidites and hemipelagites, Marnoso-arenacea Formation (Miocene), northern Apennines, Italy. *Palaios*, **6**, 281–90.

McCall, P. L. (1977) Community patterns and adaptive strategies of the infaunal benthos of Long Island Sound. *Journal of Marine Research*, **35**, 221–66.

McCall, P. L. and Tevesz, M. J. S. (1982) The effects of benthos on physical properties of freshwater sediments, in *Animal–Sediment Relations. The Biogenic Alteration of Sediments* (eds P. L. McCall and M. J. S. Tevesz), Plenum Press, New York, pp. 105–76.

McCall, P. L. and Tevesz, M. J. S. (1983) Soft-bottom succession and the fossil record, in *Biotic Interactions in Recent and Fossil Benthic Communities* (eds M. J. S. Tevesz and P. L. McCall), Plenum Press, New York, pp. 157–94.

McCave, I. N. (1988) Biological pumping upwards of the coarse fraction of deep-sea sediments. *Journal of Sedimentary Petrology*, **58**, 148–58.

McMaster, R. L. (1962) Seasonal variability of compactness in marine sediments: a laboratory study. *Geological Society of America, Bulletin*, **73**, 643–6.

Meadows, A. and Meadows, P. S. (1994) Bioturbation in deep sea Pacific sediments. *Journal of the Geological Society, London*, **151**, 361–75.

Meadows, P. S. and Meadows, A. (eds) (1991) The environmental impact of burrowing animals and animal burrows. *Symposia of the Zoological Society of London*, **63**.

Meadows, P. S., Reichelt, A. C., Meadows, A. and Waterworth, J. S. (1994) Microbial and meiofaunal abundance, redox potential, pH and shear strength profiles in deep sea Pacific sediments. *Journal of the Geological Society, London*, **151**, 377–90.

Meadows, P. S. and Tufail, A. (1994) Bioturbation, microbial activity and sediment properties in an estuarine ecosystem. *Proceedings of the Royal Society of Edinburgh*, **90B**, 129–42.

Meldahl, K. H. (1987) Sedimentologic and taphonomic implications of biogenic stratification. *Palaios*, **2**, 350–8.

Mellanby, K. (1971) *The Mole*, Collins, London.

Melville, R. V. (1979) Further proposed amendments to the International Code of Zoological Nomenclature Z.N.(G.) 182. *Bulletin of Zoological Nomenclature*, **36**, 11–14.

Mettam, C. (1969) Peristaltic waves of tubicolous worms and the problem of irrigation in *Sabella pavonina*. *Journal of Zoology, London*, **158**, 341–56.

Metz, R. (1987) Sinusoidal trail formed by a Recent biting midge (family Ceratopogonidae): trace fossil implications. *Journal of Paleontology*, **61**, 312–14.

Metz, R. (1990) Tunnels formed by mole crickets (Orthoptera: Gryllotalpidae): paleoecological implications. *Ichnos*, **1**, 139–41.

Metz, R. (1995) Ichnologic study of the Lockatong Formation (Late Triassic), Newark Basin, southeastern Pennsylvania. *Ichnos*, **4**, 43–51.

Middlemiss, F. A. (1962) Vermiform burrows and the rate of sedimentation in the Lower Greensand. *Geological Magazine*, **99**, 33–40.

Mikulás, R. (1990) The ophiuroid *Taeniaster* as a tracemaker of *Asteriacites*, Ordovician of Czechoslovakia. *Ichnos*, **1**, 133–7.

Miller, D. C. and Sternberg, R. W. (1988) Field measurements of the fluid and sediment-dynamic environment of a benthic deposit feeder. *Journal of Marine Research*, **46**, 771–96.

Miller, M. F. (1984) Distribution of biogenic structures in Paleozoic nonmarine and marine-margin sequences: an actualistic model. *Journal of Paleontology*, **58**, 550–70.

Miller, M. F. (1991) Morphology and paleoenvironmental distribution of Paleozoic *Spirophyton* and *Zoophycos*: implications for the *Zoophycos* ichnofacies. *Palaios*, **6**, 410–25.

Miller, M. F. and Collinson, J. W. (1994) Trace fossils from Permian and Triassic sandy braided stream deposits, central Transantarctic Mountains. *Palaios*, **9**, 605–10.

Miller, M. F. and Johnson, K. G. (1981) *Spirophyton* in alluvial–tidal facies of the Catskill deltaic complex: possible biological control of ichnofossil distribution. *Journal of Paleontology*, **55**, 1016–27.

Miller, W. (1988) Giant *Bathysiphon* (Foraminifera) from Cretaceous turbidites, northern California. *Lethaia*, **21**, 363–74.

Miller, W. (1990) Community replacement pathways: what do fossil sequences reveal about marine ecosystem transitions?, in *Paleocommunity Temporal Dynamics: the Long-Term Development of Multispecies Assemblies* (ed. W. Miller), *Paleontological Society Special Publications*, **5**, 262–72.

Moore, D. G. and Scruton, P. C. (1957) Minor internal structures of some recent unconsolidated sediments. *American Association of Petroleum Geologists, Bulletin*, **41**, 2723–51.

Moore, H. B. (1939) Faecal pellets in relation to marine deposits, in *Recent Marine Sediments. A Symposium* (ed. P. D. Trask), Murby, London, pp. 516–24.

Mortensen, T. (1900) Fjordens nuværende og tidligere fauna, in *Studier over Ringkøbing Fjord* (ed. S. H. A. Rambusch), Bojesen, København, pp. 49–65.

Morton, J. E. (1959) The habits and feeding organs of *Dentalium entalis*. *Journal of the Marine Biological Association, UK*, **38**, 225–38.

Morton, J. E. and Miller, M. (1968) *The New Zealand Sea Shore*, Collins, London, Auckland.

Moussa, M. T. (1970) Nematode fossil trails from the Green River Formation (Eocene) in the Uinta Basin, Utah. *Journal of Paleontology*, **44**, 304–7.

Müller, A. H. (1962) Zur Ichnologie, Taxiologie und Ökologie fossiler Tiere. *Freiberger Forschungshefte*, **C151**, 5–49.

Müller, A. H. (1982) Über Hyponome fossiler und rezenten Insekten, erster Beitrag. *Freiberger Forschungshefte*, **C366**, 7–27.

Myers, A. C. (1970) Some palaeoichnological observations on the tube of *Diopatra cuprea* (Bosc): Polychaeta, Onuphidae, in *Trace Fossils* (eds T. P. Crimes and J. C. Harper), *Geological Journal Special Issues*, **3**, 331–4.

Myers, A. C. (1972) Tube–worm–sediment relationships of *Diopatra cuprea* (Polychaeta: Onuphidae). *Marine Biology*, **17**, 350–4.

Myers, A. C. (1977a) Sediment processing in a marine subtidal sandy bottom community: I. Physical aspects. *Journal of Marine Research*, **35**, 609–32.

Myers, A. C. (1977b) Sediment processing in a marine subtidal sandy bottom community: II. Biological consequences. *Journal of Marine Research*, **35**, 633–47.

Myers, A. C. (1979) Summer and winter burrows of a mantis shrimp, *Squilla empusa*, in Narragansett Bay, Rhode Island (USA). *Estuarine Coastal Marine Science*, **8**, 87–98.

Nash, R. D., Chapman, C. J., Atkinson, R. J. A. and Morgan, P. J. (1984) Observations on burrows and burrowing behaviour of *Calocaris macandreae* (Crustacea: Decapoda: Thalassinidea). *Journal of Zoology, London*, **202**, 425–39.

Newell, R. C. (1979) *Biology of Intertidal Animals*, 3rd edn, Marine Ecological Surveys, Faversham.

Nichols, D. (1959) Changes in the chalk heart-urchin *Micraster* interpreted in relation to living forms. *Philosophical Transactions of the Royal Society, London*, **B242**, 347–437.

Nichols, F. H. (1974) Sediment turnover by a deposit-feeding polychaete. *Limnology and Oceanography*, **19**, 945–50.

Nicolaisen, W. and Kanneworff, E. (1969) On the burrowing and feeding habits of the amphipods *Bathyporeia pilosa* Lindström and *Bathyporeia sarsi* Watkin. *Ophelia*, **6**, 231–50.

Nowell, A. R. M., Jumars, P. A. and Eckman, J. E. (1981) Effects of biological activity on the entrainment of marine sediments. *Marine Geology*, **42**, 133–53.

Nowell, A. R. M., Jumars, P. A., Self, R. F. L. and Southard, J. B. (1989) The effects of sediment transport and deposition on infauna: results obtained in a specially designed flume, in *Ecology of Marine Deposit Feeders* (eds. G. Lopez, G. Taghon and J. Levinton), *Lecture Notes on Coastal and Estuarine Studies*, **31**, 247–68.

Nyholm, K.-G. and Bornö, C. (1969) The food uptake of *Echiurus echiurus* Pallas. *Zoologiska Bidrag från Uppsala*, **38**, 249–54.

Ohta, S. (1984) Star-shaped feeding traces produced by echiuran worms on the deep-sea floor of the Bay of Bengal. *Deep-Sea Research*, **31**, 1415–32.

Orr, P. J. (1994) Trace fossil tiering within event beds and preservation of frozen profiles: an example from the Lower Carboniferous of Menorca. *Palaios*, **9**, 202–10.

Oschmann, W. (1988) Upper Kimmeridgian and Portlandian marine macrobenthic associations from southern England and northern France. *Facies*, **18**, 49–82.

Oschmann, W. (1991a) Anaerobic–poikiloaerobic–aerobic: a new facies zonation for modern and ancient neritic redox facies, in *Cycles and Events in Stratigraphy* (eds G. Einsele, W. Ricken and A. Seilacher), Springer-Verlag, Berlin, pp. 565–71.

Oschmann, W. (1991b) Distribution, dynamics and palaeoecology of Kimmeridgian (Upper Jurassic) shelf anoxia in western Europe, in *Modern and Ancient Continental Shelf Anoxia* (eds R. V. Tyson and T. H. Pearson), Geological Society, London, Special Publications, **58**, 381–95.

Oschmann, W. (1993) Environmental oxygen fluctuations and the adaptive response of marine benthic organisms. *Journal of the Geological Society, London*, **150**, 187–91.

Osgood, R. G. (1970) Trace fossils of the Cincinnati area. *Palaeontographica Americana*, **6**, 281–444.

Osgood, R. G. (1975) The history of invertebrate ichnology, in *The Study of Trace Fossils* (ed. R. W. Frey), Springer-Verlag, New York, pp. 3–12.

Osgood, R. G. and Drennen, W. T. (1975) Trilobite trace fossils from the Clinton Group (Silurian) of east-central New York State. *Bulletins of American Paleontology*, **67**, 299–348.

Osgood, R. G. and Szmuc, E. (1972) The trace fossil *Zoophycos* as an indicator of water depth. *Bulletins of American Paleontology*, **62**, 1–22.

Ott, J. A. (1993) A symbiosis between nematodes and chemoautotrophic sulfur bacteria exploiting the chemocline of marine sands, in *Trends in Microbial Ecol-*

ogy (eds R. Guerrero and C. Pedrós-Alió), Spanish Society for Microbiology, Barcelona, pp. 231–4.

Ott, J. A., Fuchs, B., Fuchs, R. and Malasek, A. (1976) Observations on the biology of *Callianassa stebbingi* Borrodaille and *Upogebia litoralis* Risso and their effect upon the sediment. *Senckenbergiana Maritima*, **8**, 61–79.

Pals, G. and Pauptit, E. (1979) Oxygen binding properties of the coelomic haemoglobin of the polychaete *Heteromastus filiformis* related with some environmental factors. *Netherlands Journal of Sea Research*, **13**, 581–92.

Parkes, R. J., Cragg, B. A., Getliff, J. M. and Fry, J. C. (1993) Presence and activity of bacteria in deep sediments from marine environments, in *Trends in Microbial Ecology* (eds R. Guerrero and C. Pedrós-Alió), Spanish Society for Microbiology, Barcelona, pp. 421–6.

Pattison, S. A. J. (1992) Recognition and interpretation of estuarine mudstones (central basin mudstones) in the tripartite valley-fill deposits of the Viking Formation, central Alberta, in *Applications of Ichnology to Petroleum Exploration* (ed. S. G. Pemberton), *SEPM Core Workshop*, **17**, 223–49.

Paul, A. Z. (1977) The effect of benthic processes on the CO_2 carbonate system, in *The Fate of Fossil Fuel CO_2 in the Oceans* (eds N. R. Andersen and A. Malahoff), Plenum Press, New York, pp. 345–54.

Paul, A.Z., Thorndyke, E. M., Sullivan, L. G., Heezen, B. C. and Gerard, R. D. (1978) Observations of the deep-sea floor from 202 days of time-lapse photography. *Nature*, **272**, 812–14.

Paull, C. K., Hecker, B., Commeau, R., Freeman-Linde, R. P., Neuman, C., Corso, W. P., Golubic, S., Hook, J. E., Sikes, J. E. and Curray, J. (1984) Biological communities at the Florida Escarpment resemble hydrothermal vent taxa. *Science*, **226**, 965–7.

Pearse, A. S. (1908) Observations on the behavior of the holothurian, *Thyone briareus* (Leseur). *Biological Bulletin of the Marine Biological Laboratory, Woods Hole*, **15**, 259–88.

Pemberton, S. G. (ed.) (1992) Applications of ichnology to petroleum exploration – a core workshop. *SEPM Core Workshops*, **17**.

Pemberton, S. G. and Frey, R. W. (1982) Trace fossil nomenclature and the *Planolites-Palaeophycus* dilemma. *Journal of Paleontology*, **56**, 843–81.

Pemberton, S. G. and Frey, R. W. (1985) The *Glossifungites* ichnofacies: modern examples from the Georgia coast, U.S.A. *SEPM Special Publications*, **35**, 237–59.

Pemberton, S. G., Frey, R. W. and Saunders, T. D. A. (1990) Trace fossils, in *Palaeobiology – a Synthesis* (eds D. E. G. Briggs and P. R. Crowther), Blackwell, Oxford, pp. 355–62.

Pemberton, S. G., Flach, P. D. and Mossop, G. D. (1982) Trace fossils from the Athabasca Oil Sands, Alberta, Canada. *Science*, **217**, 825–7.

Pemberton, S. G., MacEachern, J. A. and Frey, R. W. (1992) Trace fossil facies models: environmental and allostratigraphic significance, in *Facies Models – Response to Sea Level Change* (eds R. G. Walker, and N. P. James), Geological Association of Canada, Ottawa, pp. 47–72.

Pemberton, S. G., Risk, M. J. and Buckley, D. E. (1976) Supershrimp: deep bioturbation in the Strait of Canso, Nova Scotia. *Science*, **192**, 790–1.

Pemberton, S. G. and Wightman, D. M. (1987) Brackish water trace fossil suites: examples from the Lower Cretaceous Mannville Group. *4th Symposium on Mesozoic Terrestrial Ecosystems, Short Papers* (eds P. M. Currie and E. H. Koster), 185–92.

Pemberton, S. G. and Wightman, D. M. (1992) Ichnological characteristics of brackish water deposits, in *Applications of Ichnology to Petroleum Exploration* (ed. S. G. Pemberton), *SEPM Core Workshop*, **17**, 141–69.

Pequignat, C. E. (1970) Biologie des *Echinocardium cordatum* (Pennant) de la Baie de Seine. *Forma et Functio*, **2**, 121–68.

Pervesler, P. and Dworschak, P. C. (1985) Burrows of *Jaxea nocturna* Nardo in the Gulf of Trieste. *Senckenbergiana Maritima*, **17**, 33–53.

Phillips, P. J. (1971) Observations on the biology of mudshrimps of the genus *Callianassa* (Anomura: Thalassinidae) in Mississippi Sound. *Gulf Research Reports*, **3**, 165–96.

Pianka, E. R. (1970) On *r*- and *K*-selection. *American Naturalist*, **104**, 592–7.

Pickerill, R. K. (1990) Nonmarine *Paleodictyon* from the Carboniferous Albert Formation of southern New Brunswick. *Atlantic Geology*, **26**, 157–63.

Pickerill, R. K. (1992) Carboniferous nonmarine invertebrate ichnocoenoses from southern New Brunswick, eastern Canada. *Ichnos*, **2**, 21–35.

Pickerill, R. K. (1994) Nomenclature and taxonomy of invertebrate trace fossils, in *The Palaeobiology of Trace Fossils* (ed. S. K. Donovan), Wiley, Chichester, pp. 3–42.

Pickerill, R. K., Donovan, S. K. and Dixon, H. L. (1993) The trace fossil *Dactyloidites ottoi* (Geinitz, 1849) from the Neogene August Town Formation of south-central Jamaica. *Journal of Paleontology*, **67**, 1070–4.

Pickerill, R. K. and Forbes, W. H. (1987) A trace fossil preserving its producer (*Trentonia shegiriana*) from the Trenton Limestone of the Quebec City area. *Canadian Journal of Earth Science*, **15**, 659–64.

Pickerill, R. K. and Narbonne, G. M. (1995) Composite and compound ichnotaxa: a case example from the Ordovician of Québec, eastern Canada. *Ichnos*, **4**, 53–69.

Pickerill, R. K. and Peel, J. S. (1990) Trace fossils from the Lower Cambrian Bastion formation of North-East Greenland. *Grønlands Geologiske Undersøgelse, Rapporter*, **147**, 5–43.

Picton, B. E. and Manuel, R. L. (1985) *Arachnanthus sarsi* Carlgren, 1912: a redescription of a cerianthid anemone new to the British Isles. *Zoological Journal of the Linnaean Society*, **83**, 343–9.

Peel, R. K. and Peel, (1991) *Gordia nodosa* isp. nov. and other trace fossils from the Cass Fjord formation (Cambrian) of North Greenland. *Grønlands Geologiske Undersøgelse, Rapporter*, **150**, 15–28.

Pohl, M. E. (1946) Ecological observations on *Callianassa major* Say at Beaufort, North Carolina. *Ecology*, **27**, 71–80.

Pollard, J. E. (1981) A comparison between the Triassic trace fossils of Cheshire and south Germany. *Palaeontology*, **24**, 555–88.

Pollard, J. E. (1985) *Isopodichnus*, related arthropod trace fossils and notostracans from Triassic fluvial sediments. *Transactions of the Royal Society of Edinburgh, Earth Sciences*, **76**, 273–85.

Pollard, J. E., Goldring, R. and Buck, S. G. (1993) Ichnofabrics containing *Ophiomorpha*: significance in shallow-water facies interpretation. *Journal of the Geological Society, London*, **150**, 149–64.

Powell, E. N. (1977) Particle size selection and sediment reworking in a funnel feeder, *Leptosynapta tenuis* (Holothuroidea, Synaptidae). *Internationale Revue der gesamten Hydrobiologie und Hydrographie*, **62**, 385–408.

Powell, R. R. (1974) The functional morphology of the foreguts of the thalassinid crustaceans, *Callianassa californiensis* and *Upogebia pugettensis*. *University of California Publications in Zoology*, **102**, 1–41.

Pryor, W. A. (1975) Biogenic sedimentation and alteration of argillaceous sediments in shallow marine environments. *Geological Society of America, Bulletin*, **86**, 1244–54.

Rao, K. P. (1954) Bionomics of *Ptychodera flava* Eschscholtz (Enteropneusta). *Journal of Madras University*, **B24**, 1–5.

Rasmussen, H. W. (1971) Echinoid and crustacean burrows and their diagenetic significance in the Maastrichtian-Danian of Stevns Klint, Denmark. *Lethaia*, **4**, 191–216.

Ratcliffe, B. C. and Fagerstrom, J. A. (1980) Invertebrate lebensspuren of Holocene floodplains: their morphology, origin and paleoecological significance. *Journal of Paleontology*, **54**, 614–30.

Ray, A. J. and Aller, R. C. (1985) Physical irrigation of relict burrows: implications for sediment chemistry. *Marine Geology*, **62**, 371–9.

Reichardt, W. (1978) Impact of bioturbation by *Arenicola marina* on microbial parameters in intertidal sediments. *Marine Ecology Progress Series*, **44**, 149–58.

Reichardt, W. T. (1987) Burial of Antarctic macroalgal debris in bioturbated deep-sea sediments. *Deep-Sea Research*, **34**, 1761–70.

Reichelt, A. C. (1991) Environmental effects of meiofaunal burrowing, in *The Environmental Impact of Burrowing Animals and Animal Burrows* (eds P. S. Meadows and A. Meadows), *Symposium of the Zoological Society of London*, **63**, 33–52.

Reid, D. M. (1938) Burrowing methods of *Talorchestia deshayesii* (Audouin) (Crustacea, Amphipoda). *Annals and Magazine of Natural History*, **1** (11), 155–7.

Reid, D. M. (1989) The unwhole organism. *American Zoologist*, **29**, 1133–40.

Reid, R. G. B. (1990) Evolutionary implications of sulphide-oxidizing symbiosis in bivalves, in *The Bivalvia – Proceedings of a Memorial Symposium in Honour of Sir Charles Maurice Yonge, Edinburgh* (ed. B. Morton), Hong Kong University Press, Hong Kong, pp. 127–40.

Reid, R. G. B. (1989) The unwhole organism. *American Zoologist*, **29**, 1133–40.

Reid, R. G. B. and Brand, D. G. (1986) Sufide-oxidizing symbiosis in lucinaceans: implications for bivalve evolution. *Veliger*, **29**, 3–24.

Reineck, H.-E. (1958) Wühlbau-Gefüge in Abhängigkeit von Sediment-Umlagerungen. *Senckenbergiana Letheia*, **39**, 1–56.

Reineck, H.-E. (1963) Sedimentgefüge im Bereich der südlichen Nordsee. *Senckenbergischen naturforschenden Gesellschaft, Abhandlungen*, **505**.

Reineck, H.-E. (1970) *Das Watt. Ablagerungs- und Lebensraum*. Kramer, Frankfurt am Main.

Reineck, H.-E. (1976) Zonierung von Primärgefügen und Bioturbation. *Senckenbergiana Maritima*, **8**, 155–69.

Reineck, H.-E., Dörjes, J., Gadow, S. and Hertweck, G. (1968) Sedimentologie, Faunenzonierung und Faziesabfolge vor der Ostküste der inneren Deutschen Bucht. *Senckenbergiana Letheia*, **49**, 261–309.

Reineck, H.-E., Gutmann, W. F. and Hertweck, G. (1967) Das Schlickgebiet südlich Helgoland als Beispiel rezenter Schelfablagerungen. *Senckenbergiana Letheia*, **48**, 219–75.

Reineck, H.-E. and Singh, I. B. (1971) Der Golf von Gaeta (Tyrrhenisches Meer). III. Die Gefüge von Vorstrand- und Schelfsedimenten. *Senckenbergiana Maritima*, **3**, 185–201.

Reise, K. (1979) Spatial configuration generated by motile polychaetes. *Helgoländer wissenschaftliche Meeresuntersuchungen*, **32**, 55–72.

Reise, K. (1981) High abundance of small zoobenthos around biogenic structures in tidal sediments of the Wadden Sea. *Helgoländer wissenschaftliche Meeresuntersuchungen*, **34**, 413–25.

Reise, K. and Ax, P. (1979) A meiofaunal 'thiobios' limited to the anaerobic sulfide system of marine sand does not exist. *Marine Biology*, **54**, 225–37.

Remane, A. (1940) Einführung in der zoologische Ökologie, in *Tierwelt Nord- und Ostsee* (eds Grimpe and Wagler), Leipzig. (Fide Schäfer 1962.)

Retallack, G. J. (1984) Trace fossils of burrowing beetles and bees in an Oligocene paleosol, Badlands National Park, South Dakota. *Journal of Paleontology*, **58**, 571–92.

Retallak, G. J. and Feakes, C. R. (1987) Trace fossil evidence for Late Ordovician animals on land. *Science*, **235**, 61–3.

Rhoads, D. C. (1963) Rates of sediment reworking by *Yoldia limatula* in Buzzards Bay, Massachusetts, and Long Island Sound. *Journal of Sedimentary Petrology*, **33**, 723–7.

Rhoads, D. C. (1967) Biogenic reworking of intertidal and subtidal sediments in Barnstable Harbor and Buzzards Bay, Massachusetts. *Journal of Geology*, **75**, 461–76.

Rhoads, D. C. (1974) Organism–sediment relations on the muddy sea floor. *Oceanography and Marine Biology Annual Review*, **12**, 263–300.

Rhoads, D. C. (1975) The paleoecological and environmental significance of trace fossils, in *The Study of Trace Fossils* (ed. R. W. Frey), Springer-Verlag, New York, pp. 147–60.

Rhoads, D. C., Aller, R. C. and Goldhaber, M. B. (1977) The influence of colonizing benthos on physical properties and chemical diagenesis of the estuarine seafloor, in *Ecology of Marine Benthos* (ed. B. C. Coull), University of South Carolina Press, Columbia, pp. 113–38.

Rhoads, D. C. and Boyer, L. F. (1982) The effect of marine benthos on physical properties of sediments. A successional perspective, in *Animal–sediment Relations. The Biogenic Alteration of Sediments* (eds P. L. McCall and M. J. S. Tevesz), Plenum Press, New York, pp. 3–52.

Rhoads, D. C., McCall, P. L. and Yingst, J. Y. (1978) Disturbance and production on the estuarine seafloor. *American Scientist*, **66**, 577–86.

Rhoads, D. C. and Morse, J. (1971) Evolutionary and ecologic significance of oxygen deficient marine basins. *Lethaia*, **4**, 413–28.

Rhoads, D. C. and Stanley, D. J. (1965) Biogenic graded bedding. *Journal of Sedimentary Petrology*, **35**, 956–63.

Rhoads, D. C., Webb, J. E. *et al.* (1976) Organism–sediment relationships, in *The Benthic Boundary Layer* (ed. I. N. McCave), Plenum Press, New York, pp. 273–95.

Rhoads, D. C. and Young, D. K. (1970) The influence of deposit feeding organisms on sediment stability and community trophic structure. *Journal of Marine Research*, **28**, 150–78.

Rhoads, D. C. and Young, D. K. (1971) Animal–sediment relations in Cape Cod Bay, Massachusetts. II. Reworking by *Molpadia oolitica* (Holothuroidea). *Marine Biology*, **11**, 255–61.

Rice, A. L., Billett, D. S. M., Fry, J., John, A. W. G., Lampitt, R. S., Mantoura, R. F. C. and Morris, R. J. (1986) Seasonal deposition of phytodetritus to the deep-sea floor. *Proceedings of the Royal Society of Edinburgh*, **88B**, 265–79.

Rice, A. L., Billett, D. S. M., Thurston, M. H. and Lampitt, R. S. (1991) The Institute of Oceanographic Sciences Biology Programme in the Porcupine Seabight: background and general introduction. *Journal of the Marine Biological Association, United Kingdom*, **71**, 281–310.

Rice, A. L. and Chapman, C. J. (1971) Observations on the burrows and burrowing behaviour of two mud-dwelling decapod crustaceans, *Nephrops norvegicus* and *Goneplax rhomboides*. *Marine Biology*, **10**, 330–42.

Rice, A. L. and Johnstone, A. D. F. (1972) The burrowing of the gobiid fish *Lesueurigobius friesii* (Collett). *Zeitschrift für Tierpsychologie*, **30**, 431–8.

Rice, D. L. and Rhoads, D. C. (1989) Early diagenesis of organic matter and the nutritional value of sediment, in *Ecology of Marine Deposit Feeders* (eds G. Lopez, G. Taghon and J. Levinton), *Lecture Notes on Coastal and Estuarine Studies*, **31**, 59–97.

Richards, A. F. and Parks, J. M. (1976) Marine geotechnology, in *The Benthic Boundary Layer* (ed. I. N. McCave), Plenum Press, New York, pp. 157–81.

Richter, R. (1926) Flachseebeobachtungen zur Paläontologie und Geologie XII. Bau, Begriff und paläogeographische Bedeutung von *Corophioides luniformis* (Blankenhorn, 1917). *Senckenbergiana*, **8**, 200–19.

Richter, R. (1952) Fluidal-Textur in Sediment-Gesteinen und über Sedifluktion überaupt. *Notizblatt des Hessischen Landesamtes für Bodenforschung zu Wiesbaden*, **3** (6), 67–81.

Ricketts, E. F. (1994) Mud-flat cycles, incised channels, and relative sea-level changes on a Paleocene mud-dominated coast, Ellesmere Island, arctic Canada. *Journal of Sedimentary Research*, **B64**, 211–18.

Ricketts, E. F. and Calvin, J. (1962) *Between Pacific tides*, 3rd edn, revised by J. W. Hedgpeth, Stanford University Press, Stanford.

Ride, W. D. I., Sabrosky, C. W., Bernardi, G. and Melville, R. V. (1985) *International Code of Zoological Nomenclature*, 3rd edn, International Trust for Zoological Nomenclature, London.

Rijken, M. (1979) Food and food uptake in *Arenicola marina. Netherlands Journal of Sea Research*, **13**, 406–21.

Rindsberg, A. K. (1990) Ichnological consequences of the 1985 International Code of Zoological Nomenclature. *Ichnos*, **1**, 59–63.

Rindsberg, A. K. (1994) Ichnology of the Upper Mississippian Hartselle Sandstone of Alabama, with notes on other Carboniferous Formations. *Geological Survey of Alabama Bulletin*, **158**, 1–107.

Rioult, M., Dugué, O., Jan du Chêne, R., Ponsot, C., Fily, G., Moron, J.-M. and Vail, P. R. (1991) Outcrop sequence stratigraphy of the Anglo–Paris Basin, Middle to Upper Jurassic (Normandy, Maine, Dorset). *Bulletin des Centres de Recherches, Exploration-Production, Elf Aquitaine*, **15**, 101–94.

Risk, M. J. (1973) Silurian echiuroids: possible feeding traces in the Thorold Sandstone. *Science*, **180**, 1285–7.

Risk, M. J. and Tunnicliffe, V. J. (1978) Intertidal spiral burrows: *Paraonis fulgens* and *Spiophanes wigleyi* in the Minas Basin, Bay of Fundy. *Journal of Sedimentary Petrology*, **48**, 1287–92.

Robbins, J. A. (1986) A model for particle-selective transport of tracers in sediments with conveyor belt deposit feeders. *Journal of Geophysical Research*, **91**, 8542–58.

Roberts, H. H., Wiseman, W. J. and Suchanek, T. H. (1981) Lagoon sediment transport: the significant effect of *Callianassa* bioturbation. *Proceedings of the 4th International Coral Reef Symposium, Manila*, **1**, 459–65.

Roček, Z. and Rage, J.-C. (1994) The presumed amphibian footprint *Notopus petri* from the Devonian: a probable starfish trace fossil. *Lethaia*, **27**, 241–344.

Röder, H. (1971) Gangsysteme von *Paraonis fulgens* Levinsen 1883 (Polychaeta) in ökologischer, ethologischer und aktuopaläontologischer Sicht. *Senckenbergiana Maritima*, **3**, 3–51.

Rollins, H. B., West, R. R. and Busch, R. M. (1990) Hierarchical genetic stratigraphy and marine paleoecology, in *Paleocommunity Temporal Dynamics: the Long-Term Development of Multispecies Assemblies* (ed. W. Miller), *Paleontological Society Special Publications*, **5**, 273–308.

Romano, M. and Whyte, M. A. (1987) A limulid trace fossil from the Scarborough formation (Jurassic) of Yorkshire; its occurrence, taxonomy and interpretation. *Proceedings of the Yorkshire Geological Society*, **46**, 85–95.

Romero-Wetzel, M. B. (1987) Sipunculans as inhabitants of very deep, narrow burrows in deep-sea sediments. *Marine Biology*, **96**, 87–91.

Romero-Wetzel, M. B. (1989) Branched burrow-systems of the enteropneust *Ste-*

reobalanus canadensis (Spengel) in deep-sea sediments of the Vöring-Plateau, Norwegian Sea. *Sarsia*, **74**, 85–9.

Ronan, T. E. (1977) Formation and paleontologic recognition of structures caused by marine annelids. *Paleobiology*, **3**, 389–403.

Ronan, T. E. (1978) Food-resources and the influence of spatial pattern on feeding in the phoronid *Phoronopsis viridis*. *Biological Bulletin of the Marine Biological Laboratory, Woods Hole*, **154**, 472–84.

Ronan, T. E., Miller, M. F. and Farmer, J. D. (1981) Organism–sediment relationships on a modern tidal flat, Bodega Harbor, California. *Annual Meeting, Pacific Section of the SEPM, Field Trip* **3**, 15–31.

Root, R. B. (1967) The niche exploitation pattern of the blue-grey gnatcatcher. *Ecological Monographs*, **37**, 317–50.

Ruddiman, W. F. and Glover, L. K. (1972) Vertical mixing of ice-rafted volcanic ash in North Atlantic sediments. *Geological Society of America, Bulletin*, **83**, 2817–36.

Sageman, B. B., Wignall, P. B. and Kauffman, E. G. (1991) Biofacies models for oxygen-deficient facies in epicontinental seas: tool for paleoenvironmental analysis, in *Cycles and Events in Stratigraphy*, Springer-Verlag, Berlin, pp. 542–64.

Sandberg, C. A. and Gutschick, R. C. (1984) Distribution, microfauna and source-rock potential of Mississippian Della Phosphatic Member of Woodman Formation and equivalents, Utah and adjacent states, in *Hydrocarbon Source Rocks of the Greater Rocky Mountain Region* (eds J. Woodward, F. F. Meissner and J. L. Clayton), Rocky Mountain Association of Geologists, Denver, pp. 135–78.

Sanders, H. L. (1958) Benthic studies in Buzzards Bay. I. Animal–sediment relationships. *Limnology and Oceanography*, **3**, 245–58.

Sarjeant, W. A. S. (1979) Code for trace fossil nomenclature. *Palaeogeography, Palaeoclimatology Palaeoecology*, **28**, 147–66.

Sarjeant, W. A. S. and Kennedy, W. J. (1973) Proposal for a code for the nomenclature of trace fossils. *Canadian Journal of Earth Science*, **10**, 460–75.

Sassaman, C. and Mangum, C. P. (1972) Adaptations to environmental oxygen levels in infaunal and epifaunal sea anemones. *Biological Bulletin*, **143**, 657–78.

Saunders, T. D. A. and Pemberton, S. G. (1990) On the palaeoecological significance of the trace fossil *Macaronichnus*. *Abstracts, Papers, International Sedimentological Congress, Nottingham*, **13**, 416.

Savrda, C. E. (1986) Development and evaluation of a trace fossil model for the reconstruction of paleo-oxygenation in marine environments. PhD thesis, University of Southern California, Los Angeles.

Savrda, C. E. (1991) Ichnology in sequence stratigraphic studies: example from the Lower Paleocene of Alabama. *Palaios*, **6**, 39–53.

Savrda, C. E. (1992) Trace fossils and benthic oxygenation, in *Trace Fossils* (eds C. G. Maples and R. R. West), *Short Courses in Paleontology*, **5**, 172–96, Paleontological Society, Knoxville.

Savrda, C. E. (1993) Ichnosedimentologic evidence for a noncatastrophic origin of Cretaceous–Tertiary boundary sands in Alabama. *Geology*, **21**, 1075–8.

Savrda, C. E. (in press) Ichnologic applications in paleoceanographic, paleoclimatic and sea-level studies. *Palaios*.

Savrda, C. E. and Bottjer, D. J. (1986) Trace fossil model for reconstruction of paleo-oxygenation in bottom waters. *Geology*, **14**, 3–6.

Savrda, C. E. and Bottjer, D. J. (1987a) The exaerobic zone, a new oxygen-deficient marine biofacies. *Nature*, **327**, 54–6.

Savrda, C. E. and Bottjer, D. J. (1987b) Trace fossils as indicators of bottom-water redox conditions in ancient marine environments, in *New Concepts in the*

Use of Biogenic Sedimentary Structures for Paleoenvironmental Interpretation (ed. D. J. Bottjer), SEPM Pacific Section, Los Angeles, pp. 3–26.

Savrda, C. E. and Bottjer, D. J. (1994) Ichnofossils and ichnofabrics in rhythmically bedded pelagic/hemi-pelagic carbonates: recognition and evaluation of benthic redox and scour cycles. *International Association of Sedimentologists, Special Publications*, **19**, 195–210.

Savrda, C. E., Bottjer, D. J. and Gorsline, D. S. (1984) Development of a comprehensive oxygen-deficient marine biofacies model: evidence from Santa Monica, San Pedro, and Santa Barbara Basins, California borderland. *American Association of Petroleum Geologists, Bulletin*, **68**, 1179–92.

Savrda, C. E., Bottjer, D. J. and Seilacher, A. (1991) Redox-related benthic events, in *Cycles and Events in Stratigraphy* (eds G. Einsele, W. Ricken and A. Seilacher), Springer-Verlag, Berlin, Heidelberg, pp. 524–41.

Savrda, C. E. and Ozalas, K. (1993) Preservation of mixed-layer ichnofabrics in oxygenation-event beds. *Palaios*, **8**, 609–13.

Schäfer, W. (1956) Wirkungen der Benthos-Organismen auf den jungen Schichtverband. *Senckenbergiana Letheia*, **37**, 183–263.

Schäfer, W. (1962) *Aktuo-Paläontologie nach Studien in der Nordsee*, Kramer, Frankfurt am Main.

Schembri, P. J. and Jaccarini, V. (1977) Locomotory and other movements of the trunk of *Bonellia viridis* (Echiura, Bonelliidae). *Journal of Zoology, London*, **182**, 477–94.

Schimper, W. P. and Schenk, A. (1890) *Palaeophytologie*, in *Handbuch de Palaeontologie, Teil II* (ed. K. A. von Zittel), Oldenbourg, München and Leipzig.

Schink, D. R. and Guinasso, N. L. (1977) Modeling the influence of bioturbation and other processes on calcium carbonate dissolution at the sea floor, in *The Fate of Fossil Fuel CO_2 in the Oceans* (eds N. R. Andersen and A. Malahoff), Plenum Press, New York, pp. 375–99.

Scholle, P. A., Arthur, M. A. and Ekdale, A. A. (1983) Pelagic environment. *American Association of Petroleum Geology, Memoirs*, **33**, 619–91.

Scoffin, T. P. (1992) Taphonomy of coral reefs: a review. *Coral Reefs*, **11**, 57–77.

Seilacher, A. (1951) Der Röhrenbau von *Lanice conchilega* (Polychaeta). Ein Beitrag zur Deutung fossiler Lebensspuren. *Senckenbergiana*, **32**, 267–80.

Seilacher, A. (1953a) Studien zur Palichnologie. I. Über die Methoden der Palichnologie. *Neues Jahrbuch für Geologie und Paläontologie, Abhandlungen*, **96**, 421–52.

Seilacher, A. (1953b) Studien zur Palichnologie. II. Die fossilen Ruhespuren (Cubichnia). *Neues Jahrbuch für Geologie und Paläontologie, Abhandlungen*, **98**, 87–124.

Seilacher, A. (1955) Spuren und Lebensweise der Trilobiten. *Abhandlungen der Akademie der wissenschaften und der Literatur, Mainz, mathematisch-naturwissenschaftliche Klasse*, Jahrgang 1955, 342–72.

Seilacher, A. (1957) An-aktualistisches Wattenmeer? *Paläontologische Zeitschrift*, **31**, 198–206.

Seilacher, A. (1959) Vom Leben der Trilobiten. *Naturwissenschaften*, **46**, 389–93.

Seilacher, A. (1960) Lebensspuren als Leitfossilien. *Geologische Rundschau*, **49**, 41–50.

Seilacher, A. (1962a) Form und Function des Trilobiten-Daktylus. *Paläontologische Zeitschrift, Sonderausgabe, Festband Hermann Schmidt*, 218–27.

Seilacher, A. (1962b) Paleontological studies on turbidite sedimentation and erosion. *Journal of Geology*, **70**, 227–34.

Seilacher, A. (1964) Biogenic sedimentary structures, in *Approaches to Paleoecology* (eds J. Imbrie and N. Newell), Wiley, New York, pp. 296–316.

Seilacher, A. (1967a) Bathymetry of trace fossils. *Marine Geology*, **5**, 413–28.

Seilacher, A. (1967b) Fossil behavior. *Scientific American*, **217**, 72–80.

Seilacher, A. (1968) Sedimentationsprozesse in Ammonitenhäusen. *Akademie der Wissenschaften und der Literatur, Abhandlungen der mathematisch-naturwissenschaftlichen Klasse*, Jahrgang 1967, 191–203.

Seilacher, A. (1970) *Cruziana* stratigraphy of 'non-fossiliferous' Palaeozoic sandstones, in *Trace Fossils* (eds T. P. Crimes and J. C. Harper), *Geological Journal Special Issues*, **3**, 447–76.

Seilacher, A. (1972) Divaricate patterns in pelecypod shells. *Lethaia*, **5**, 325–43.

Seilacher, A. (1974) Flysch trace fossils: evolution of behavioural diversity in the deep-sea. *Neues Jahrbuch für Geologie und Paläontologie, Monatshefte,* Jahrgang 1974, 233–45.

Seilacher, A. (1977) Pattern analysis of *Paleodictyon* and related trace fossils, in *Trace Fossils 2* (eds T. P. Crimes and J. C. Harper), *Geological Journal Special Issues*, **9**, 289–334.

Seilacher, A. (1978) Use of trace fossil assemblages for recognizing depositional environments, in *Trace Fossil Concepts* (ed. P. B. Basan), *SEPM Short Courses*, **5**, 167–81.

Seilacher, A. (1985) Trilobite palaeobiology and substrate relationships. *Transactions of the Royal Society of Edinburgh*, **76**, 231–7.

Seilacher, A. (1989) *Spirocosmorhaphe*, a new graphoglyptid trace fossil. *Journal of Paleontology*, **63**, 116–17.

Seilacher, A. (1990) Aberrations in bivalve evolution related to photo- and chemosymbiosis. *Historical Biology*, **3**, 289–311.

Seilacher, A. (1992) An updated *Cruziana* stratigraphy of Gondwanian Palaeozoic sandstones, in *The Geology of Libya, Part 8* (ed. M. J. Salem), Elsevier, Amsterdam, pp. 1565–80.

Seilacher, A. (1993) Problems of correlation in the Nubian Sandstone facies, in *Geoscientific Research in Northwest Africa* (eds U. Thorweihe and H. Schandelmeier), Balkema, Rotterdam, pp. 329–33.

Seilacher, A. and Meischner, D. (1964) Fazies-Analyse im Paläozoikum des Oslo-Gebietes. *Geologische Rundschau*, **54**, 596–619.

Seilacher, A. and Seilacher, E. (1994) Bivalvian trace fossils: a lesson from actuopaleontology. *Courier Forschungsinstitut Senckenberg*, **169**, 5–15.

Sellwood, B. W. (1971) A *Thalassinoides* burrow containing the crustacean *Glyphaea udressieri* (Meyer) from the Bathonian of Oxfordshire. *Palaeontology*, **14**, 589–91.

Shick, J. M., Edwards, K. C. and Dearborn, J. H. (1981) Physiological ecology of the deposit-feeding sea star *Ctenodiscus crispatus*: ciliated surfaces and animal–sediment interactions. *Marine Ecology – Progress Series*, **5**, 165–84.

Shinn, E. A. (1968) Burrowing in recent time sediments of Florida and the Bahamas. *Journal of Paleontology*, **42**, 879–94.

Shourd, M. L. and Levin, H. L. (1976) *Chondrites* in the Upper Plattin Subgroup (Middle Ordovician) of eastern Missouri. *Journal of Paleontology*, **50**, 260–8.

Silva, A. J. (1974) Marine geomechanics: overview and projections, in *Deep-Sea Sediments: Physical and Mechanical Properties* (ed. A. L. Inderbitzen), Plenum Press, New York, pp. 45–76.

Simpson, S. (1957) On the trace fossil *Chondrites*. *Quarterly Journal of the Geological Society, London*, **107**, 475–99.

Skoog, S. Y., Venn, C. and Simpson, E. L. (1994) Distribution of *Diopatra cuprea* across modern tidal flats: implications for *Skolithos*. *Palaios*, **9**, 188–201.

Smith, A. B. and Crimes, T. P. (1983) Trace fossils formed by heart urchins – a study of *Scolicia* and related traces. *Lethaia*, **16**, 79–92.

Smith, C. R., Jumars, P. A. and DeMaster, D. J. (1986) *In situ* studies of megafaunal mounds indicate rapid sediment turnover and community response at the deep-sea floor. *Nature*, **323**, 251–3.

Smith, L. S. (1961) Clam-digging behavior in the starfish, *Pisaster brevispinus* (Stimpson 1857). *Behaviour*, **18**, 148–51.

Smith, N. D. and Hein, F. J. (1971) Biogenic reworking of fluvial sediments by staphylinid beetles. *Journal of Sedimentary Petrology*, **41**, 598–602.

Smith, R. M. H. (1987) Helical burrow casts of therapsid origin from the Beaufort Group (Permian) of South Africa. *Palaeogeography, Palaeoclimatology, Palaeoecology*, **60**, 155–70.

Smith, R. M. H. (1993) Vertebrate taphonomy of Late Permian floodplain deposits in the southwestern Karoo Basin of South Africa. *Palaios*, **8**, 45–67.

Southward, A. J. and Dando, P. R. (1988) Distribution of Pogonophora in canyons of the Bay of Biscay: factors controlling abundance and depth range. *Journal of the Marine Biological Association, UK*, **68**, 627–38.

Southward, A. J., Southward, E. C., Dando, P. R., Barrett, R. L. and Ling, R. (1986) Chemoautotrophic function of bacterial symbionts in small Pogonophora. *Journal of the Marine Biological Association, UK*, **66**, 415–37.

Southward, E. C. (1979) Horizontal and vertical distribution of Pogonophora in the Atlantic Ocean. *Sarsia*, **64**, 51–5.

Stanley, S. M. (1970) Relation of shell form to life habits of the Bivalvia (Mollusca). *Geological Society of America, Memoirs*, **125**.

Stanley, S. M. (1972) Functional morphology and evolution of byssally attached bivalve mollusks. *Journal of Paleontology*, **46**, 165–212.

Sternberg, K. M. von (1833) *Versuch einer geognostisch-botanischen Darstellung der Flora der Vorwelt*, **5–6**, Fleischer, Leipzig, Prague.

Stevens, B. A. (1929) Ecological observations on Callianassidae of Puget Sound. *Ecology*, **10**, 399–405.

Stiasny, G. (1910) Zur Kenntnis der Lebensweise von *Balanoglossus clavigerus* Delle Chiaje. *Zoologischer Anzeiger*, **35**, 561–5, 633.

Suchanek, T. H. (1983) Control of seagrass communities and sediment distribution by *Callianassa* (Crustacea, Thalassinidea) bioturbation. *Journal of Marine Research*, **41**, 281–98.

Suchanek, T. H. (1985) Thalassinid shrimp burrows: ecological significance of species-specific architecture. *Proceedings of the 5th International Coral Reef Congress*, 205–10.

Suchanek, T. H. and Colin, P. L. (1986) Rates and effects of bioturbation by invertebrates and fishes at Enewetak and Bikini Atolls. *Bulletin of Marine Science*, **38**, 25–34.

Suchanek, T. H., Colin, P. L., McMurtry, G. M. and Suchanek, C. S. (1986) Bioturbation of sediment radionuclides in Enewetak Atoll lagoon by callianassid shrimp: biological aspects. *Bulletin of Marine Science*, **38**, 144–54.

Suchanek, T. H., Williams, S. L., Ogden, J. C., Hubbard, D. K. and Gill, I. P. (1985) Utilization of shallow-water seagrass detritus by Caribbean deep-sea macrofauna: delta-^{13}C evidence. *Deep-Sea Research*, **32**, 201–14.

Surlyk, F. (1991) Sequence stratigraphy of the Jurassic–Lowermost Cretaceous of East Greenland. *American Association of Petroleum Geologists, Bulletin*, **75**, 1468–88.

Surlyk, F. and Noe-Nygaard, N. (1991) Sand bank and dune facies architecture of a wide intercratonic seaway: Late Jurassic–Early Cretaceous Raukelv Formation, Jameson Land, East Greenland, in *The Three-Dimensional Facies Architecture of Terrigenous Clastic Sediments and its Implication for Hydrocarbon Discovery and Recovery* (eds A. D. Miall and N. Tyler), *SEPM, Cencepts in Sedimentology and Paleontology* **3**, 261–76.

Swedmark, B. (1964) The interstitial fauna of marine sand. *Biological Review*, **39**, 1–42.

Swinbanks, D. D. (1981a) Sediment reworking and the biogenic formation of clay laminae by *Abarenicola pacifica. Journal of Sedimentary Petrology*, **51**, 1137–45.

Swinbanks, D. D. (1981b) Sedimentology photo. *Journal of Sedimentary Petrology*, **51**, 1146.

Swinbanks, D. D. and Luternauer, J. L. (1987) Burrow distribution of thalassinidean shrimp on a Fraser Delta tidal flat, British Columbia. *Journal of Sedimentary Petrology*, **61**, 315–32.

Swinbanks, D. D. and Murray, J. W. (1981) Biosedimentological zonation of Boundary Bay tidal flats, Fraser River delta, British Columbia. *Sedimentology*, **28**, 201–37.

Swinbanks, D. D. and Shirayama, Y. (1984) Burrow stratigraphy in relation to manganese diagenesis in modern deepsea carbonates. *Deep-Sea Research*, **31**, 1197–223.

Szulc, J. (1990) Ichnological indicators of the sedimentary environment fluctuation, in *International Workshop – Field Seminar: the Muschelkalk – Sedimentary Environments, Facies and Diagenesis* (eds J. Szulc *et al.*), International Association of Sedimentologists, Krakow, Opole, pp. 23–5.

Taghon, G. L. (1989) Modeling deposit feeding, in *Ecology of Marine Deposit Feeders* (eds G. Lopez, G. Taghon and J. Levinton), *Lecture Notes on Coastal and Estuarine Studies*, **31**, 223–46.

Taghon, G. L., Nowell, A. R. M. and Jumars, P. A. (1980) Induction of suspension feeding in spionid polychaetes by high particulate fluxes. *Science*, **210**, 562–4.

Taghon, G. L., Self, R. F. L. and Jumars, P. A. (1978) Predicting particle selection by deposit feeders: a model and its implications. *Limnology and Oceanography*, **23**, 752–9.

Tamaki, A. (1988) Effects of the bioturbating activity of the ghost shrimp *Callianassa japonica* Ortmann on migration of a mobile polychaete. *Journal of Experimental Marine Biology and Ecology*, **120**, 81–95.

Taylor, A. M. and Gawthorpe, R. L. (1993) Application of sequence stratigraphy and trace fossil analysis to reservoir description: examples from the Jurassic of the North Sea, in *Petroleum Geology of Northwest Europe: Proceedings of the 4th Conference* (ed. J. R. Parker), Geological Society, London, pp. 317–35.

Taylor, A. M. and Goldring, R. (1993) Description and analysis of bioturbation and ichnofabric. *Journal of the Geological Society, London*, **150**, 141–8.

Tedesco, L. P. and Wanless, H. R. (1991) Generation of sedimentary fabrics and facies by repetitive excavation and storm infilling of burrow networks, Holocene of south Florida and Caicos Platform, B.W.I. *Palaios*, **6**, 326–43.

Tendal, O. S. (1972) A monograph of the Xenophyophoria (Rhizopodea, Protozoa). *Galathea Reports*, **12**.

Tendal, O. S. (1980) Xenophyophores from the French expeditions 'INCAL' and 'BIOVEMA' in the Atlantic Ocean. *Cahiers de Biologie Marine*, **21**, 303–6.

Tendal, O. S. and Hessler, R. R. (1977) An introduction to the biology and systematics of Komokiacea (Textulariina, Foraminiferida). *Galathea Reports*, **14**, 165–94.

Tevesz, M. J. S. and McCall, P. L. (1982) Geological significance of aquatic nonmarine trace fossils, in *Animal–Sediment Relations. The Biogenic Alteration of Sediments* (eds P. L. McCall and M. J. S. Tevesz), Plenum Press, New York, pp. 257–85.

Thackray, G. D. (1994) Fossil nest of sweat bees (Halictinae) from a Miocene paleosol, Rusinga Island, western Kenya. *Journal of Paleontology*, **68**, 795–800.

Thamdrup, H. M. (1935) Beiträge zur Ökologie der Wattenfauna auf experimenteller Grundlage. *Meddelelser fra Kommissionen for Danmarks Fiskeri og Havundersøgelse, Serie Fiskeri*, **10** (2).

Thayer, C. W. (1979) Biological bulldozers and the evolution of marine benthic communities. *Science*, **203**, 458–61.

Thayer, C. W. (1983) Sediment-mediated biological disturbance and the evolution of marine benthos, in *Biotic Interactions in Recent and Fossil Benthic Communities* (eds M. J. S. Tevesz and P. L. McCall), Plenum Press, New York, pp. 475–625.

Thistle, D. (1981) Natural physical disturbances and communities of marine soft bottoms. *Marine Ecology Progress Series*, **6**, 223–8.

Thistle, D., Yingst, J. Y. and Fauchald, K. (1985) A deep-sea benthic community exposed to strong near-bottom currents on the Scotian Rise (western Atlantic). *Marine Geology*, **66**, 91–112.

Thompson, B. E. (1980) A new bathyal sipunculan from southern California, with ecological notes. *Deep-Sea Research*, **27A**, 951–7.

Thompson, L. C. and Pritchard, A. W. (1969) Osmoregulatory capacities of *Callianassa* and *Upogebia* (Crustacea, Thalassinidea). *Biological Bulletin of the Marine Biological Laboratory, Woods Hole*, **136**, 114–29.

Thompson, R. K. (1972) Functional morphology of the hind-gut gland of *Upogebia pugettensis* (Crustacea, Thalassinidea) and its role in burrow construction. PhD thesis, University of California, Berkeley.

Thompson, R. K. and Pritchard, A. W. (1969) Respiratory adaptations of two burrowing crustaceans, *Callianassa californiensis* and *Upogebia pugettensis* (Decapoda, Thalassinidea). *Biological Bulletin of the Marine Biological Laboratory, Woods Hole*, **136**, 274–87.

Thomsen, E. and Vorren, T. O. (1984) Pyritization of tubes and burrows from Late Pleistocene continental shelf sediments off north Norway. *Sedimentology*, **31**, 481–92.

Thomson, J. and Wilson, T. R. S. (1980) Burrow-like structures at depth in a Cape Basin red clay core. *Deep-Sea Research*, **27A**, 197–202.

Thorson, G. (1968) Havbundens dyreliv. Infaunaen, den jævne havbunds dyresamfund, in *Danmarks Natur 3, Havet* (eds T. W. Böcher, C. O. Nielsen and A. Schou), Politikens Forlag, København, pp. 82–166.

Trewin, N. H. (1976) *Isopodichnus* in a trace fossil assemblage from the Old Red Sandstone. *Lethaia*, **9**, 29–37.

Trewin, N. H. (1994) A draft system for the identification and description of arthropod trackways. *Palaeontology*, **37**, 811–23.

Trewin, N. H. and Welsh, W. (1976) Formation and composition of a graded estuarine shell bed. *Palaeogeography, Palaeoclimatology, Palaeoecology*, **19**, 219–30.

Trueman, E. R. (1968a) Burrowing habit and the early evolution of body cavities. *Nature*, **218**, 96–8.

Trueman, E. R. (1968b) The burrowing process of *Dentalium* (Scaphopoda). *Journal of Zoology, London*, **154**, 19–27.

Trueman, E. R. (1968c) The mechanism of burrowing of some naticid gastropods in comparison with that of other molluscs. *Journal of Experimental Biology*, **48**, 663–78.

Trueman, E. R. (1971) The control of burrowing and the migratory behaviour of *Donax denticulatus* (Bivalvia: Tellinacea). *Journal of Zoology, London*, **165**, 453–69.

Trueman, E. R. (1975) *The Locomotion of Soft-bodied Animals*, Arnold, London.

Trueman, E. R. and Ansell, A. D. (1969) The mechanisms of burrowing into soft substrata by marine animals. *Marine Biology Annual Review*, **7**, 315–66.

Tuck, I. D., Atkinson, R. J. A. and Chapman, C. J. (1994) The structure and seasonal variability in the spatial distribution of *Nephrops norvegicus* burrows. *Ophelia*, **40**, 13–25.

Tudhope, A. W. and Scoffin, T. P. (1984) The effects of *Callianassa* bioturbation on the preservation of carbonate grains in Davis Reef Lagoon, Great Barrier Reef, Australia. *Journal of Sedimentary Petrology*, **54**, 1091–6.

Tyler, J. C. and Smith, C. L. (1992) Systematic significance of the burrow form of seven species of garden eels (Congridae: Heterocongrinae). *American Museum Novitates*, **3037**.

Tyson, R. V. and Pearson, T. H. (1991) Modern and ancient continental shelf anoxia: an overview, in *Modern and Ancient Continental Shelf Anoxia* (eds R. V. Tyson and T. H. Pearson), *Geological Society, London, Special Publications*, **58**, 1–24.

Tyszka, J. (1994) Paleoenvironmental implications from ichnological and microfaunal analysis of Bajocian spotty carbonates, Pieniny Klippen Belt, Polish Carpathians. *Palaios*, **9**, 175–87.

Uchman, A. (1991a) Zróżnicowanie batymetryczne i facjalne skamieniałości śladowych w południowej części płaszczowiny magurskiej (polskie Karpaty zewnętrzne). *Kwartaknik Geologiczny*, **35**, 437–48.

Uchman, A. (1991b) 'Shallow water' trace fossils in Paleogene flysch of the southern part of the Magura Nappe, Polish Outer Carpathians. *Annales Societatis Geologorum Poloniae*, **61**, 61–75.

Uchman, A. (1991c) Diverse tiering patterns in Paleogene flysch trace fossils, Magura nappe, Carpathian Mountains, Poland. *Ichnos*, **1**, 287–92.

Uchman, A. (1992a) Ichnogenus *Rhizocorallium* in the Paleogene flysch (outer Western Carpathians, Poland). *Geologica Carpathica*, **43**, 57–9.

Uchman, A. (1992b) An opportunistic trace fossil assemblage from the flysch of the Inoceramian beds (Campanian–Palaeocene), Bystrica Zone of the Magura Nappe, Carpathians, Poland. *Cretaceous Research*, **13**, 539–47.

Underwood, C. J. (1993) The position of graptolites within Lower Palaeozoic planktic systems. *Lethaia*, **26**, 189–202.

Van der Horst, C. J. (1934) The burrow of an enteropneust. *Nature*, **134**, 852.

Van der Land, J. (1970) Systematics, zoogeography, and ecology of the Priapulida. *Zoologische Verhandelingen*, **112**.

van Straaten, L. M. J. U. (1952) Biogene textures and the formation of shell beds in the Dutch Wadden Sea. *Koninklijke Nederlandse Akademie van Wetenschappen, Proceedings*, **B55**, 500–16.

van Straaten, L. M. J. U. (1956) Composition of shell beds formed in tidal flat environment in the Netherlands and in the Bay of Arcachon (France). *Geologie en Mijnbouw (N.S.)*, **18**, 209–26.

Vaugelas (1989) to follow

Veldhuizen, H. D. van and Phillips, D. W. (1978) Prey capture by *Pisaster brevispinus* (Asteroidea: Echinodermata) on soft substrate. *Marine Biology*, **48**, 89–97.

Vogel, S. (1978) Organisms that capture currents. *Scientific American*, **239** (2), 108–17.

Voigt, E. (1959) Die ökologische Bedeutung der Hartgründe ('Hardgrounds') in der oberen Kreide. *Paläontologische Zeitschrift*, **33**, 129–47.

Voigt, E. (1974) Über die Bedeutung der Hartgründe ('Hardgrounds') für die Evertebratenfauna der Maastrichter Tuffkreide. *Natuurhistorisch Maandblad*, **63**, 32–9.

Voorhies, M. R. (1975) Vertebrate burrows, in *The Study of Trace Fossils* (ed. R. W. Frey), Springer-Verlag, New York, pp. 325–50.

Vossler, S. M. and Pemberton, S. G. (1988a) Superabundant *Chondrites*: a response to storm buried organic material? *Lethaia*, **21**, 94.

Vossler, S. M. and Pemberton, S. G. (1988b) *Skolithos* in the Upper Cretaceous Cardium Formation: an ichnofossil example of opportunistic ecology. *Lethaia*, **21**, 351–62.

Vossler, S. M. and Pemberton, S. G. (1989) Ichnology and paleoecology of offshore siliciclastic deposits in the Cardium Formation (Turonian, Alberta, Canada). *Palaeogeography, Palaeoclimatology, Palaeoecology*, **74**, 217–39.

Waldron, J. W. F. (1988) Determinination of finite strain in bedding surfaces using sedimentary structures and trace fossils: a comparison of techniques. *Journal of Structural Geology*, **10**, 273–81.

Walker, E. F. (1985) Arthropod ichnofauna of the Old Red Sandstone at Dunure and Montrose, Scotland. *Transactions of the Royal Society of Edinburgh: Earth Sciences*, **76**, 287–97.

Walker, K. R. (1972) Trophic analysis: a method for studying the function of ancient communities. *Journal of Paleontology*, **46**, 82–93.

Walker, K. R. and Diehl, W. W. (1986) The effect of synsedimentary substrate modification on the composition of paleocommunities: paleoecologic succession revisited. *Palaios*, **1**, 65–74.

Walker, K. R. and Laporte, L. F. (1970) Congruent fossil communities from Ordovician and Devonian carbonates of New York. *Journal of Paleontology*, **44**, 928–44.

Wallace, M. W. (1987) The role of internal erosion and sedimentation in the formation of stromatactis mudstones and associated lithologies. *Journal of Sedimentary Petrology*, **57**, 697–700.

Walter, H. (1980) Zur Kenntnis der Ichnia limnisch-terrestrischer Arthropoden des Rotliegenden. *Freiberger Forschungshefte*, **C357**, 61–8.

Walter, H. (1982) Zur Ichnologie der Oberen Hornburger Schichten des östlichen Harzvorlandes. *Freiberger Forschungshefte*, **C366**, 45–63.

Walter, H. (1983) Zur Taxonomie, Ökologie und Biostratigraphie der Ichnia limnisch-terrestrischer Arthropoden der mitteleuropäischen Jungpaläozoikums. *Freiberger Forschungshefte*, **C382**, 146–93.

Wanless, H. R., Tedesco, L. P. and Tyrrell, K. M. (1988) Production of subtidal tubular and surficial tempestites by Hurricane Kate, Caicos Platform, British West Indies. *Journal of Sedimentary Petrology*, **58**, 739–50.

Warme, J. E. (1967) Graded bedding in the recent sediment of Mugu Lagoon, California. *Journal of Sedimentary Petrology*, **37**, 540–7.

Watkin, E. E. (1939) The swimming and burrowing habits of some species of the amphipod genus *Bathyporeia*. *Journal of the Marine Biological Association, UK*, **23**, 457–65.

Watkin, E. E. (1940) The swimming and burrowing habits of the amphipod *Urothoë marina* (Bate). *Proceedings of the Royal Society of Edinburgh*, **60**, 271–80.

Watling, L. (1991) The sedimentary milieu and its consequences for resident organisms. *American Zoologist*, **31**, 789–96.

Watson, A. T. (1927) Observations on the habits and life-history of *Pectinaria (Lagis) koreni*, Mgr. *Proceedings and Transactions of the Liverpool Biological Society*, **42**, 25–59.

Weaver, P. P. E. and Schultheiss, P. J. (1983) Vertical open burrows in deep-sea sediments 2 m in length. *Nature*, **301**, 329–31.

Webb, J. E. (1969) Biologically significant properties of submerged marine sands. *Proceedings of the Royal Society of London*, **B174**, 355–402.

Webb, J. E. and Hill, M. B. (1958) The ecology of Lagos Lagoon IV. On the reactions of *Branchiostoma nigeriense* Webb to its environment. *Philosophical Transactions of the Royal Society of London*, **B241**, 355–91.

Weimer, R. J. and Hoyt, J. H. (1964) Burrows of *Callianassa major* Say, geologic indicators of littoral and shallow neritic environments. *Journal of Paleontology*, **38**, 761–7.

Wells, G. P. (1944) Mechanism of burrowing in *Arenicola marina* L. *Nature*, **154**, 396.

Wells, G. P. (1945) The mode of life of *Arenicola marina* L. *Journal of the Marine Biological Association, UK*, **26**, 170–207.

Wells, G. P. (1948) Thixotropy, and the mechanics of burrowing in the lugworm (*Arenicola marina* L.). *Nature*, **162**, 652–3.

Werner, F. and Wetzel, A. (1982) Interpretation of biogenic structures in oceanic sediments. *Actes Colloque International CNRS, Bordeaux, Sept. 1981. Bulletin de l'Institute Géologique du Bassin d'Aquitaine 1982*, **31**, 275–88.

West, R. R. and Ward, E. L. (1990) *Asteriacites lumbricalis* and a protasterid ophiuroid, in *Evolutionary Paleobiology of Behavior and Coevolution* (ed. A. J. Boucot), Elsevier, Amsterdam, pp. 321–7.

Westall, F. and Rincé, Y. (1994) Biofilms, microbial mats and microbe–particle interactions: electron microscope observations from diatomaceous sediments. *Sedimentology*, **41**, 147–62.

Wetzel, A. (1981) Ökologische und stratigraphische Bedeutung biogener Gefüge in quartären Sedimenten am NW-afrikanischen Kontinentrand. *'Meteor' Forschungs-Ergebnisse*, **C34**, 1–47.

Wetzel, A. (1983a) Biogenic structures in modern slope to deep-sea sediments in the Sulu Sea Basin (Philippines). *Palaeogeography, Palaeoclimatology, Palaeoecology*, **42**, 285–304.

Wetzel, A (1983b) Biogenic sedimentary structures in a modern upwelling region: northwest African continental margin, in *Coastal Upwelling and its Sediment Record: Part B, Sedimentary Records of Ancient Coastal Upwelling* (eds J. Thiede and E. Suess), Plenum Press, New York, pp. 123–44.

Wetzel, A. (1984) Bioturbation in deep-sea fine-grained sediments: influence of sediment texture, turbidite frequency and rates of environmental change, in *Fine-Grained Sediments: Deep Water Processes and Facies* (eds D. A. V. Stow and D. J. W. Piper), Geological Society, London, pp. 595–608.

Wetzel, A. and Aigner, T. (1986) Stratigraphic completeness: tiered trace fossils provide a measuring stick. *Geology*, **14**, 234–7.

Wetzel, A. and Bromley, R. G. (1994) *Phycosiphon incertum* revisited: *Anconichnus horizontalis* is its junior subjective synonym. *Journal of Paleontology*, **68**, 1396–402.

Wetzel, A. and Werner, F. (1981) Morphology and ecological significance of *Zoophycos* in deep-sea sediments off NW Africa. *Palaeogeography, Palaeoclimatology, Palaeoecology*, **32**, 185–212.

Wheatcroft, R. A. (1989) Comment and reply on 'Characteristic trace-fossil associations in oxygen-poor sedimentary environments'. *Geology*, **17**, 674–5.

Wheatcroft, R. A. and Jumars, P. A. (1987) Statistical re-analysis for size dependency in deep-sea mixing. *Marine Geology*, **77**, 157–63.

Whitlatch, R. B. (1974) Food-resource partitioning in the deposit feeding polychaete *Pectinaria gouldii. Biological Bulletin of the Marine Biological Laboratory, Woods Hole*, **147**, 227–35.

Whitlatch, R. B. (1977) Seasonal changes in the community structure of the macrobenthos inhabiting the intertidal sand and mud flats of Barnstable Harbor, Massachusetts. *Biological Bulletin of the Marine Biological Laboratory, Woods Hole*, **152**, 275–94.

Whitlatch, R. B. (1980) Pattern of resource utilization and coexistence in marine intertidal deposit-feeding communities. *Journal of Marine Research*, **38**, 743–65.

Whitlatch, R. B. (1981) Animal–sediment relationships in intertidal marine benthic habitats: some determinants of deposit-feeding species diversity. *Journal of Experimental Marine Biology and Ecology*, **53**, 31–45.

Whittington, H. B. (1980) Exoskeleton, moult stage, appendage morphology, and habits of the Middle Cambrian trilobite *Olenoides serratus*. *Palaeontology*, **23**, 171–204.

Whittington, H. B. (1985) *Tegopelte gigas*, a second soft-bodied trilobite from the Burgess Shale, Middle Cambrian, British Columbia. *Journal of Paleontology*, **59**, 1251–74.

Wightman, D. M., Pemberton, S. G. and Singh, C. (1987) Depositional modelling of the Upper Mannville (Lower Cretaceous), east central Alberta: implications for the recognition of brackish water deposits. *SEPM Special Publications*, **40**, 189–220.

Wignall, P. B. (1990) Observations on the evolution and classification of dysaerobic communities, in *Paleocommunity Temporal Dynamics: the Long-Term Development of Multispecies Assemblies* (ed. W. Miller), *Paleontological Society Special Publications*, **5**, 99–111.

Wignall, P. B. (1991) Dysaerobic trace fossils and ichnofabrics in the Upper Jurassic Kimmeridge Clay of southern England. *Palaios*, **6**, 264–70.

Wignall, P. B. (1993) Distinguishing between oxygen and substrate control in fossil benthic assemblages. *Journal of the Geological Society, London*, **150**, 193–6.

Wignall, P. B. (1994) *Black Shales*, Clarendon Press, Oxford.

Wignal, P. B. and Pickering, K. T. (1993) Palaeoecology and sedimentology across a Jurassic fault scarp, NE Scotland. *Journal of the Geological Society, London*, **150**, 323–40.

Wikander, P. B. (1980) Biometry and behaviour in *Abra nitida* (Müller) and *A. longicallus* (Scacchi) (Bivalvia: Tellinacea). *Sarsia*, **65**, 255–68.

Wilke, D. E. (1952) Beobachtungen über den Bau und die Funktion des Röhren- und Kammersystems der *Pectinaria koreni* Malmgren. *Helgoländer wissenschaftliche Meeresuntersuchungen*, **4**, 130–7.

Williams, D. D., Williams, N. E. and Hynes, H. B. N. (1974) Observations on the life history and burrow construction of the crayfish *Cambarus fodiens* (Cottle) in a temporary stream in southern Ontario. *Canadian Journal of Zoology*, **52**, 365–70.

Wilson, M. A. and Palmer, T. J. (1992) Hardgrounds and hardground faunas. *Publications of the Institute of Earth Studies, University of Wales, Aberystwyth*, **9**.

Wilson, W. H. (1980) A laboratory investigation of the effect of a terebellid polychaete on the survivorship of nereid polychaete larvae. *Journal of Experimental Marine Biology and Ecology*, **46**, 73–80.

Wilson, W. H. (1981) Sediment-mediated interactions in a densely populated infaunal assemblage: the effects of the polychaete *Abarenicola pacifica*. *Journal of Marine Research*, **39**, 735–48.

Wohlenberg, E. (1939) Die Wattenmeer-Lebensgemeinschaften im Königshafen von Sylt. *Helgoländer wissenschaftliche Meeresuntersuchungen*, **1**, 1–92.

Woodin, S. A. (1977) Algal 'gardening' behavior by nereid polychaetes: effects on soft-bottom community structure. *Marine Biology*, **44**, 39–42.

Woodin, S. A. (1978) Refuges, disturbance, and community structure: a marine soft-bottom example. *Ecology*, **59**, 274–84.

Woodin, S. A. (1982) Browsing: important in marine sedimentary environments? Spionid polychaete examples. *Journal of Experimental Marine Biology and Ecology*, **60**, 35–45.

Woodin, S. A. (1991) Recruitment of infauna: positive or negative cues? *American Zoologist*, **31**, 797–807.

Woodin, S. A. and Jackson, J. B. C. (1979) Interphyletic competition among marine benthos. *American Zoologist*, **19**, 1029–43.

Woodin, S. A. and Martinelli, R. (1991) Biogenic habitat modification in marine sediments: the importance of species composition and activity, in *The Environmental Impact of Burrowing Animals and Animal Burrows* (eds P. S. Meadows and A. Meadows), *Symposium of the Zoological Society of London*, **63**, 231–50.

Woodin, S. A., Walla, M. D. and Lincoln, D. E. (1987) Occurrence of brominated compounds in soft-bottom benthic organisms. *Journal of Experimental Marine Biology and Ecology*, **107**, 209–17.

Wyatt, T. D. and Foster, W. A. (1991) Intertidal invaders: burrow design in marine beetles, in *The Environmental Impact of Burrowing Animals and Animal Burrows* (eds P. S. Meadows and A. Meadows), *Symposium of the Zoological Society of London*, **63**, 281–96.

Yeo, R. K. and Risk, M. J. (1981) The sedimentology, stratigraphy, and preservation of intertidal deposits in the Minas Basin system, Bay of Fundy. *Journal of Sedimentary Petrology*, **51**, 245–60.

Yingst, J. Y. and Aller, R. C. (1982) Biological activity and associated sedimentary structures in HEBBLE-area deposits, western North Atlantic. *Marine Geology*, **48**, M7–M15.

Yingst, J. Y. and Rhoads, D. C. (1980) The role of bioturbation in the enhancement of bacterial growth rates in marine sediments, in *Marine Benthic Dynamics* (eds K. R. Tenore and B. C. Coull), University of South Carolina Press, Durham, pp. 407–21.

Yonge, C. M. (1939) The protobranchiate Mollusca: a functional interpretation of their structure and evolution. *Philosophical Transactions of the Royal Society of London*, **B230**, 79–147.

Yonge, C. M. (1949) On the structure and adaptations of the Tellinacea, deposit-feeding Eulamellibranchia. *Philosophical Transactions of the Royal Society of London*, **B234**, 29–76.

Yonge, C. M. and Thompson, T. E. (1976) *Living Marine Molluscs*, Collins, London.

Young, D. K. (1971) Effects of infauna on the sediment and seston of a subtidal environment. *Vie et Milieu, Supplement*, **22**, 557–71.

Young, D. K., Jahn, W. H., Richardson, M. D. and Lohanick, A. W. (1985) Photographs of deep-sea Lebensspuren: a comparison of sedimentary provinces in the Venezuela Basin, Caribbean Sea. *Marine Geology*, **68**, 269–301.

Young, D. K. and Rhoads, D. C. (1971) Animal–sediment relations in Cape Cod Bay, Massachusetts. I. A transect study. *Marine Biology*, **25**, 203–19.

Yurewicz, D. A. (1977) Sedimentology of Mississippian basin-facies carbonates, New Mexico and west Texas – the Rancheria Formation. *SEPM Special Publications*, **25**, 203–19

Ziegelmeier, E. (1952) Beobachtungen über den Röhrenbau von *Lanice conchilega* (Pallas) im Experiment und am naturlichen Standort. *Helgoländer wissenschaftliche Meeresuntersuchungen*, **4**, 107–29.

Ziegelmeier, E. (1969) Neue Untersuchungen über die Wohnröhren-Bauweise von *Lanice conchilega* (Polychaeta, Sedentaria). *Helgoländer wissenschaftliche Meeresuntersuchungen*, **19**, 216–29.

Zijlstra, J. J. P. (1994) Sedimentology of the late Cretaceous and early Tertiary (tuffaceous) chalk of northwest Europe. *Geologica Ultraiectina. Mededelingen van de Faculteit Aardwetenschappen Universiteit Utrecht*, **119**.

Glossary

active fill Burrow fill directly emplaced by the burrower.
advection In **bioturbation**, fluid or particle advection involves bulk transport of material, in contrast to **diffusion**.
aedificichnia Trace fossil edifices constructed on as opposed to within the substrate (section 9.2.10; Bown and Ratcliffe 1988).
aerobic biofacies Biofacies dependent on an **oxic environment** (Tyson and Pearson 1991).
agrichnion (Pl. agrichnia); permanent dwelling burrow used to trap organisms for food, or to culture them for food (**gardening**) or to employ them in **chemosymbiosis** (Ekdale *et al.* 1984a).
amensalism Exclusion of one species or group of species by the life activities of another.
anaerobic biofacies Biofacies dependent on an **anoxic environment** (Tyson and Pearson 1991).
anoxic environment Environment lacking free oxygen.
backfill Material actively emplaced by an animal posteriorly as it progresses through sediment.
bed-junction preservation Burrow fill consisting of material that contrasts with that of the matrix, owing to piping down of sediment from above a bed junction at which a corresponding change in sediment has taken place (Simpson 1957).
benthic boundary layer Biologically active zone around the sediment–water interface.
benthos Flora and fauna of the sea or lake floor.
biodeformation structure Sediment structure arising from **eddy diffusion** where an organism moves through soupy or unconsolidated sediment (Schäfer 1956).
biodeposition Entrapment, during suspension feeding, of fine particles in suspension mode and their incorporation in and deposition as sand-sized pellets.
bioglyph Ornament on a burrow wall produced by the life activity of the occupant (Bromley *et al.* 1984).
bioturbation Process by which the primary consistency and structure of a sediment are modified by the activity of organisms living within it. Sediment mixing by animals.
biowinnowing Resuspension of fine-grained sediment fractions through biological activity. Currents may remove this suspended material from the area.
body fossil Fossilized bodily remains of the skeletal or soft parts of an organism. Cf. **trace fossil**.
boxwork Burrow network of interconnection shafts and galleries in a three-dimensional configuration (Ekdale *et al.* 1984a). Cf. **maze**.
browsing predation Browsing off parts of animals without killing the prey, allowing for regeneration of the lost parts. Also known as harvesting.
burrow Space within sediment occupied and maintained by an animal.
burrow boundary Limit of the **burrow**. The interface between the **burrow fill** or the **lining** and the substrate. Cf. **wall**.
burrow fill Material that fills a burrow. May be active (**backfill**) or passive (gravitational).
burrow structure Biogenic sedimentary structure created in sediment by the activity of a burrowing animal.

calichnia Trace fossil structures made for breeding purposes (section 9.2.11; Genise and Bown 1994a).

casting medium The more resistant substance in which a trace fossil is preserved in a lithologically heterogeneous deposit (i.e. sand in an alternating sand–mud heterolithite) (Fig. 9.1.).

chemosymbiosis A commensal association between an animal host and autolithotrophic bacteria. The bacteria oxidize various reduced compounds using oxygen provided by the host, thereby allowing fixation of carbon and the production of carbohyrates and enzymes (Reid 1989).

chimney Upward extension of a burrow **lining** or **tube** above the substrate surface (section 3.7).

cleavage relief Seilacher's (1964) term for **trace fossil** preservation in **semirelief**, as expressed on cleavage surfaces of a fissile rock.

climax community Community that has been allowed to develop a mature population structure, owing to environmental stability. Normally comprises a high diversity of **equilibrium species** occupying numerous narrow niches (see **equilibrium strategy**).

climax trace fossil Ichnotaxon that is characteristic of **ichnocoenoses** produced by **climax communities**.

colonization window The time available for colonization of the substrate (section 12.2.3; Pollard *et al.* 1993).

compression Burrowing process that involves the compaction of a plastic substrate so as to allow the passage of an animal, and creating an open space.

conveyor Animal that causes by its feeding process a continual upward **advection** of grains, i.e. from a deeper level within the substrate to the surface. Cf. **reverse conveyor**. Abbreviated from 'conveyor-belt feeder' (Rhoads 1974).

coprolite Fossilized excrement. If less than or about 1 mm in size, may be referred to as microcoprolite or faecal pellet.

core In ichnological sense, the central part of a **zoned fill**. Cf. **mantle**.

cubichnion (Pl. cubichnia); temporary resting trace (Seilacher 1953b).

cryptobioturbation Bioturbation by interstitial **meiofauna** and small macrofauna displacing grains very short distances. Cryptobioturbation causes softening or blurring of the fabric and leads to homogeneity (Howard and Frey 1975).

deposit feeder Animal that obtains nutriment from particulate organic matter and microbes incorporated within the sediment (Levinton 1972).

detritus feeder Deposit feeder that feeds on the nutrient-rich topmost layer of the substrate.

diffusion In bioturbation, particle diffusion refers to particle displacement at a grain-to-grain scale, as opposed to the bulk transport of **advection**.

dilatant Of sediment that, under the application of a force, increases in strength. See **thixotropic**.

domichnion (Pl. domichnia); permanent dwelling structure constructed within the substrate (Seilacher 1953a). See **aedificichnia**.

double-anchor technique Sediment-penetration technique involving anchorage at two points alternately (Trueman 1968a). A terminal anchor secures the fore end of the animal while the hind end is drawn up; then a penetration anchor braces the hind end while the fore end intrudes into the substrate ahead. Produces push-and-pull structures (Seilacher and Seilacher 1994).

dysoxic Conditions of very low oxygen tension (section 12.1.1; Fig. 12.1; Tyson and Pearson 1991).

eddy diffusion Sediment mixing that results from turbulent flow of the substrate, where there is a bulk motion of small volumes that retain their identity for a period of time (Hanor and Marshall 1971).

elite structure Biogenic structure within sediment that attracts more than its spatial share of activity on account of high oxygen or organic content (Reise 1981).

elite trace fossil Ichnotaxon that receives special treatment from diagenesis (e.g. colour enhancment, concretionary development) on account of some feature of its structure, and is thereby rendered more conspicuous than other ichnotaxa in a given assemblage.

endobenthic Pertaining to endobenthos, the organisms that live within the sediment of the sea or lake floor.

endobiont Animal living within the sea or lake floor, a member of an **endobenthic** community.

epibenthic Pertaining to the epibenthos, living upon the sea or lake floor as opposed to within it. Cf. **endobenthic**.

epirelief Seilacher's (1953a, 1964) term for a trace fossil preserved in either 'concave' or 'convex' semirelief on the upper surface of a **casting medium** (usually sandstone: Fig. 9.1).

equilibrichnion (Pl. equilibrichnia); same as **equilibrium structure**.

equilibrium species Species capable of using K- or **equilibrium strategy**.

equilibrium strategy Population strategy used by organisms in mature and **climax communities**, involving narrow specialization for specific niches.

equilibrium structure Sediment structure produced by an animal adjusting its position relative to the sea floor in response to gradual or minor aggredation or degredation of the sediment surface. Also equilibrium trace. Cf. **escape trace**.

escape trace Sediment structure produced by an animal escaping from threat of sudden burial, erosion or predation.

exaerobic biofacies Biofacies developed in dysoxic regions on a generally anoxic substrate, colonized by shelled chemosymbiotic fauna (Fig. 12.1; Savrda and Bottjer 1987a).

excavation Burrowing technique that involves loosening of compacted substrate and transporting it out of the system, either to the sea floor or to another part of the burrow, so as to create an open space.

exogenic Biogenic sediment structures produced at the depositional interface. Also exogene.

false branching Branching or apparent branching in a trace fossil that has arisen from the shifting of an unbranched original burrow (Fig. 8.8; Bromley and Frey 1974).

firmground Substrate composed of stiff but uncemented sediment. Cf. **soupground** and **softground** (Fürsich 1978).

fodinichnion (Pl. fodinichnia); sediment structure produced by a non-vagile **deposit feeder** (Seilacher 1953a).

fossilization barrier The rigorous taphonomic screen between the activity of living **endobenthic** animals and **trace fossils**.

fugichnion (Pl. fugichnia); same as **escape trace** (Frey 1973).

full relief Seilacher's (1953a) term for a trace fossil preserved in three dimensions and contained within a rock unit, i.e. not along a bedding plane. Cf. **semirelief**.

funnel feeding Detritus trapping by means of a funnel developed at one or both burrow apertures.

gallery more or less horizontal burrows or parts of burrow systems. Cf. **shaft**.

gardening Culturing of microbes for food.

graphoglyptid Complex horizontal tunnel system that apparently served as both a permanent dwelling and a garden or trap for obtaining food. See agrichnion.

guild Group of species that exploit the same class of environmental resource by similar means. Species occupying neighbouring ecological niches (Root 1967).

halo burrows Burrows that have been rendered conspicuous by chemical diagenesis in the sediment immediately surrounding them, to create a discoloured halo (Chamberlain 1975).

hardground Syn-sedimentarily cemented sea floor. Intergranular cement produces a rigid substrate impenetrable by burrowing animals (Voigt 1959).

historical layer Zone of the substrate beneath the deepest tier of burrowers (Berger *et al.* 1979).

hyporelief Seilacher's (1953a) term for a trace fossil preserved in either 'concave' or 'convex' **semirelief** on the sole of a bed.

ichnoclast Reworked trace fossil (Goldring 1991).

ichnocoenosis (Pl. ichnocoenoses); an ecologically pure assemblage of trace fossils, i.e. deriving from the work of a single **endobenthic** community. Palaeoichnocoenosis is more correct for fossil material.

ichnodiversity Number of ichnotaxa in an **ichnocoenosis**. Cf. diversity, which concerns the number of taxa of organisms in a biocoenosis.

ichnofabric All aspects of the texture and internal structure of a sediment that result from bioturbation at all scales (Bromley and Ekdale 1986).

ichnogenus Genus-group name formally assigned to trace fossils. Name is italicized and capitalized. Abbreviated as **igen.** or ichnogen.

ichnoguild Grouping of **ichnospecies** that express a similar sort of behaviour, belong to the same **trophic group** and occupy a similar tier or location within the substrate.

ichnospecies Species-group name formally assigned to trace fossils. Name is italicized. Abbreviated as **isp.** or ichnosp. (Pl. ispp. or ichnospp.).

ichnotaxobase Morphological feature of a trace fossil that is considered a valid basis for ichnotaxonomy.

ICZN International Code of Zoological Nomenclature (Ride *et al.* 1985); the established rules for classifying living and fossil animals.

igen. Abbreviation of **ichnogenus**.

intrusion Burrowing technique that involves the temporary displacement of sedimentary grains by **eddy diffusion** to allow the passage of an animal through watery substrate. The grains flow back behind the animal to form a **biodeformation structure**, and no cavity is produced.

inverted conveyor See **reverse conveyor**.

isp. Abbreviation of **ichnospecies**.

key bioturbator Animal species that causes a particularly high rate of sediment processing or reworking so that the results of its activity come to dominate the fabric of the substrate.

laminated-to-scrambled Of a bed or rock unit that is generally laminated, but is bioturbated at the top (Fig. 10.14).

lining Material applied to the burrow wall by the occupant. Passively accumulated material may be included, which adhered to the wall while the burrow was open (Fig. 1.7).

looseground Unconsolidated sandy substrate, arenaceous equivalent of **softground** (Goldring 1995a).

mantle Outer zone of a zoned **backfill**; see **core**. (Heinberg 1970.)

maze Burrow **network** restricted to one horizon. Cf. **boxwork**.

meiofauna Size-class between macro- and microfauna. Endobenthic meiofauna is the interstitial fauna that lives within the porosity of the sediment.

meniscus structure A form of **backfill** composed of dish-shaped packages of sediment (crescent-shaped in axial section).

mixed layer Uppermost layer of the substrate, totally bioturbated and homogenized (Berger *et al.* 1979). Cf. **transition** and **historical layers**.

network Burrow system of interconnected **shafts** and **galleries**. See **boxwork** and **maze**.

omission suite Trace fossils emplaced within sediment beneath an omission surface and dating from the period of non-deposition (Bromley 1975).

opportunistic species Species that are capable of using r- or **opportunistic strategy**.

opportunistic strategy Population strategy used by organisms that can rapidly colonize vacant niches, e.g. a habitat depopulated by a temporary environmental change. r-strategists.

oxic environment One in which free oxygen is fully available (Fig. 12.1; Tyson and Pearson 1991).

palimpsesting Overprinting of the work of one endobenthic community onto that of another (11.1.2). Cf. **time averaging**.

pascichnion (Pl. pascichnia); structures produced by vagile **detritus** and **deposit feeders** (Seilacher 1953a).

passive fill Material that entered an open burrow gravitationally, by physical sedimentation. Cf. **active fill**.

penetration anchor See **double-anchor technique**.

pioneer species **Opportunistic species** that are the first to colonize newly available habitats.

post-depositional trace fossil Used particularly for assemblages associated with turbidites and storm beds. A structure that was emplaced after an erosive or depositional event and which may cut the sole interface. See **pre-depositional trace fossil**.

post-event trace fossil See **post-depositional trace fossil**.

post-lithification suite Trace fossils (borings) emplaced in an omission surface subsequent to its cementation as a **hardground** (Bromley 1975).

post-omission suite Trace fossils emplaced within sediment beneath an omission surface, but dating from after renewal of sedimentation and the burial of the omission surface (Bromley 1975).

praedichnion (Pl. praedichnia); structures produced as a result of predation (Ekdale 1985).

pre-depositional trace fossil Used particularly for semirelief assemblages on soles of turbidites and storm beds. Trace fossil that was emplaced in a pre-event sediment prior to sudden erosion and deposition; the burrow is cast on the sole of the post-event sand bed (Fig. 6.4).

pre-event trace fossil See **pre-depositional trace fossil**.

pre-lithification suite In the case of hardgrounds, these are trace fossils of the **omission suite** that were emplaced in sediment beneath an omission surface prior to its lithification (Bromley 1975).

pre-omission suite Trace fossils immediately beneath an omission surface but dating from before the depositional hiatus (Bromley 1975). Cf. **omission suite** and **post-omission suite**.

protrusive Of a **spreite** produced by distalward movement, i.e. away from the apertures.

repichnion (Pl. repichnia); a trace fossil expressing locomotion, including some **backfills**, trails and trackways.

retrusive Of a **spreite** produced by proximalward movement, i.e. toward the apertures.

reverse conveyor Animal of which the feeding activity conveys material downwards from the sediment surface and emplaces it deeply within the substrate. Cf. **conveyor**.

rind burrows Burrows the wall materials of which have been rendered conspicuous through chemical alteration (Chamberlain 1975; Ekdale 1977). See **elite trace fossil**.

RPD Redox potential discontinuity. The layer within an aquatic substrate at which oxidizing processes become displaced by reducing processes (Fenchel and

Riedl 1970). Boundary between the oxic 'yellow' sediment above and the anoxic 'black', chemically reducing sediment below (Fig. 5.7).

semirelief Originally Seilacher's (1953a) term for a trace fossil preserved at a bedding plane by erosion-and-fill processes. Thus, a cast of the lower half of the trace, containing no internal (biogenic) structure, a **pre-depositional trace fossil**. Commonly used also for trace fossils emplaced along pre-existing bedding planes; but these **post-depositional trace fossils** contain intrinsic internal structure and, although appearing as semirelief structures, are actually preserved in full relief.

seston Organic and inorganic particles in suspension in the water column.

shaft Vertical burrow or vertical element of a burrow system (Frey 1973).

shear Mixing or deformation of sediment by laminar flow (Fig. 1.1b).

softground Soft, unconsolidated sedimentary substrate intermediate between **firmground** and **soupground**. Cf. **looseground**.

soupground Substrate composed of water-laden sediment having a fluid consistency (Ekdale *et al.* 1984a).

spreite Laminated biogenic structure composed of closely spaced successive tunnel walls produced as a burrow is shifted laterally (broadside-on) through the sediment. Cf. **backfill**.

suite Subset of the heterogeneous trace fossil assemblage representing an approach towards the ecologically pure ichnocoenosis (Bromley 1975).

suspension feeder Animal that traps and feeds on seston (Levinton 1972).

terminal anchor See **double-anchor technique**.

thixotropic Of sediment that, under the application of a force, decreases in strength and becomes fluid. Cf. **dilatant**.

tiering Vertical partitioning of a community. In **endobenthic** communities, different types of sediment processing occurring at several levels beneath the sea floor (Bottjer and Ausich 1982).

time-averaging Crowding of time-lines, resulting in the combining of several successive communities in a single assemblage (section 5.3.39; Fig. 10.2b). Cf. **palimpsesting**.

trace fossil Fossilized structure produced in unlithified sediment, sedimentary rock or other substrate by the activity or growth of organisms.

tracemaker The animal responsible for the trace fossil.

transition layer Zone just beneath the **mixed layer** of the substrate in which heterogeneous mixing results from the activity of deeper-burrowing animals (Berger *et al.* 1979).

trophic group High-ranking term embracing species having similar feeding behaviour, e.g. **deposit feeders, suspension feeders**.

true branching Branching of a trace fossil that represents branching of the original causative burrow (Fig. 8.8; Bromley and Frey 1974). Cf. **false branching**.

tube Well developed **burrow lining**.

undertracks Biogenic structure produced by deformation of sediment laminae directly beneath a track and usually preserved in cleavage relief (Osgood 1970).

wall Constructional features at the **burrow boundary**, including compression and disturbance zones within the adjacent substrate. Cf. **lining** (Figs 1.7 and 8.1). The wall may show sculpture or ornament, resulting from scratching and digging activity of the tracemaker (cf. Keighley and Pickerill 1994).

worm Used colloquially here in a functional sense for soft, elongate animals. No taxonomic position is implied.

zoned fill Burrow fill that consists of two or more discrete sediment types arranged concentrically; a **mantle** surrounding a **core**.

Index

Figures appearing in **bold** refer to sections and subsections. Figures appearing in *italics* refer to illustrations.